# Industrial Engineering and Management

Edited by **Jeff Hansen**

**NY** RESEARCH
P R E S S

New York

Published by NY Research Press,
23 West, 55th Street, Suite 816,
New York, NY 10019, USA
www.nyresearchpress.com

**Industrial Engineering and Management**
Edited by Jeff Hansen

International Standard Book Number: 978-1-63238-501-7 (Hardback)

Printed in the United States of America.

# Contents

# Preface

Every book is a source of knowledge and this one is no exception. The idea that led to the conceptualization of this book was the fact that the world is advancing rapidly; which makes it crucial to document the progress in every field. I am aware that a lot of data is already available, yet, there is a lot more to learn. Hence, I accepted the responsibility of editing this book and contributing my knowledge to the community.

Industrial engineering specifically focuses on improving quality and productivity. It utilizes a combination of disciplines such as system engineering, manufacturing engineering, operations research, management science and safety engineering to design and optimize complex systems and processes. This branch of engineering tries to reduce or eliminate unproductive processes. Conventionally industrial engineering was used to set up machines and assembly lines for factories and manufacturing units, but now along with setting up a manufacturing unit it also helps in streamlining the procedures. This book elucidates the concepts and innovative models around prospective developments with respect to this field. Those with an interest in the area of industrial engineering would find this book helpful. This book consists of contributions made by international experts which unravel the recent studies and futuristic aspects of industrial engineering.

While editing this book, I had multiple visions for it. Then I finally narrowed down to make every chapter a sole standing text explaining a particular topic, so that they can be used independently. However, the umbrella subject sinews them into a common theme. This makes the book a unique platform of knowledge.

I would like to give the major credit of this book to the experts from every corner of the world, who took the time to share their expertise with us. Also, I owe the completion of this book to the never-ending support of my family, who supported me throughout the project.

**Editor**

# A new heuristic algorithm for the planar minimum covering circle problem

Hasan Hosseini Nasab*, Mahdi Tavana and Mohsen Yousefi

*Industrial Engineering Department , Faculty of Engineering, Yazd University, Yazd, Islamic Republic of Iran*

The minimum covering circle problem is widely utilized in the studies of single or multiple facility location problems. It may be employed to locate the essential locations of factories, schools, fire departments, hospital establishments, and other facilities, which could be considered as a point in a plane theoretically. In this paper, a new geometrical algorithm is presented which determines the minimum covering circle of all points on a plane in four steps. The model validity was considered by studying the coordinates of points with random numbers and different distributions. In order to show the accuracy of the proposed algorithm, numerical experiments were carried out and compared with other studies in the field. The results show that the proposed algorithm extremely outperforms other examined algorithms.

**Keywords:** minimum covering circle; facility location; geometrical algorithm

## 1. Introduction

A wide range of location problems effect today's firm and organizational objectives and decision-makings. The interest and service level, economic profits, quality of service, security, environmental protection, availability, optimum performance are some of the many criteria and objectives of the facility location problem; and hence, different methods and approaches have been employed in the literature to maximize the achievement of these objectives.

The facility location problem may be generally categorized in the two groups of multiple or single facility location, however the multi-facility location may be assumed as a combination of several single facility problems. The importance of the single facility location problem is clear as it provides a good insight into multi-facility problems and may even present suitable initial solutions. Hence, various researchers have focused on developing the single facility location problem, as further described in the literature review. Another approach to categorize the facility location problem includes grouping the solving process, for example discrete or continuous. The continuous category itself may be further grouped in different approaches i.e. the minimum covering circle solution, which is the focus of this paper.

The history of the classical minimum covering circle tracks back to Sylvester, who introduced this problem in 1857. Also known as the Euclidean 1-center and the smallest enclosing circle, the minimum covering circle problem seeks the smallest circle which

---

*Corresponding author. Email: hhn@yazd.ac.ir

covers or encloses all existingd points on a plane. In other words, the goal is to find the exact location of a facility with the minimum distance from the farthest point of interest. Sylvester (1857) and Chrystal (1885) proposed primal approaches to solve this problem, which start from a large circle and reduce the radius until it is enclosed by 2 or 3 points, as further described below:

Let point $p_i = (a_i, b_i)$ be a member of the set $P$ which contains $n$ number of points $p_1, p_2, \ldots, p_n$ in a $R^2$ space. Let $f = (x, y)$ be the center of an assumed circle. It is clear that the distance between any of the existing points and the center of the circle may be calculated from Equation (1):

$$d(f, p_i) = \sqrt{(x - a_i)^2 + (y - b_i)^2} \tag{1}$$

In order to achieve the minimum covering circle, Equation (1) has to be minimized for all existing points. Therefore, the objective formulates are as below:

$$\min_{f \in R^2} \quad \max_{p_i \in P} \quad d(f, p_i). \tag{2}$$

Similar to the model of Dearing and Zeck (2009) the objective function may be redefined as below:

$$\min z \tag{3}$$

$$\text{st: } z \geq \sqrt{(x - a_i)^2 + (y - b_i)^2}, \quad p_i \in P \tag{4}$$

where '$\text{Min}_{f \in R^2} \max_{p_i \in P} d(f, p_i)$' is considered as variable '$Z$' to simplify the sentence.

The minimum covering circle problem is applicable for the following fields and problems:

- To find an appropriate place to facilitate an area by a radio station.
- To find an appropriate place to construct a fire station.
- To find an appropriate place for an emergency helicopter station to service all nodes in minimal time.

The various existing approaches in the minimum covering circle problem contain two major setbacks: first, the difficult perception of each method which requires a high degree of visualization; and second, the complexity of the algorithms which are time-consuming even with the use of modern computer wizards.

## 2.  Literature of study

The smallest circle problem or minimum covering circle problem is a mathematical problem of computing the smallest circle that contains all of a given set of points in the Euclidean plane. The corresponding problem in n-dimensional space, the smallest bounding sphere problem, is to compute the smallest n-sphere that contains all of a given set of points. The smallest circle problem was initially proposed by the English mathematician Sylvester in 1857. Nair and Chandrasekaran (1971) developed a new method for the minimum covering circle problem using the polynomial quadratic method. A minimum covering circle problem with Euclidean distance was considered by Elzinga and Hearn (1972). The first linear algorithm was proposed by Megiddo (1983a) that was established based on Voronoi diagram. He showed that any fixed dimension problem can be modeled as a convex quadratic programming and be solved in an $O(n)$ time. In another paper, Megiddo (1983b) investigated a weighted problem in the plane and presented an $O(n(\log n)^3 (\log \log n)^2)$ algorithm for finding the minimum covering circle. An iterative algorithm based on Voronoi diagram was posed by Skyum (1991). The algorithm was appropriate for only medium size problem and cannot be used to obtain an optimum solution for large size problems. Welzl (1991) solved minimum covering circle based on Seidel's LP algorithm in $d \geq 2$ and showed some efficiency of a heuristic algorithm. He considered 1000 and 5000 points with 2, 3, 5, and 10 dimensions to prove the advantages of the algorithm.

Minimum covering problem was solved with $d \geq 2$ by Hopp and Reeve (1996), Gartner (1999), Fisher, Gartner, and Kutz (2003), and Zhou, Tohemail, and Sun (2005). In addition Gartner (1999) and Fisher et al. (2003) presented a robust algorithm based on the pivoting approach that solves the problem in a short time for $d \leq 20$.

Fuzzy approach is also used for minimum covering circle problem. Li, Kabadi, and Nair (2002, 2005) presented two fuzzy models that were based on possibility constraints and necessary constraints. In both algorithms, the coordination of points was considered using fuzzy variables. The algorithms were able to find the minimum covering circle in a fuzzy method.

For the large size problem, Zhou et al. (2005) presented two new algorithms. Its computational results showed that for a large size problem such as 512,000 points with $d = 100$, the result can be gained in a timescale less than an hour. Two other algorithms were proposed by Yıldırım (2007) which was structured with the aim of Frank–Wolfe approach.

Shamos and Hoey (1975) proposed an $O(n \log n)$ time algorithm for the problem based on the observation that the center of the smallest enclosing circle must be a vertex of the farthest-point Voronoi diagram of the input point set.

Dearing and Zeck (2009) used the dual algorithm and solved the problem with ($d = 32$ and $d = 64$). El-Tamimi and Al-Zahrani (2012) developed a new geometrical algorithm based on the quadratic modeling, originally proposed by Nair and Chandrasekaran (1971). Their computational result was examined for the dimension of ($d = 32$).

The minimum covering circle problem is a mathematical problem of computing the smallest circle that contains all of a given set of points in the Euclidean plane. The minimum covering circle of a set $S$ can be determined by at most three points in $S$ which lie on the boundary of the circle. In such cases we need to investigate $(n-1)(n-2)/3$ solutions. However, the minimum enclosing circle can be found in linear time, but it is still time-consuming especially when the number of points increases. Several exact and heuristic algorithms are proposed for finding the minimum covering circle but most of them perform in many steps and need a lot of CPU time to reach the final circle. As discussed in literature of study, many approaches exist, however the difficult perception of each method, which needs a high degree, or the time-consuming overall processes due to the complexity of existing algorithms present much room for improvement. In this paper, a new geometrical algorithm is presented which determines the minimum covering circle of all points on a plane in four simple steps. At the result, in respect to other methods, the required mathematical calculations and the CPU time to find the minimum covering circle decrease. The proposed algorithm is described below.

## 3. Achieving the minimum covering circle

The proposed method in this paper consists of four consecutive steps, as presented below.

In typical assumption of the problem it is implied that there are specified points which can be customers of a department store, applicants for wireless communication service, patients or citizens who want to take advantage from services. Coordinate of each point is determined with respect to a hypothetical source. The purpose is, to determine the minimum covering circle or in other words to determine the center and radius of the circle.

Defining the center can help decision-makers determine the site of the facility and the radius indicates the consideration of the power and capacities of service. For example, in a communication antenna problem and/or radio mast, in addition to determining the construction site of the facility, decision-makers need to define the power and capacity for the facility.

The proposed method of this paper to determine the minimum covering circle is defined in the following four steps:

*Step 1*: Measure the distances between each pair of points and sort them in a chronological order. Choose the two points with the maximum distance. These points will be referred to as $P_1$ and $P_2$ (Figure 1). Using these two points form a circle with the diameter of the measured maximum distance. If all available points are covered with the drawn circle, the smallest covering circle is achieved, otherwise go to step 2.

$$d(p_i, p_j) = \sqrt{(a_i - a_j)^2 + (b_i - b_j)^2} \qquad (5)$$

where $(a_i, b_i)$ and $(a_j, b_j)$ are the coordinates of points $p_i$ and $p_j$, respectively.

*Step 2*: Measure the distance of all points from the center of the drawn circle in step 1, and determine the farthest uncovered point. Refer to this point as $P_3$. Draw a unique circle using the three points of $P_1$, $P_2$, and $P_3$ which are calculated using the following formula:

For any three points $P_1$, $P_2$, and $P_3$ with coordinates $(a, b)$, $(c, d)$, and $(e, f)$, respectively, the center of circle, $(X, Y)$ and the radius $R$, can be calculated as:

$$X = 1/2((a^2 + b^2) \times (f - d) + (c^2 + d^2) \times (b - f) + (e^2 + f^2) \times (d - b))/(a \times (f - d) + c \times (b - f) + e \times (d - b))$$

$$Y = 1/2((a^2 + b^2) \times (e - c) + (c^2 + d^2) \times (a - e) + (e^2 + f^2) \times (c - a))/(b \times (e - c) + d \times (a - e) + f \times (c - a))$$

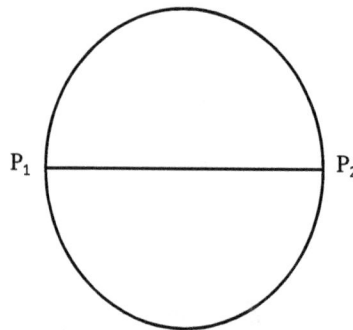

Figure 1.   The first circle which is drawn by $P_1$ and $P_2$.

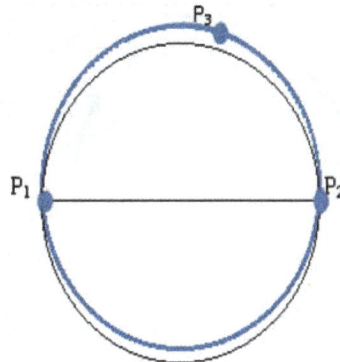

Figure 2.    Second circle which is drawn by $P_1$, $P_2$, and $P_3$.

$$R = \sqrt{(X - a)^2 + (Y - b)^2}$$

The achieved circle drawn by these three points (Figure 2) may be optimum.

It is necessary to note that $P_3$ must only be located in the two areas of $R_1$ or $R_2$ as shown in yellow color in Figure 3 otherwise the distance of this point from one of the two previous points will exceed the maximum calculated distance in step 1.

If all points have been covered by the drawn circle, this circle is optimum otherwise go to step 3.

*Step 3*: Repeat the process of finding the farthest point from the center of the last drawn circle in step 2 and refer to it as $P_4$. Using the three points, that is $P_3$, $P_4$ and either one of the points $P_1$ or $P_2$, draw the final circle Figure 4. Based on our observations, one of these two circles will be the minimum covering circle. Meaning that no points will exist outside of the drawn circle, in addition to that the drawn circle is the minimum circle that may be formed according to the following: whenever an arc is bigger than half of the circumference of the circle which no points are located on, the circle has the potential to be improved, otherwise the existing circle is the minimum circle (Elzinga & Hearn, 1972). To choose between $P_1$ and $P_2$ go to step 4.

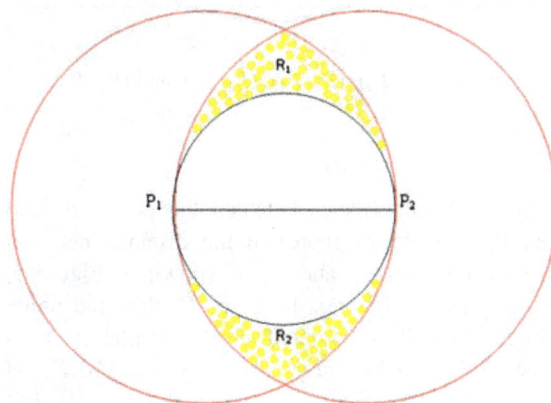

Figure 3.    Possible zone for $P_3$.

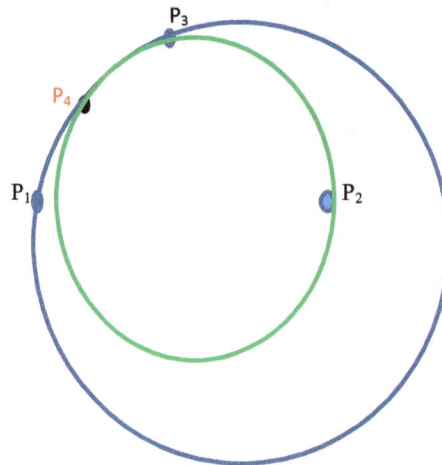

Figure 4.   Possible zone for $P_3$.

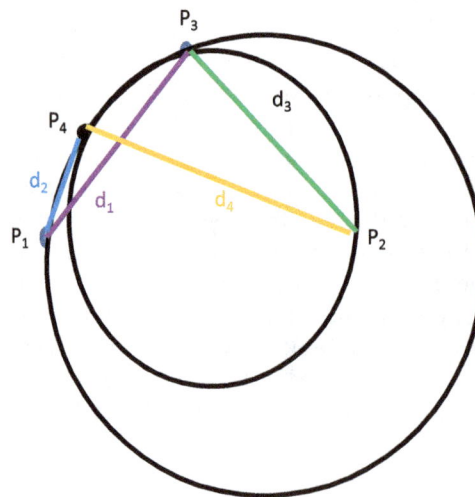

Figure 5.   The distance between $(P_1, P_3)$, $(P_1, P_4)$, $(P_2, P_3)$, and $(P_2, P_4)$.

*Step 4*: Let $d_1$ represent the distance between the points $P_1$ and $P_3$, $d_2$ represent the distance between $P_1$ and $P_4$, $d_3$ represent the distance between $P_2$ and $P_3$, and $d_4$ represent the distance between $P_2$ and $P_4$ as shown in Figure 5. If the minimum value among $P_1$, $P_2$, $P_3$, and $P_4$ belongs to $P_1$ or $P_2$ then the point $P_1$ will be omitted otherwise the point $P_2$ will be omitted. For example, in Figure 5, $d_2$ has the smallest value, so the circle will be drawn using $P_2$, $P_3$, and $P_4$ which is the smallest covering circle.

In the next two sections, the validation procedure and the comparison of the proposed method with other existing methods are presented.

## 4. Validation of the proposed algorithm

The minimum covering circle can be drawn by many algorithms proposed in the literature or by drawing the circle crossing any three points which are not located in a direct line. But the important issue is the necessary steps to find the minimum covering circle that has effect on the calculation time. Therefore, the required time for reaching the final solution of the proposed algorithm is compared with the existing algorithms. Point scatterings and the total number of points in the study are also considered as two important factors with probabilistic impact on the speed and accuracy of the optimum solution. Therefore, it should be ensured that the proposed algorithm is capable of achieving optimum solutions based on any numbers of points and scattering distribution. The other aspect of attention is the impacts of the number of points and their distribution on the required computational time. Generally, it is expected that increasing the number of the points will lead to an increase in CPU time, which remains to be observed.

In this regard, four tests based on various probability distributions were executed with a number of 100, 1000, and 10,000 points each. Each test defines the distribution of points on a plane as presented below:

Test 1: $x, y \sim \text{Uniform}(0, 100)$,
Test 2: $x, y \sim \exp(\beta = 50)$,
Test 3: $x, y \sim \text{Normal}(\mu = 100, \sigma = 50)$,
Test 4: $x, y \sim \text{Normal}(\mu = \exp(\beta = 50), \sigma = \exp(\beta = 100))$.

### 4.1. Assessing the CPU computational time

A statistical two-tailed hypothesis test was designed to examine the impacts of point scattering and the number of points on the CPU computational time.

The independent variables are classified into two categories: (1) the total number of points and (2) the points' scattering distribution; the response variable is determined as the CPU time. The below two null hypotheses are considered:

(1)  $H_0$: The point scattering distribution does not have an impact on CPU time.
(2)  $H_0$: The total number of points does not have an impact on CPU time.

In the designed hypothesis test, two numeric results (CPU time) are obtained for each intersection of independent variables which are shown in Table 1. In relation to the numeric results of Table 1, 12 figures which describe the point scattering and drawn circle are exhibited in Appendix 1.

Based on the numeric results presented in Table 1, the hypothesis test is considered at a 5% significant level. The SPSS 18 software was used for the variance analysis. As

Table 1.   CPU time.

| Number of points | Scattering (distribution) | | | |
| --- | --- | --- | --- | --- |
| | Test 1 | Test 2 | Test 3 | Test 4 |
| 100 | .0029 | .0028 | .003 | .0023 |
| | .0023 | .0027 | .0029 | .0024 |
| 1000 | .122 | .122 | .1291 | .1312 |
| | .1252 | .1298 | .1261 | .1267 |
| 10,000 | 16.493 | 16.148 | 16.417 | 16.587 |
| | 16.273 | 16.289 | 16.793 | 16.886 |

Table 2.   Analysis of the variance based on the data in Table 1.

| Dependent variable: CPU time | Type III sum of squares | df | Mean square | $F$ | Sig. |
|---|---|---|---|---|---|
| Corrected model | 1438.579 | 11 | 130.780 | 1.051E4 | .000 |
| Intercept | 736.171 | 1 | 736.171 | 5.913E4 | .000 |
| Distribution | .108 | 3 | .036 | 2.882 | .080 |
| No. of point | 1438.261 | 2 | 719.130 | 5.777E4 | .000 |
| Distribution × no. of point | .211 | 6 | .035 | 2.821 | .060 |
| Error | .149 | 12 | .012 | | |
| Total | 2174.899 | 24 | | | |
| Corrected total | 1438.729 | 23 | | | |

shown in Table 2, for the first independent variable 'distribution' the $p$-value is equal to .08, which means the first null hypothesis (H: the point scattering distribution does not have an impact on CPU time) is not rejected. In other words points' scattering distribution does not have any impact on CPU time. In another case, for the 'number of points' variable the $p$-value equals .000, which shows an extreme impact on CPU time. Therefore, the second null hypothesis (H: the total number of points does not have an impact on CPU time) is rejected. As mentioned before, it was predictable that an increase in the number of points will lead to an increase in CPU time.

### 4.2.   Assessing the number of circles

According to the proposed algorithm, the solution can be obtained by drawing one, two, or three circles. In this, the impact of the two independent variables (refer to 4.1) on the number of drawn circles which lead to the final solution is examined. The final solution here is the minimum covering circle. Hence, two null hypotheses were defined as below:

(1) $H_0$: Point scattering does not have an impact on the number of drawn circles.
(2) $H_0$: The total number of points does not have an impact on the number of drawn circles.

The results of studying the impacts of the points' scattering distribution and their total number are provided in Table 3.

As shown in Table 4, the $p$-value is .044 for the 'distribution' factor which means the first hypothesis (H: point scattering does not have an impact on the number of drawn circles) is rejected. In other words altering the points' scattering distribution has a

Table 3.   The number of drawn circles.

| Number of points | Scattering (distribution) | | | |
|---|---|---|---|---|
| | Test 1 | Test 2 | Test 3 | Test 4 |
| 100 | 3 | 3 | 1 | 1 |
| | 1 | 2 | 2 | 1 |
| 1000 | 2 | 2 | 2 | 1 |
| | 1 | 2 | 2 | 1 |
| 10,000 | 3 | 2 | 2 | 1 |
| | 1 | 3 | 2 | 1 |

Table 4.   Analysis of variance based on the data in Table 3.

| Response variable: drawn circle | Type III sum of squares | df | Mean square | $F$ | Sig. |
|---|---|---|---|---|---|
| Corrected model | 6.500 | 11 | .591 | 1.182 | .388 |
| Intercept | 73.500 | 1 | 73.500 | 147.000 | .000 |
| Distribution | 5.500 | 3 | 1.833 | 3.667 | .044 |
| No. of point | .250 | 2 | .125 | .250 | .783 |
| Distribution × no. of point | .750 | 6 | .125 | .250 | .950 |
| Error | 6.000 | 12 | .500 | | |
| Total | 86.000 | 24 | | | |
| Corrected total | 12.500 | 23 | | | |

significant effect on the number of drawn circles. For the second hypothesis in this section (H: the total number of points does not have an impact on the number of drawn circles), the $p$-value equals .783 which shows that the number of points does not have any impact on the number of drawn circles.

Based on Sections 4.1 and 4.2, the validity of the proposed algorithm is confirmed. For further verification and validation the power of the proposed algorithm is compared with a recent proposed algorithm for finding minimum covering circle in Section 5.

## 5.   Comparing computational results

In this study, the proposed algorithm was coded by the MATLAB 10 software, and the experiments were performed by a PC with an Intel Pentium Dual-Core (2.2 GHz) processor under Windows 7 using 3 GB of RAM. One of the factors in the proposed algorithm's capability is the computational speed in comparison with the other studies in this field. Therefore, the recent study was chosen for comparison. The results of the comparison are provided in Table 5. The position of the experimented points is also provided in Figure A4.

The results show that our algorithm extremely outperforms other examined algorithms. For more perception, the difference in CPU time with El-Tamimi and Al-Zahrani is visualized in Figure 6.

Based on the CPU time results, the CPU computational time was fitted using a function which may be utilized to predict the CPU time in different numbers of points. The fitting process was performed using the SPSS 18 software. Figure 7 shows the resultant polynomial function.

Table 5.   CPU time (seconds), comparing the results of the proposed algorithm with some algorithms in the literature.

| Algorithms | Number of points | | | |
|---|---|---|---|---|
| | 500 | 1000 | 3000 | 5000 |
| Proposed algorithm | .0311 | .1246 | 1.0990 | 3.2231 |
| El-Tamimi and Al-Zahrani (2012) | 43.08 | 172.11 | 1085.39 | 4240.23 |
| Patel (1995) | 39.224 | 83.234 | 301.45 | 1389.45 |
| Sarkar and Chaudhuri (1996) | 34.180 | 60.111 | 200.01 | 1004.87 |
| Das, Chakraborti, and Chaudhuri (2001) | 24.255 | 80.569 | 349.99 | 1012.2 |

**Estimated Marginal Means of TIME**

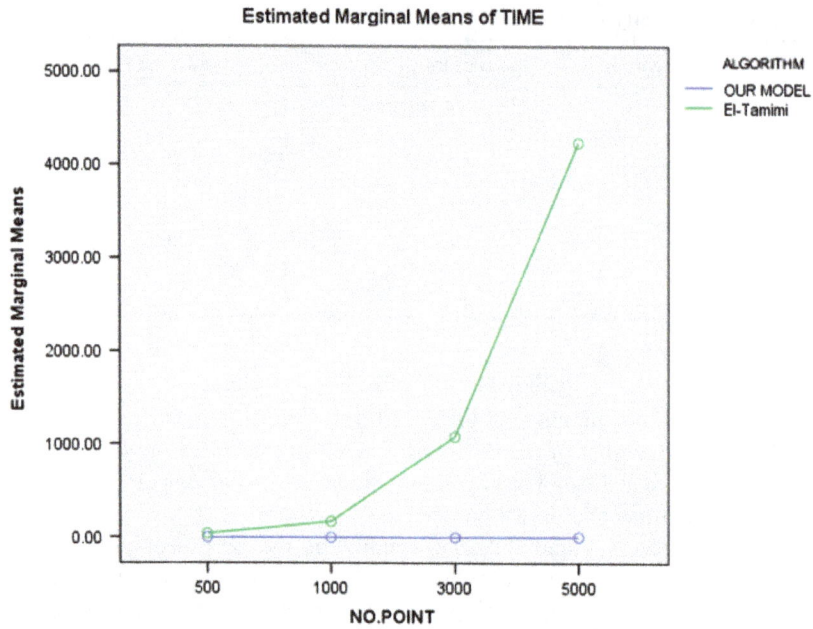

Figure 6.    Visualized CPU time (comparing algorithms with El-Tamimi and Al-Zahrani).

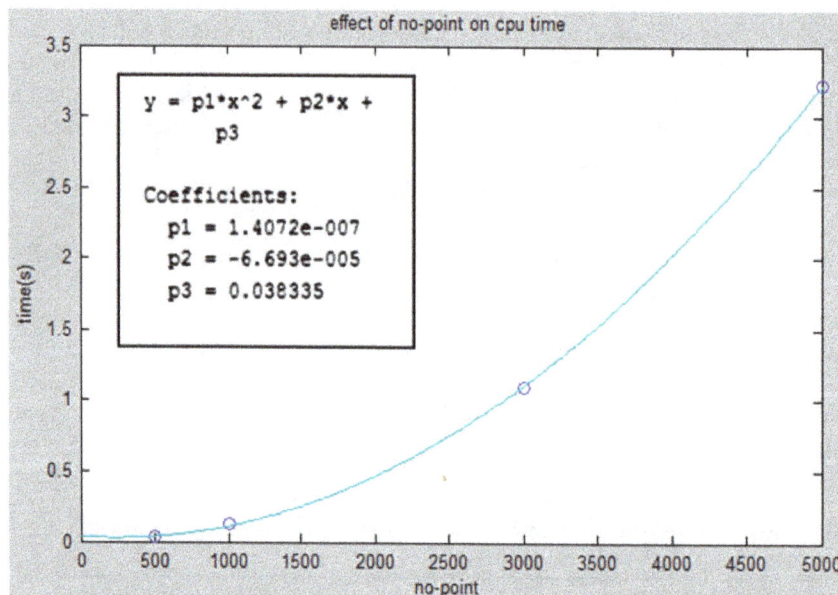

Figure 7.    Fitted polynomial function and function coefficient.

## 6.   Conclusion and suggestions for future research

The minimum covering circle is one of the most important problems in common location problems and optimization scopes. Minimum covering circle solutions may be employed to locate the essential locations of facilities such as hospital establishments or fire departments.

In this scope, many approaches exist; however, the difficult perception of each method, which need a high degree or the time-consuming overall processes due to the complexity of existing algorithms present much room for improvement.

In this paper, a simple method for determining the minimum covering circle is presented, which is based on geometric approaches. The model validity was considered by studying the coordinates of points with random numbers and different distributions.

The observations indicated that an increase in the scattering of points led to the achievement of the final solution in the first or second step. The distribution of points scattering was also assessed by the means of ANOVA which resulted in the conclusion that changes the number of total points which will have a significant effect on the computational speed in reaching the solution. In contrast, the points' scattering distribution (Test 1–Test 4) has no impact on the solution speed.

Compared with the recent proposed methods, the proposed algorithm of this paper has a much higher computational speed and less sensitivity to an increase in the total number of points.

The proposed method is designed for achieving the minimum covering circle in a two-dimensional space, regarding various applications, this algorithm can be developed for three- or more than three-dimensional spaces. Also, using the concept of reliability and the service level considering the available budget may be a new concept of study for further research.

## References

Chrystal, G. (1885). On the problem to construct the minimum circle enclosing n given points in the plane. *Proceedings of the Edinburgh Mathematical Society, 3*, 30–33.

Das, P., Chakraborti, N. R., & Chaudhuri, P. K. (2001). Spherical minimax location problem. *Computational Optimization and Applications, 18*, 311–326.

Dearing, P. M., & Zeck, C. R. (2009). A dual algorithm for the minimum covering ball problem in $R^n$. *Operations Research Letters, 37*, 171–175.

El-Tamimi, A. A., & Al-Zahrani, K. (2012). An improved quadratic program for unweighted Euclidean 1-center location problem. *Journal of King Saud University – Engineering, 25*, 161–165. Retrieved from http://dx.doi.org/10.1016/j.jksues.2012.04.003

Elzinga, J., & Hearn, D. W. (1972). Geometrical solutions for some minimax location problems. *Transportation Science, 6*, 379–394.

Fisher, K., Gartner, B., & Kutz, M. (2003). Fast smallest-enclosing-ball computation in high dimensions. In *Algorithms – ESA, volume 2832 of Lecture Notes in Computer Science* (pp. 630–641).

Gartner, B. (1999). Fast and robust smallest enclosing balls. In *Proceedings of the 7th Annual European Symposium on Algorithms, ESA*. In *Lecture Notes in Computer Science* (pp. 325–338). Zurich: Springer-Verlag.

Hopp, T. H., & Reeve, C. P. (1996). *An algorithm for computing the minimum covering sphere in any dimension* (Technical Report NISTIR 5831). Gaithersburg, MD: National Institute of Standards and Technology.

Li, L., Kabadi, S. N., & Nair, K. P. K. (2002). Fuzzy versions of the covering circle problem. *European Journal of Operational Research, 137*, 93–109.

Li, L., Kabadi, S. N., & Nair, K. P. K. (2005). Fuzzy disk for covering fuzzy points. *European Journal of Operational Research, 160*, 560–573.

Megiddo, N. (1983a). The weighted Euclidean 1-center problem. *Mathematics of Operations Research, 8*, 498–504.

Megiddo, N. (1983b). Linear-time algorithms for linear programming in R3 and related problems. *SIAM Journal on Computing, 12*, 759–776.

Nair, K. P. K., & Chandrasekaran, R. (1971). Optimal location of a single service center of certain types. *Naval Research Logistics Quarterly, 18*, 503–510.

Patel, M. H. (1995). Spherical minimax location problem using the Euclidean norm: Formulation and optimization. *Computational Optimization and Applications, 4*, 79–90.

Sarkar, A. K., & Chaudhuri, P. K. (1996). Solution of an equiweighted minimax location problem on a hemisphere. *Computational Optimization and Applications, 6*, 73–82.

Shamos, M. I., & Hoey, D. (1975). Closest point problems. *Proceedings of 16th Annual IEEE Symposium on Foundations of Computer Science*, 151–162. Berkeley, CA. doi:10.1109/SFCS.1975.8

Skyum, S. (1991). A simple algorithm for computing the smallest enclosing circle. *Information Processing Letters, 37*, 121–125.

Sylvester, J. J. (1857). A question in the geometry of situation. *Quarterly Journal of Mathematics, 1*, 79–86.

Welzl, E. (1991). Smallest enclosing disks (balls and ellipsoids). In H. Maurer (Ed.), *New results and new trends in computer science*. In *Lecture Notes in Computer Science* (Vol. 555, pp. 359–370). Berlin: Springer.

Yıldırım, E. A. (2007). Two algorithms for the minimum enclosing ball problem. *SIAM Journal on Optimization, 17*, 621–641.

Zhou, G., Tohemail, K. C., & Sun, J. (2005). Efficient algorithms for the smallest enclosing ball problem. *Computational Optimization and Applications, 30*, 147–160.

## Appendix 1

The test figures of Tables 1 and 3 are provided below. Figures A1–A3 show the algorithm's general solution processes when the number of points are 100, 1000, and 10,000, respectively. The blue, red, and green circles represent the first, second, and third (final) circle, respectively.

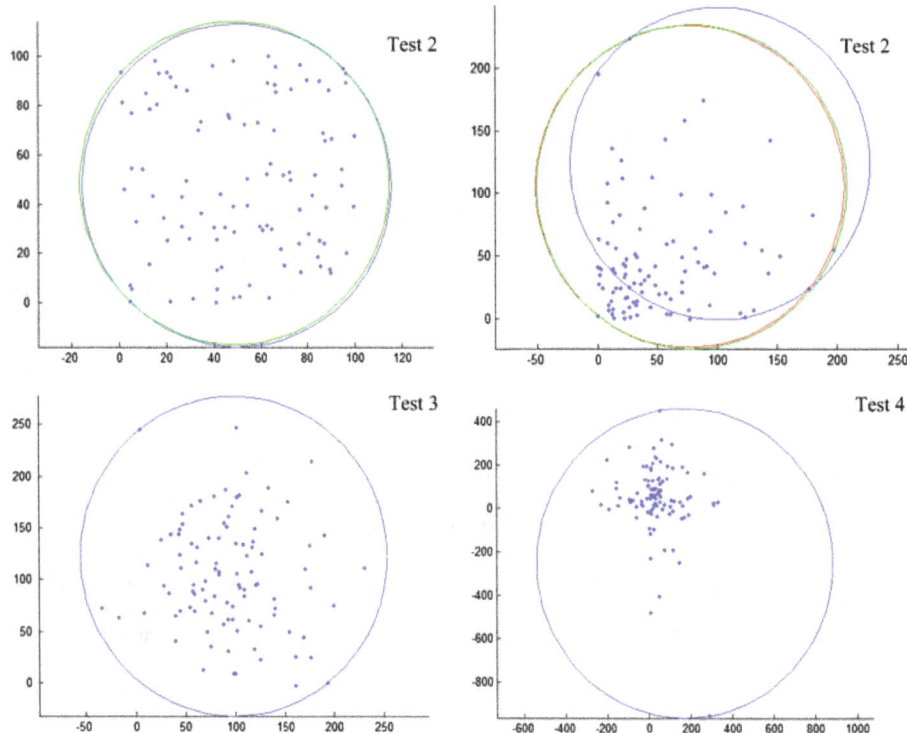

Figure A1.    Point scattering of the four executed tests with the total number of points of 100.

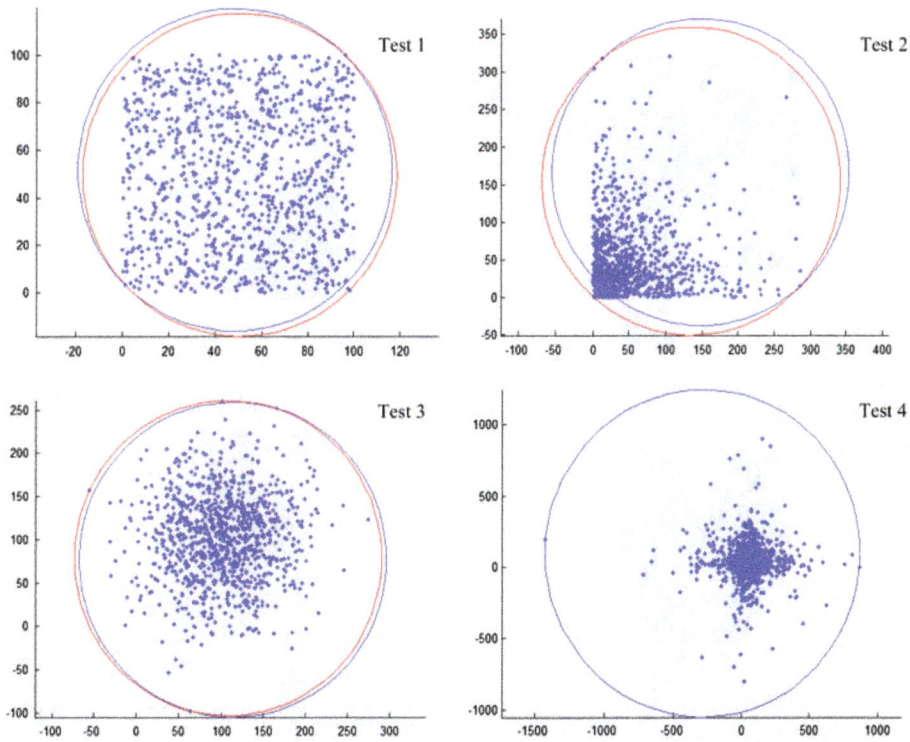

Figure A2.   Point scattering of the four executed tests with the total number of points of 1000.

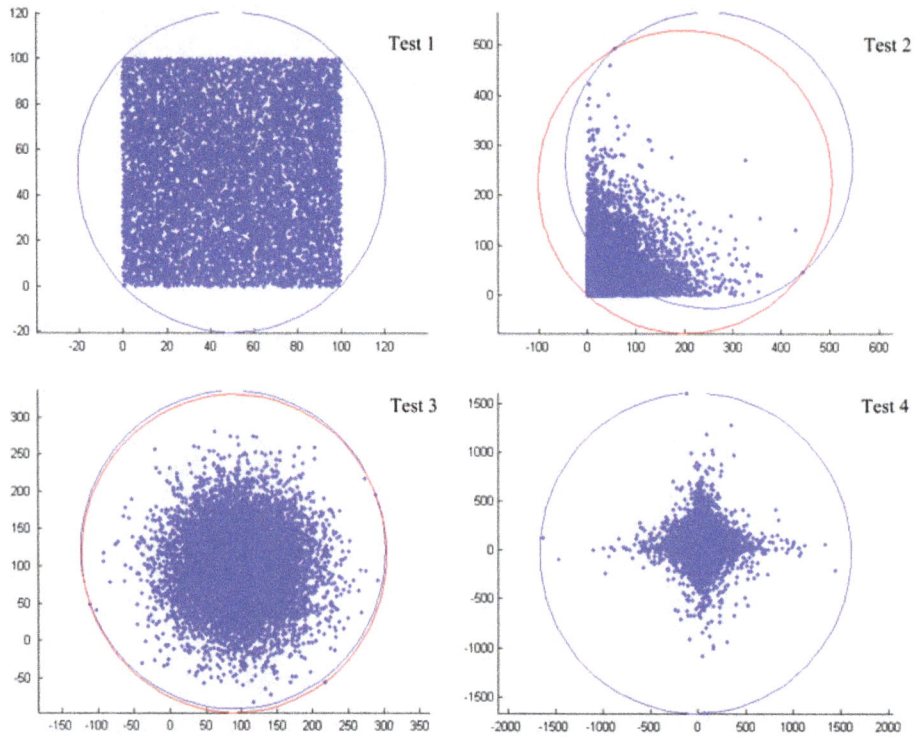

Figure A3.    Point scattering of the four executed tests with the total number of points of 10,000. Visual results are shown in Figure A4 in accordance to Section 5.

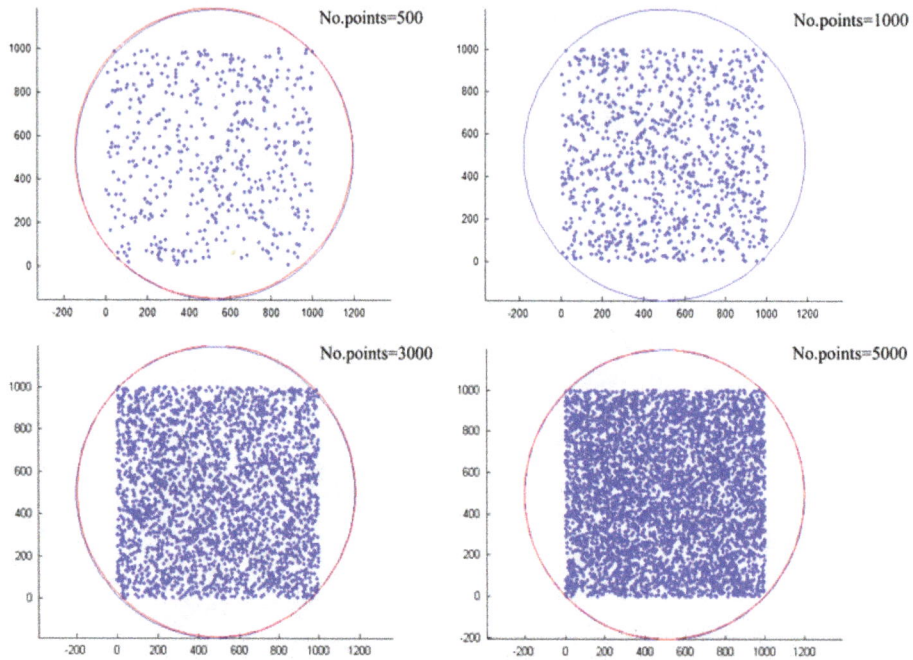

Figure A4.    The proposed algorithm results, in accordance to the comparison tests.

# 2

# Accelerated life tests as an integrated methodology for product design, qualification and production control: a case study in household appliances

M. Tucci*, F. De Carlo, O. Borgia and N. Fanciullacci

*IBIS Lab, Department of Industrial Engineering, University of Florence, Viale Morgagni 40, Florence 50134, Italy*

Cost reduction and product quality are key factors in the present competitive market. Product reliability is strongly correlated with customer satisfaction. Accelerated life tests (ALTs) represent a methodology able to investigate product reliability performance in a short time with respect to the classical testing methods in the design. The aim of this paper is the proposal and development of an integrated procedure based on ALTs in order to evaluate the reliability performance of a new product; the use of such testing methods is useful to systematically support the design and qualification phases and to help the service demand forecasting before the product launch. The smart choice of the acceleration parameter and the knowledge developed during the upstream phase in the product lifecycle can be useful and cost effective for its utilization in the downstream lifecycle, as an acceptable substitute of specific advanced final inspection methodologies. The methodology has been tested during the development of a new model of washing machine. The application of ALTs was able to stress one of the most important failure modes of the product, returning important validation data for the design phase and qualification phase and giving good expectation for its fruitful utilization for final inspection in production.

**Keywords:** accelerated life test; reliability performance; product development; final inspection

## 1. Introduction and literature review

The household appliances market has reached a very high level of competitiveness. The main reasons can be identified as the increase in the number of companies operating in this business area and in the expansion of emerging countries. The concentration of all these companies in the same market sector leads them to invest a lot of resources and time in product research and development (R&D) in order to persuade customers to choose their products. Continuous technological development and increase of capital investments are key requirements to follow customer needs and to keep pace with competitors (Minderhoud, 1999). Because of the competition, companies try to develop products with increasing performance and new features. Moreover, they try to reduce the time to market, to introduce innovation and to avoid losing their competitive advantage over competitors. As customer satisfaction remains the key point for a successful product, one of the main issues of the R&D activities is to design products with a

*Corresponding author. Email: mario.tucci@unifi.it

longer life far beyond the warranty expiration, ensuring long availability and proper functioning of the goods. Reliability and durability (Silvestri, Falcone, De Felice, & Di Bona, 2010) are important quality features of a new product which are, unlike static attributes, measurable on the bench with immediate results; they are dynamic and they must be estimated in two ways. The first approach is the reliability data analysis of similar products, applicable when there are similarities among them. For example, it is possible to estimate the assembly reliability using a reliability block diagram combined with a Weibull analysis (Kim, 2011; Zhai, Lu, Liu, Li, & Vachtsevanos, 2013).

The second way to evaluate reliability is through very long tests. New and innovative product reliability performances are commonly estimated through experimental tests, to simulate environmental life stress. Usually, such tests are time consuming and need dedicated resources. In the industrial setting, furthermore, the approach to these tests is not rigorous and scientifically robust. As a consequence, most of them don't provide the expected results and are not cost effective (O'Connor, 2001). Overlooked or poorly designed testing phases can lead to a product of poor quality, because some unapparent failure modes could go undetected (Yadav, Singh, & Goel, 2006).

In the household appliances field, and especially for washing machines, R&D achieved a very high level of reliability as the standard. Therefore, testing procedures of products have become very lengthy, since these are not prone to fail, while manufacturing companies need to estimate reliability performance more rapidly under the pressure of shorter time to market. Traditional reliability performance tests are no longer efficient enough and solutions are needed to accelerate them while not losing significance and correspondence with user experience. Already in the 70s, Nelson (1974) introduced the theory of accelerated life tests (ALTs) and showed its validity. ALTs are a category of test methodology able to assess reliability performance of a product or component in a short time, through the use of stress condition higher than the that in real time (Nelson, 2004). These tests are based on the hypothesis that there must be a relationship between the stress level and the product life. Hence, starting from a parametric form of this relation, a series of tests are performed to experimentally estimate it. Escobar and Meeker (2006) presented a review of ALTs and of their various applications. They are grouped based on: methods of acceleration, type and number of overstresses and life–stress relations. Several studies regarding different issues of ALTs are available in literature: Meeker (1984) compares compromise test plan and optimum test plan, Nelson (1980) used a step-stress procedure to further accelerate tests, while Fard and Li (2009) determined the optimum duration for each step. Yang (1994) deals with test durations using different censoring times for each level of overstress while Zhu and Elsayed (2013) propose an approach to ALT design with multiple stresses depending on different objectives and constraints. Yang and Zaghati (2006) increase the usage rate of a compact relay to accelerate its failure mechanism. ALTs can be used also to predict product field reliability using a relationship between ALT results and field returns as shown by Meeker, Escobar, and Hong (2008). There are also some ALTs' pitfalls, investigated by Meeker and Escobar (1998).

According to the presented studies, ALTs are typically implemented in the product lifecycle after the design phase and they integrate with the other production stages, as shown in Figure 1.

The diagram reported in Figure 1 shows the main stages that a new product goes through before it is put into production. ALTs are usually carried out after the final design step. The results of the experiments are mainly used for the finalization of the

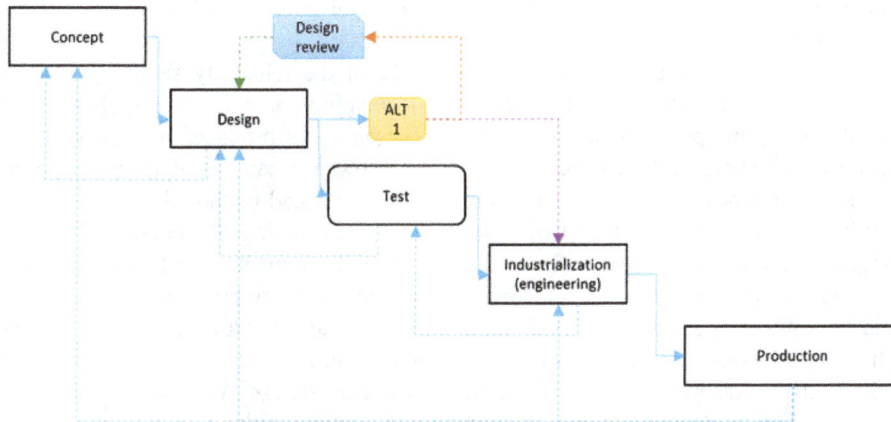

Figure 1.   Integration of ALTs in the product design and development stages. The solid lines indicate a material flow, while the dashed lines represent the flow of information. ALT1 indicates the traditional application of accelerated testing.


design, highlighting the critical issues that the normal tests fail to reveal. The experimental evidence can also be useful in the production stage, going hand in hand with the usual tests on prototypes.

Although a lot has been written in literature on ALTs, at the moment they do not represent the standard test methodology when long-term product reliability must be investigated. Even though ALTs are more efficient than the traditional tests, they involve risks and technical difficulties. In fact, they need a lot of resources for design and implementation; this often worries the producer who sees in ALTs a risky methodology, also because of the high cost and for the chance of getting wrong results. Actually, to obtain correct and accurate results, the analyst must thoroughly investigate each interesting failure mode and pay attention on the sample size and on the number and type of overstresses applied. The bigger the sample size, the more accurate the results are. On the other hand, wider sample sizes would increase the costs, since more items and operators should be involved and the total test time will be greater.

The definition of the reliability performance must not be limited only to the design phase, but it should be updated even during production. In fact, during its life cycle, a product might be subjected to productive modifications (suppliers change, production layout modification, assembly sequence change, etc.). All these specific factors could bring the process out of control affecting product reliability. Thus they must be discovered, investigated and revised in order to elaborate corrective actions and to bring the process back under control. Some authors faced these problems: Meeker and Hamada (1995) presented in a tutorial the most suitable statistical tools for each phase of product lifecycle, pointing out how ALT is mainly suitable for the upstream phase (namely, concept, design, test and industrialization) and only for unavoidable degradation failures and known infant mortality, while other methodologies are able to capture unexpected infant mortality, known accidental failures or unexpected accidental failures. For the latter, the early field data and the warranty return data are obviously important, but if the company waits until the availability of these data, that means the economical and image damage was already done. Hobbs (2000) proposes an ad hoc methodology, the highly accelerated stress screens (HASS) just for such kind of failures both in the design and production phases.

Recently, Isaic-Maniu and VodĂ (2009) reviewed the reliability sampling plans to be used as final inspection of the production, developing some specific applications of ISO 2859 sampling plans to survival performance on items produced and inspected by means of accelerated tests. Hence, as far as we know, we can say that in the normal practice white goods industry does not employ such advanced methodologies.

With so much effort devoted to the development and tuning of the ALTs for a new product, it appears convenient to try to extend their deployment also to the control of the production process. Limiting their use only to the reliability performance assessment could be a mistake, while they could be suitable to make other kind of defects emerge, such as production problems, defective quality controls and so on (O'Connor & Kleyner, 2011; Raheja & Gullo, 2012). To this extent, design ALTs are not the most appropriate tool in themselves, but they can become useful if the choice of the accelerated parameters is modified according to those employed in HASS (Hobbs, 2000): vibrations and temperature variations.

The aim of this study is to increase the efficiency of ALTs to support the production stage, choosing conveniently the accelerated parameters, so to produce vibrations and temperature increase. In this paper a new methodological approach to ALTs is proposed, which essentially consists of applying them to three steps of the product life: testing, engineering and production. In particular, ALTs will be used not only as presented in Figure 1 (where these tests are useful for a limited period of the product lifecycle) but also during production and even in the wider after-sale phase (service, warranty, warehouse management, etc.).

As a case study, the reliability performances of a domestic washing machine were investigated using ALTs during the production phase.

The remainder of the present paper is as follows: after a description of the methodology in the methods section, the following results section presents a brief introduction on the case study, the experimental procedure adopted and the experimental outcomes. The discussion section follows, analysing the main achievements and showing a further possible area for ALTs deployment. Finally, some conclusions and interesting future developments are shown with a suggestion of a new possible way to use ALTs together with final inspections for statistical quality control (SQC).

## 2. Methods

ALT is a methodology based on the assumption that there is a relationship between the stress level and product life. The modelling function depends on two main factors: the component type under analysis and the stress type. Typically, the lifetime of a mechanical component is affected by operating parameters such as temperature, vibrations, environmental conditions and load. Literature review suggests the 'Inverse Power Law' model as the most suitable and flexible model for mechanical components if the stress is non-thermal. Escobar and Meeker (2006), Yang (2007) indicate the Inverse Power Law as the most suitable relationship that describes the dependence of the washing machine life $L$ on mechanical stress.

$$L = \frac{A}{S^B} \tag{1}$$

where $L$ is the nominal life, $S$ is the stress (in this case a mechanical load) while $A$ and $B$ are constants, dependent on material properties, product design, etc.

The independent variable chosen for the reliability model was the number of standard washing cycles, even if in the following formulas it is referred to as $t$, recalling its link to the lifetime. As it will be explained in the next section, the tests were 'Type I censoring' with a censoring time fixed at 500 cycles. The estimation of the reliability function at the user level, obtained from the overstress values, was carried out with the software ALTA© (ReliaSoft Corporation, Tucson, Arizona, USA). To assess the reliability of the machines at the user level (normal operating conditions), it was necessary to find the probability function that best represents the data collected at each level of overstress; in this case, a two-parameters Weibull function was chosen. The probability density function (p.d.f.) is shown in Equation (2):

$$f(t) = \frac{\beta}{\alpha^\beta} t^{\beta-1} e^{\left[\left(\frac{t}{\alpha}\right)^\beta\right]}, \quad t > 0 \tag{2}$$

where $\alpha$ is the scale parameter, also called characteristic life (it is the life at which 63.2% of the population will be failed) and $\beta$ is the shape parameter. When $\beta > 1$ ($\beta < 1$), the distribution has an increasing (decreasing) hazard rate, while for $\beta = 1$ it has a constant hazard rate.

The probability that a washing machine will fail during an interval $[0, t]$ is defined by the cumulative distribution function $F(t)$. The latter is obtained by integrating the p.d.f. within the interval $[0, t]$:

$$F(t) = \int_0^t f(t)dt = 1 - e^{-\left[\left(\frac{t}{\alpha}\right)^\beta\right]} \tag{3}$$

As a consequence, the reliability function $R(t)$, derived by the cumulative function, is given by:

$$R(t) = 1 - F(t) = e^{-\left[\left(\frac{t}{\alpha}\right)^\beta\right]} \tag{4}$$

In some cases, when performing ALTs, we can define a useful parameter called accelerating factor. It is defined as (5):

$$A_f = \frac{t_p}{t_p'} \tag{5}$$

where $t_p$ is the 100th percentile of the life distribution at user stress level and $t_p'$ is the 100th percentile of the life distribution at the overstress level. Using $A_f$, it is possible to calculate user-level reliability performance like test hours (cycles), life percentiles, etc. from those obtained with higher stress levels. For example, known the test hours with high stress level, $A_f$ can be used to obtain the equivalent test hours at the user stress level. The choice of an acceleration factor is based on the hypothesis that the reliability function for all stress levels has the same shape parameter; because we are using a Weibull function, it means that $\beta$ must be the same for all the stress levels. As a consequence, once that $\beta$ is calculated for a certain level of stress, the scale parameter of the Weibull function for each stress level is the only missing parameter to be estimated.

The two Weibull parameters ($\alpha$ and $\beta$) were calculated using the maximum likelihood estimation (MLE) as it is the most suitable method with several censored data (Nelson & Meeker, 1978). The chosen unilateral lower confidence interval is 97%.

The reliability function at user stress level was obtained using data from tests implemented with three different mechanical overstress levels. As previously stated, to find the relationship between life and stress, the Inverse Power Law correlation function (1) was used.

In this case, the acceleration factor between two stress levels is,

$$A_f = \left(\frac{S'}{S}\right)^B \tag{6}$$

where $S'$ is the highest stress level.

## 3. Results

As already mentioned in the introduction, the aim of the present study is to show how ALTs can support the industrialization phase, providing useful information about the long-term product reliability. These results are critical, for example, for a more correct initial sizing of inventories, for the prediction of the extents of the after-sales service, for understanding the issue of service calls management under warranty. The possible representation of the interaction of ALTs with the life cycle of the product is shown in Figure 2.

Figure 2 shows how this new ALTs application allows to interact directly on the production process, as a feedback. The comparison between the results of ALT1 and ALT2 allows you to assess the long-term performance of production with those foreseen in the design. For any discrepancy, you will be able to undertake a control and corrective action.

In addition, the predictions of the ALT2 can be compared with the actual performance of the product, rising from the field. So you will be able to promptly detect abnormalities and you might intervene with measures at different levels: from design to engineering, even up to the redesign of ALTs, if problems arise in the experimental phase of the accelerated testing.

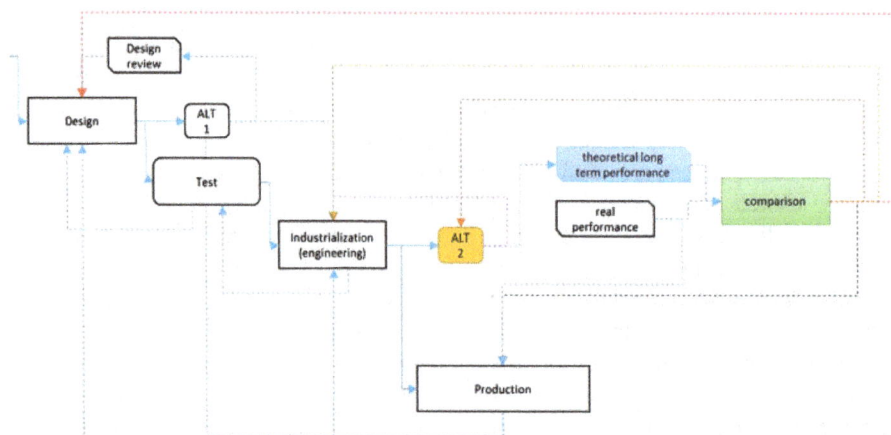

Figure 2. Possible interactions between industrialization ALTs and the other phases of product development. The second application of ALTs (ALT2) can have many interactions (dashed lines) with the other stages of development, making it possible to improve the process as a whole.

### 3.1.  Case study

To verify the goodness of fit of ALTs as a greater support to the design, engineering and production steps, we applied them to a real case, to sustain the process of developing a new model of washing machine. The manufacturer decided to launch onto the market a new model developed on the basis of an existing top product with important innovations regarding the enhancement of the maximum drum capacity without altering the case size. This new specification had some relevant consequences since it involved the redesign of one of the most important components of the product: the oscillating system.

Reliability performance of the unchanged components was previously studied by the manufacturer, and a lot of information coming from warranty data analysis was available. So the real challenge was to assess the reliability performance of the new oscillating system during the industrialization phase, but the classical tests were unable to highlight any sources of criticality.

The manufacturer was interested in this issue for several reasons. First of all there was the need to validate the design and industrialization steps through an experimental approach in order to discover the real limit of the new solution and, if necessary, redesign the new component.

Another important goal was to adopt a methodology in order to keep the quality of the manufacturing process under control. Line set-up, suppliers changes, process capability demonstrated in the past to be important issues that could influence the company's product reliability. Therefore, some deviations or drifts should be evaluated as quickly as possible in order to avoid consequences on customer satisfaction due to any increase in faults. Moreover, an early and fast reliability performance assessment can be always very useful in the viewpoint of warranty analysis and service demand forecasting and planning.

The application of ALT to this case study is not trivial because, according to international scientific publications, no study was published regarding the specific application of ALTs to the oscillating system of a washing machine. Only a paper was presented by Park, Park, Kim, and Cho (2006) regarding the implementation of ALTs on the drain pump of a washing machine. Nevertheless, the research expertise of the authors in the context of accelerated degradation and testing as in De Carlo, Borgia, and Tucci (2014) and the comparison between in-house and field reliability have allowed us to design and fulfil the tests, see Borgia, De Carlo, Tucci, and Fanciullacci (2013). Also the ability to extend the applications of known methods and to apply new approaches to existing methods falls within the competence of the authors, as shown in De Carlo, Borgia, and Tucci (2011).

### 3.2.  Experimental procedure

As said, the most important role for the identification of the product reliability performance using ALT is played by the testing phase. Before performing tests, it is fundamental to plan and determine the characteristic of the sample and the stressscheme. Planning activities were performed according to the specifications, the budget and the time constraints imposed by the manufacturer. These led to: (1) the choice of the component to be tested; (2) the failure mode to be stressed and the different levels of overstress; and (3) the test length and the number of samples tested at each level.

ALTs focused on the new mechanical oscillating group (formed by the tub and the drum) being the main innovation introduced in the new model, with an unknown behaviour. The new mechanical group was bigger than the previous one, in order to contain a larger quantity of clothes. The improvements were: a new material, a decreased thickness of the drum and of the tub and several new reinforcement ribs. The design changes implied also a lower gap between the tub and the drum both in static and dynamic conditions. It is important to analyse the latter parameter because it influences the probability that the two components collide, causing a failure: the wider the distance between the two components, the lower the probability that the drum comes into attrition with the tub. According to the root cause analysis of the field repairs, the contact between these two components is one of the most common failure modes. Hence, we decided to perform tests with only one overstressing parameter. The mechanical overstress chosen was a load imbalance placed inside the drum, since it intensifies the drum deformation and its attitude to get in contact with the tub. It also overloads the bearing group, producing heat and possibly deforming its polymeric housing.

The drum imbalance is normally regulated by the electronic control unit that stops the washing cycle if it exceeds a threshold value decided by the manufacturer. The maximum user-allowable imbalance was the baseline. Three equally spaced levels of overstress for testing the washing machines were elected: 165, 200 and 237% of the baseline. These values came from the results of a previous FEM analysis and were confirmed by the designer of the assembly.

The three overstress levels were applied during a programmed washing cycle that, after a marketing survey, was identified to be the most common washing cycle performed by the standard user. It consists of two phases: 90 min of low speed cycles and 20 min of spin cycle. The experimental measures acquisition was carried out only during the spin phase, when the group reaches its most severe dynamic conditions (maximum rotating speed).

It should be added that, according to the results of some preliminary tests, it was found that the usage rate does not affect the failure mode considered in this study. Hence, cycles carried out at low speed can be neglected without altering the results.

For each level of overstress, eight washing machines were used. The sample sizes were chosen accordingly to the planning sets proposed by Yang (1994, 2007), and Yang and Jin (1994) with the 'compromise test plans' and to the requirement of the company in terms of resources and time dedicated, while overstress levels were imposed by designer knowledge. With the 'compromise test plan' methodology, test variables were chosen to minimize the asymptotic variance of the estimate of a life percentile at the user stress level. Given as an input the higher stress level, the censoring time (500 cycles), the total number of machine to be tested (24) and the shape ($\beta$) and scale ($\alpha$) parameter at the user and at the highest stress level (derived from historical data), we could calculate the asymptotic variance of the MLE of the mean life of the Weibull distribution at the user stress:

$$\text{Var} = \frac{\sigma^2}{n} V \tag{7}$$

where $\sigma = 1/\beta$, $n$ is the total number of units tested and $V$ is the standardized variance defined by Yang and Jin (1994). Hence, because $\sigma$ and $n$ are predetermined, we can calculate the lower and the intermediate stress levels and the number of units to be allocated at each stress level by minimizing $V$. Yang (2007), using a numerical method, presents a table where, for various sets of input parameters, the optimum value of $V$ is calculated.

The sample sizes obtained mathematically were also dynamically compared with the experimental evidences obtained with previous tests.

Tests were designed as right censored by imposing 500 cycles as censor 'time', which corresponds to an average of 2 years of use. For each test, every 50 cycles, a cycle called 'characterization' was performed. During this cycle, the load was set at the user level; afterwards, the machine was loaded again with the imbalance already established. The aim of the characterization cycle will be explained in the next paragraph.

### 3.3. Experimental set-up

A laboratory with three positions for the simultaneous testing of washing machines was set up in order to implement ALTs. In each test station, the washing machines were put into metal cages for safety reasons. In fact, in case of contact between the drum and the tub, especially with high imbalances, the drum coupling could break and could be thrown outside the washing machine by its inertia, with severe risks for the safety of workers.

During the tests, three different types of sensors were set on each washing machine in order to monitor those parameters useful for a following degradation analysis. This would be performed using also the data acquired during the 'characterization' cycle.

The measurement system consisted of three types of sensors, each one with its hardware and software acquisition equipment. To measure the drum–tub gap, four sensors were used: three longitudinally placed on the top of the tub, and the fourth at the back side.

Vibrations were measured with three sensors which monitored the acceleration along the axes of a Cartesian tern with origin on the top of the tank, $z$-axis vertically upwards, $x$-axis parallel to the axis of the drum and directed anteriorly. The vibration acquisition system consisted of a data logger with signal-processing capabilities connected to the three accelerometers.

The bearing temperature was measured by two thermocouples in contact with the bearing. The two values were then averaged to obtain a more reliable value.

Test activities were preceded by a calibration phase in which the measuring instruments and their software were set to ensure a proper operation. Finally, the data obtained from experimental tests were appropriately post-processed in order to make them suited for an eventual degradation analysis.

### 3.4. Experimental outcomes

ALTs were performed on 24 machines equally distributed over the three overstress levels according to the standard methodology. The first important outcome was the identification of a new (hidden) failure mode (in addition to those already known by the not accelerated tests and experience) that never occurred before. It involved the new oscillating system and it occurred seven times out of a total of eight failures. Although this fact is not a proof of the equivalence of ALT and HASS, it is an empirical evidence of some similarities, at least for industrial product with significant mechanical components, where the vibration and heat produced by the functioning of the product itself can be used as stress factors for other components, even revealing hidden failure modes.

The results obtained from the three overstress levels are shown in Table 1; next to the number of cycles performed by the machine, in brackets, it is indicated if the test finished without any failure (C – censored) or if a failure occurred (F – failed).

Table 1.  Outcomes of the 24 tests (censored C, failure F). The test is right censored at 500 cycles (corresponding to 2 years of use).

| Machine | Load imbalance | | |
| | 65% | 80% | 95% |
|---|---|---|---|
| 1 | 500 (C) | 500 (C) | 50 (F) |
| 2 | 500 (C) | 500 (C) | 350 (F) |
| 3 | 500 (C) | 500 (C) | 250 (F) |
| 4 | 500 (C) | 500 (C) | 300 (F) |
| 5 | 500 (C) | 350 (F) | 300 (F) |
| 6 | 500 (C) | 500 (C) | 350 (F) |
| 7 | 500 (C) | 500 (C) | 500 (C) |
| 8 | 500 (C) | 500 (C) | 400 (F) |

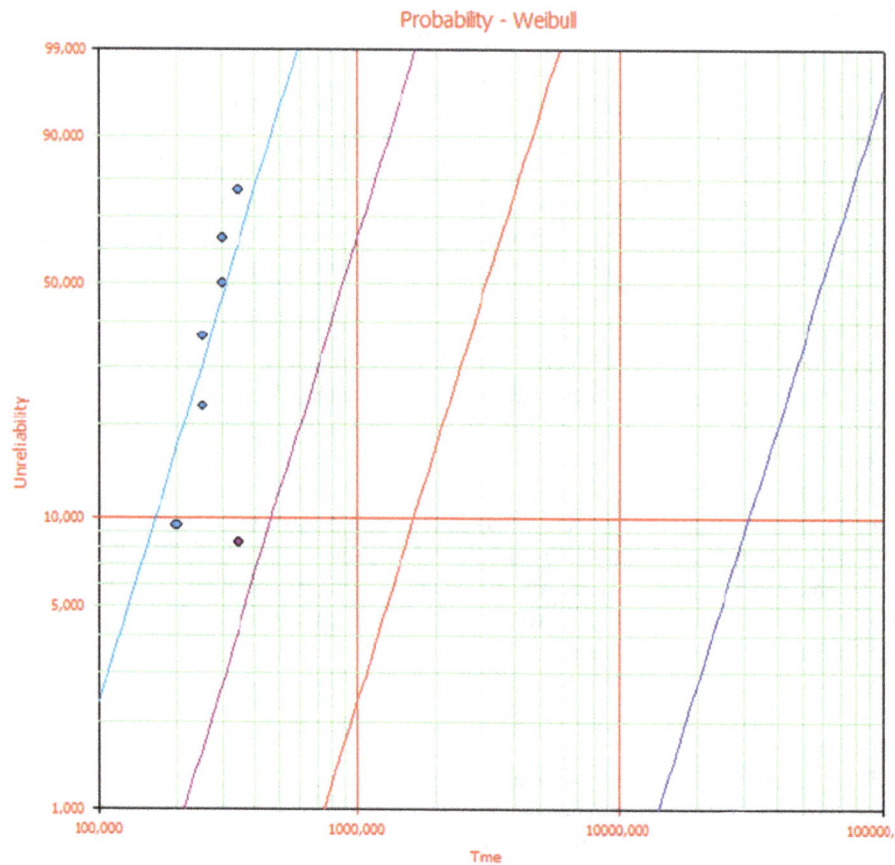

Figure 3.  Weibull chart of the four life data distribution. From the left to the right you can see 237, 200, 165 and 100% distributions. Parallel lines indicate that the distributions have the same shape factor.

Data represented in Table 1 are used to estimate the reliability distribution at the user stress level. As already mentioned, a two-parameters Weibull distribution was used to fit the life data obtained during testing activities at the three overstress levels.

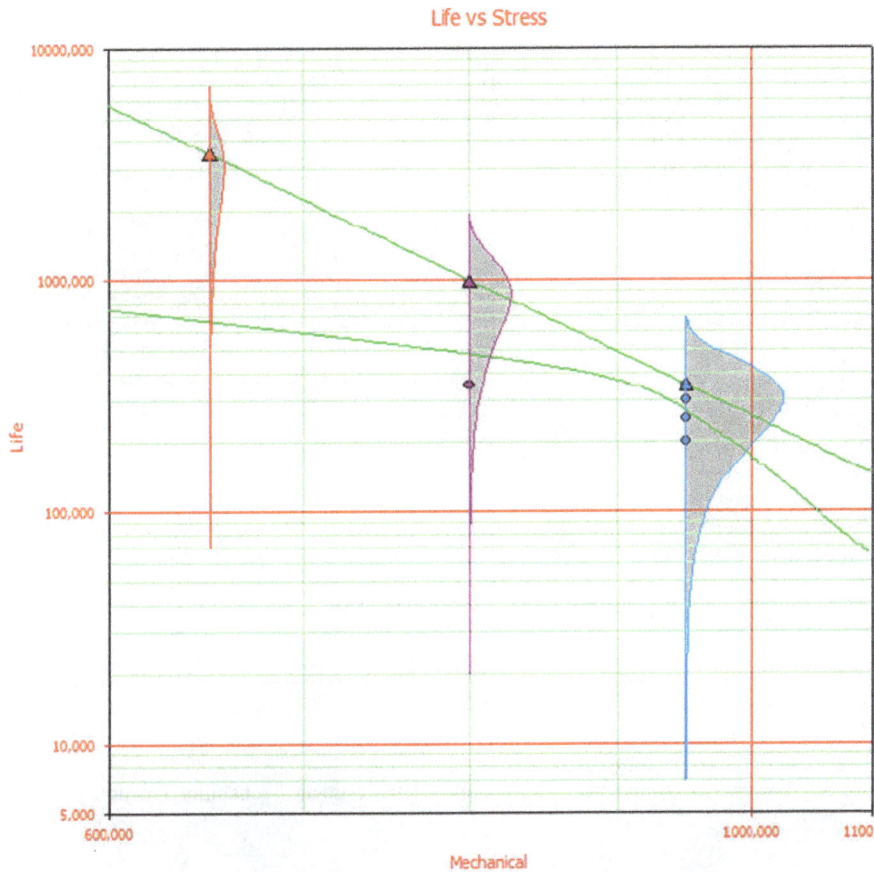

Figure 4.  Correlation between the three different levels of overstress (abscissa) and the component life (ordinate) with the 97% lower confidence interval of the frequency distributions.

Based on the maximum likelihood method, the shape parameter $\beta$ was calculated. Its value is 2.97 and it is equal in every distribution at all levels of overstress (165, 200 and 237%). The assumption of constant shape parameter is evident from the graphical representation of the four straight parallel lines on the Weibull chart (Figure 3).

Hence, the reliability at the user stress level follows a Weibull distribution with $\beta = 2.97$ and $\alpha = 53,166$ cycles. This distribution is derived from the correlation between life and mechanical stress obtained by the inverse power law (Figure 4).

Fundamental for service planning were the reliability performances at 500 and 1250 cycles; in fact, 500 cycles correspond to two years of standard usage – the normal warranty period – while 1250 cycles represent an extended warranty (5 years of usage). As regards these values, the estimated reliability performance was very close to one. It is important to underline that the values found were the average ones; as a precautionary measure, we have adopted the lower limit of the confidence interval, because it indicates the pessimistic reliability values for each number of cycle. Hence with the unilateral confidence level chosen (97%) there was a lower reliability of 99.2% for 500 cycles and 89.78% for 1250 cycles (Figure 5). Furthermore the user level was set to maximum

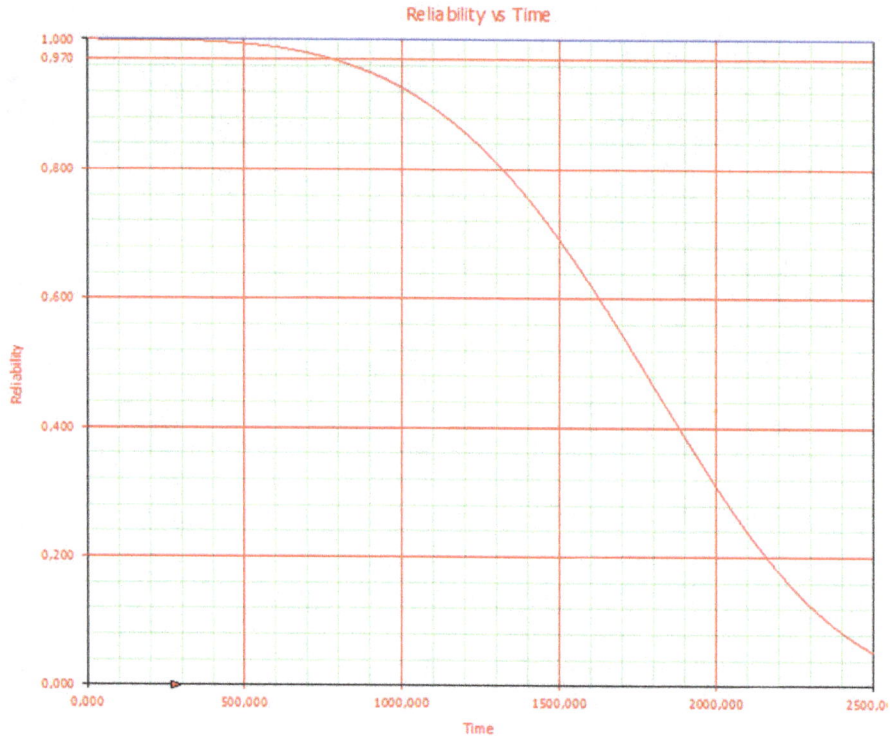

Figure 5. Lower 97% confidence interval limit of the estimated reliability function at the user stress level.

unbalance allowed by the control unit, which is not the average unbalance that loads the product during its life.

The value of 97% was chosen because it is the best reliability performance level to obtain a good trade-off between production costs and service costs. In accordance with the manufacturer, reliability performance specification was then used to calculate the number of cycles that ensures a reliability performance of 97%: 800 cycles that correspond to 3 years and 2 months. The mean life was equal to 1248 cycles (about 5 years).

As a further elaboration of the test results, it was possible to obtain the reliability surface as a function of the stress level and of the number of the performed cycles (Figure 6).

## 4.  Discussion

The aim of the study was to develop a test methodology able to simultaneously identify the reliability performances of a domestic washing machine and to use it to sustain the industrialization phase. This procedure was carried out fulfilling deadlines and tight budget constraints.

By applying ALTs in an innovative manner (after the engineering phase and in parallel to the production), the experimental results identified some critical events related to the production phase of the new mechanical assembly. Satisfying the test duration requirements, ALTs gave results in less than 30% of the traditional test time. During the

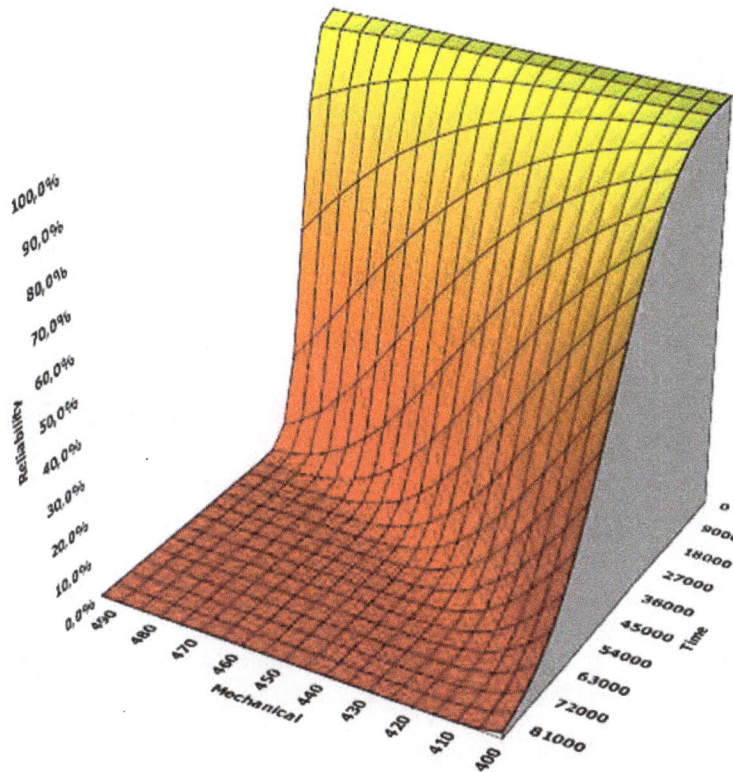

Figure 6.　Reliability surface. In a three-dimensional space, the reliability values depend on the number of cycles performed and on the level of stress applied. By this graphical representation, it was simple to assess the reliability performance of the washing machine defining the stress level and the number of cycles.

FMEA analysis, particular attention was dedicated to determinate a realistic overstress that could lead to a truthful failure mode (Lu, Loh, Brombacher, & Ouden, 2000). In fact, although imbalance values that cause failures were higher than the maximum value that was normally allowed, their systematic occurrence should led the manufacturer to well investigate the causes of such phenomena. The failure mode found was catastrophic because it could stop the functioning of the washing machine and cause severe damages. Although reliability performance was not alarming, they were useful to give indications to production managers and designers about the root cause of such a problem which was identified in a supplier change.

ALTs have, therefore, demonstrated their potential in identifying production problems, but might they also be integrated with field data? As a matter of fact, it may happen that the reliability values obtained through the life test performed in the laboratories inside the company (in-house) are different from those that come from field (estimated using the service data). This difference can be due to the fact that ALTs were designed incorrectly because they do not exactly represent the actual product use conditions. Comparing the in-field and the in-house reliability data could give a quick validation about the goodness of the test design that took place in the early product stages, giving the opportunity to the company for continual improvement of these tools along the lifecycle of the product.

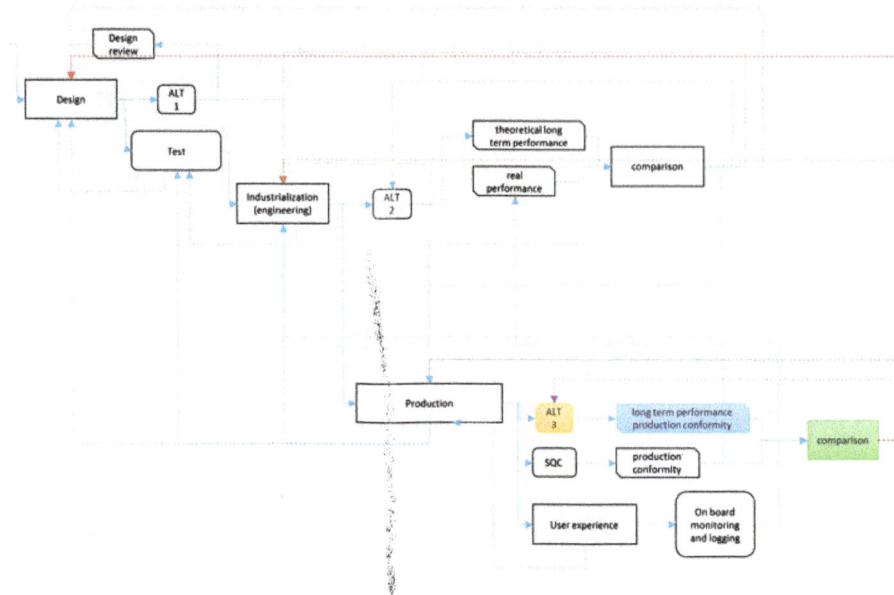

Figure 7. A possible third application of ALTs (ALT3). In the figure is shown how the experimental results of accelerated tests can be integrated in to the SQC to predict the long-term performance of the goods produced.

The favourable results and the demonstrated ability of these tests to reveal hidden failure modes as well, prompted us to think that ALTs could be a tool to be used even after the industrialization phase. Their results could be fundamental for the ultimate estimation of the optimal warranty period and for the full sizing of the supply system involved with the service. If the warranty period is known, the manufacturer could plan and manage all the activities necessary for the service, reducing costs while ensuring a high quality of the service offered. Consider, for example, the ability to plan the production and distribution of all the spare parts and their stocks or the possibility to extend the warranty period.

To sum up, in the previous paragraphs we introduced the typical use of ALTs, which reflects the ideas on which they were originally developed (Figure 1) and the use of ALTs as they have been implemented in the present work by the authors (Figure 2). The results of the study are very encouraging and have also suggested to the authors to propose an additional application area of ALTs to support industrial production. This third mode includes the steps of production and of use by end-users, as shown in Figure 7.

The idea behind this application is that the accelerated testing provides an insight into the long-term performance of the goods under investigation. This investigation, usually made on prototypes or first series' items, can be also extended to the production stage, in which it could stand beside the widespread techniques of SQC. This deployment is feasible because of the small sample size, compatible with the nature of a destructive test performed on production batches; it is also effective because they are able to reveal hidden failure modes, differently from traditional SQC in final testing. Finally, differently from the field data returned analysis, the production ALTs (ALT3 in Figure 7) make it possible to intervene in the production more promptly. In this way the manufacturer is able to proactively develop and maintain the production under control, contrasting negative production trends. As a further implementation, if the product is equipped with a system of on-board data logging and monitoring, the information downloaded from the assets during technical assistance may provide a benchmark for comparing the ALTs results. Ultimately, the possibilities of integration of ALTs during the phases of design, engineering and production are many and allow you to fully support each of these phases.

The use of ALTs as a tool to be applied in parallel with SQC is still in progress in the case study, as the production of the new model of washing machine has just started yet. The experience done in the design phase and the hidden defects discovered and corrected, albeit not related to the failure mode studied, gives an excellent outlook on its effectiveness for the planned downstream utilization.

ALTs showed their potential but also highlighted some critical issues and limitations. For example it is essential to have a dedicated laboratory for the experimental measurements and to have scrupulously investigated the aspects of all the possible failure modes (e.g. by means of an FMEA). After finding all possible failure modes, before running ALTs, you need to identify the most frequent or catastrophic ones through a deep Pareto analysis. It should analyse and classify all the data coming from the service reports. It is therefore clear which enormous effort must be taken, even before the execution of the tests. A great work across all the departments involved in the product development (design, materials, structural, etc.) is needed. Consequently, a considerable amount of resources and time is necessary to develop the preparation phase of the tests, in addition to their execution. Companies that have invested in this direction will be able to easily overcome the problems related to implementation and evaluation of reliability performance such as determination of threshold values and the choice of parameters. The results obtained have wide confidence intervals; it would be interesting to increase sample sizes for the three overstress levels in order to obtain more confident outcomes. In particular, it would be interesting to increase the number of machines tested at the lowest level of stress (165%). Another alternative could be the implementation of unequally spaced overstress levels, increasing only the lower stress level; in this way we could obtain more failure data without incrementing the number of washing machines tested.

A future development of the case study will be to characterize the standard functioning of the washing machine using the data acquired with the sensors and logged in the control unit; these can be also used to compare, at any time, the standard performance with those of machines just produced. Any difference found will be the issue of investigations aimed at finding non-compliances of production process.

## References

Borgia, O., De Carlo, F., Tucci, M., & Fanciullacci, N. (2013). Service demand forecasting through the systemability model: A case study. *International Journal of Engineering & Technology, 5*, 4312–4319. ISSN:0975-4024.

De Carlo, F., Borgia, O., & Tucci, M. (2011). Risk-based inspections enhanced with Bayesian networks. *Proceedings of the Institution of Mechanical Engineers, Part O: Journal of Risk and Reliability, 225*, 375–386.

De Carlo, F., Borgia, O., & Tucci, M. (2014). Accelerated degradation tests for reliability estimation of a new product: A case study for washing machines. *Proceedings of the Institution of Mechanical Engineers, Part O: Journal of Risk and Reliability*, in press.

Escobar, L. A., & Meeker, W. Q. (2006). A review of accelerated test models. *Statistical Science, 21*, 552–577.

Fard, N., & Li, C. (2009). Optimal simple step stress accelerated life test design for reliability prediction. *Journal of Statistical Planning and Inference, 139*, 1799–1808.

Hobbs, G. K. (2000). *Accelerated reliability engineering: HALT and HASS*. Chichester: Wiley.

Isaic-Maniu, A., & VodĂ, V. G. (2009). Reliability sampling plans: A review and some new results. *Journal of Applied Quantitative Methods, 4*, 190–201.

Kim, M. C. (2011). Reliability block diagram with general gates and its application to system reliability analysis. *Annals of Nuclear Energy, 38*, 2456–2461.

Lu, Y., Loh, H. T., Brombacher, A. C., & Ouden, E. (2000). Accelerated stress testing in a time-driven product development process. *International Journal of Production Economics, 67*, 17–26.

Meeker, W. Q. (1984). A comparison of accelerated life test plans for Weibull and lognormal distributions and type I censoring. *Technometrics, 26*, 157–171.

Meeker, W. Q., & Escobar, L. A. (1998). Pitfalls of accelerated testing. *IEEE Transactions on Reliability, 47*, 114–118.

Meeker, W. Q., Escobar, L. A., & Hong, Y. (2008). Using accelerated life tests results to predict product field reliability. *Technometrics, 51*, 146–161.

Meeker, W. Q., & Hamada, M. (1995). Statistical tools for the rapid development and evaluation of high-reliability products. *IEEE Transactions on Reliability, 44*, 187–198.

Minderhoud, S. (1999). Quality and reliability in product creation – Extending the traditional approach. *Quality and Reliability Engineering International, 15*, 417–425.

Nelson, W. B. (1974). A survey of methods for planning and analyzing accelerated tests. *IEEE Transactions on Electrical Insulation, EI-9*, 12–18.

Nelson, W. B. (1980). Accelerated life testing – Step-stress models and data analyses. *IEEE Transactions on Reliability, R-29*, 103–108.

Nelson, W. B. (2004). *Accelerated testing: Statistical models, test plans, and data analyses*. Hoboken, NJ: Wiley-Interscience. Retrieved from www.wiley.com

Nelson, W. B., & Meeker, W. Q. (1978). Theory for optimum accelerated censored life tests for Weibull and extreme value distributions. *Technometrics, 20*, 171–177.

O'Connor, P. D. T. (2001). *Test engineering: A concise guide to cost-effective design, development, and manufacture*. Chichester: Wiley.

O'Connor, P., & Kleyner, A. (2011). *Practical reliability engineering*. Chichester: Wiley.

Park, S. J., Park, S. D., Kim, K. S., & Cho, J. H. (2006). Reliability evaluation for the pump assembly using an accelerated test. *International Journal of Pressure Vessels and Piping, 83*, 283–286.

Raheja, D. G., & Gullo, L. J. (2012). *Design for reliability*. Hoboken, NJ: Wiley.

Silvestri, A., Falcone, D., De Felice, F., & Di Bona, G. (2010). New reliability allocation methodology: The integrated factors method. *International Journal of Operations and Quantitative Management, 16*, 67–85. ISSN:1082-1910.

Yadav, O. P., Singh, N., & Goel, P. S. (2006). Reliability demonstration test planning: A three dimensional consideration. *Reliability Engineering & System Safety, 91*, 882–893.

Yang, G. (1994). Optimum constant stress accelerated life test plans. *IEEE Transactions on Reliability, 43*, 575–581.

Yang, G. (2007). *Life cycle reliability engineering*. Hoboken, NJ: Wiley.

Yang, G., & Jin, L. (1994). Best compromise test plans for Weibull distributions with different censoring times. *Quality and Reliability Engineering International, 10*, 411–415.

Yang, G., & Zaghati, Z. (2006). *Accelerated life tests at higher usage rates: A case study*. In *Proceedings of the Reliability and Maintainability Symposium* (pp. 313–317). Philadelphia, PA: IEEE.

Zhai, L.-Y., Lu, W.-F., Liu, Y., Li, X., & Vachtsevanos, G. (2013). Analysis of time-to-failure data with Weibull model in product life cycle management. In A. Y. C. Nee, B. Song, & S.-K. Ong (Eds.), *Re-engineering manufacturing for sustainability* (pp. 699–703). Singapore: Springer.

Zhu, Y., & Elsayed, E. A. (2013). Design of accelerated life testing plans under multiple stresses. *Naval Research Logistics (NRL), 60*, 468–478.

**3**

# An investigation on the roll force and torque fluctuations during hot strip rolling process

Mahdi Bagheripoor* and Hosein Bisadi

*Department of Mechanical Engineering, Iran University of Science and Technology, Tehran, Iran*

Accurate prediction of the roll separating force and roll torque is critical to assuring the quality of the product. Fluctuation of these parameters during the rolling process is even more important because of its effects on rolling setups, geometrical accuracy (especially flatness in strip rolling), and the uniformity of mechanical and microstructural properties of the rolled material. The main approach of the present study is a precision analysis of the roll force and roll torque and their instabilities during the rolling process. In doing so, a coupling multi-variable simulation model for hot strip rolling was built using non-linear thermo-viscoplastic finite element method. Fluctuations of roll force and torque about the steady-state operating point are derived and the effects of main process parameters such as rolling speed and strip reduction on these instabilities are investigated. To check the validity of the employed model, predicted results are compared with experimental data.

**Keywords:** hot rolling process; finite element method; roll force; roll torque; aluminum alloys

## 1. Introduction

Considering the market requirements for flat-rolled aluminum products, an increasing demand can be observed for tight dimension tolerance, especially for thickness distribution along the strip length and profile (thickness distribution along the strip width). At the same time, the increased running speed brings out possible vibration problems in the rolling process, especially in a cold strip mill. If the thickness variation of the hot-rolled aluminum strip can be reduced, it will be possible to increase the production speed of the cold strip mill. These claims set new demands on the acceptable rolling forces and torques which, together with the roll pass design, have a significant influence on gage variations, the shape of the product, and power consumption.

Important pieces of information such as separating force, required torque, their time history and variations during the rolling process, and causes of these variations can be utilized for developing and optimizing pass schedules and providing setup data to the mills renting powerful process tools to satisfy the requirement of product development.

Throughout this century, much research has been devoted to the roll force and torque modeling and prediction using different techniques. Conventional models which are mostly based on thermo-mechanical theory were aimed at relating the roll force, roll torque, and power or energy consumption to the rolling parameters such as the roll diameter, initial temperature of the strip, strip thickness, reduction ratio, number of

*Corresponding author. Email: m_bagheripoor@mecheng.iust.ac.ir (M. Bagheripoor)

passes, speed of rolling, and interface friction (Ford & Alexander, 1964; Sims, 1954). Arnold and Whitton in 1975 proposed a formula for roll-separating force based on Sims' (1954) hot flat rolling theory, which included modifications for projected area of contact and empirical factors (Biswas, 2003). Joun and Hwang (1992) presented a new method of roll force prediction which was based on approximate solutions to the velocity, strain rate and stress distributions in the roll gap. Said, Lenard, Ragab, and Elkhier (1999) determined the roll separating forces and torques via low carbon steel rolling as a function of the area reduction at entry temperatures of 900, 950, 1000, 1050, and 1100 °C. They examined four empirical, mathematical models of the process for their ability to predict the roll separating forces. Yanagimoto, Morimoto, Kurahashi, and Chikushi (2002) proposed a mathematical model for prediction of rolling force and microstructure evolution in hot strip rolling (HSR), by Orowan's theory, FDM analysis for temperature, and incremental theory for the evolution of microstructure. Bayoumil (2007) developed a kinematic analytical approach for the prediction of rolling force, rolling torque, and forward slip in the hot strip continuous rolling, based on formulating a velocity field in the roll bite zone that expresses the effect of interfacial friction on the distribution of axial velocity and longitudinal stresses across the strip thickness. Chen, Zhang, Sun, Wang, and Song (2012) developed an online rolling force model for tandem cold rolling mill by numerical integration method, whereby the plastic deformation zone was divided into small units perpendicular to the motion of the strip.

The use of the finite element (FE) technique to predict roll forces and torque has been a popular approach for many researchers, and many commercial FE software packages now exist. Kim, Lee, Shin, and Shivpuri (1992) investigated shape rolling process using computer software, TASKS, based on the finite slab-element method and their predicted roll force and torque indicated reasonable agreement with other published results. Combining the FE and boundary element methods, Shangwu, Rodrigues, and Martins (1999) carried out the three-dimensional (3-D) modeling of hot rolling process of flat strips. They predicted rolling force, rolling torque, and contact pressure on the roll for both rigid and flexible roll cases. Kwak, Lee, Hwang, and Kim (2002) used FE simulation to investigate the effect of HSR process variables on some selected non-dimensional parameters characterizing the thermo-mechanical behavior of the strip and based on those parameters derived an online model for the precise prediction of roll force and roll power. Duan and Sheppard (2002) simulated the hot flat rolling of aluminum alloy 3003 by the commercial FEM program FORGE3. They adopted an inverse analysis method to match the calculated rolling force and torque with the measured rolling force and torque by treating the friction law and the friction coefficient as free parameters. A large-deformation constitutive model applicable to the calculation of roll force and torque in heavy-reduction rolling was presented by Byon, Kim, and Lee (2004). 3-D FE analysis coupled with the proposed constitutive models has been carried out to calculate work-piece deformation and the rolling force in their investigation. An approximate model for predicting roll force and torque in plate rolling was proposed by Moon and Lee (2008). Peening effect which unavoidably occurs in plate rolling due to a small ratio of work roll radius over mean thickness of slab being deformed has been considered in their model. Wang, Peng, Xu, and Liu (2010) investigated comprehensive influences of normal stress and shear stress in longitudinal, transverse direction, and altitude to improve the accuracy of the calculation of rolling force. In their study, the rolling force was solved in 3-D finite differential method. A new model for the prediction of the roll force and tension profiles was presented by Kim, Kwak, Shin, and Hwang (2010). Their approach was based on an approximate 3-D theory of plasticity, and considered the effect of the pre-deformation of the strip. They used a rigid-plastic FE model to examine the prediction accuracy of the proposed model. Byon, Na, and Lee (2013) coupled FE method with a modified constitutive model in which the flow stress equation has been modified according to the working temperature and strain rate at four stands (passes) of the finishing block mill of an actual rod mill, and showed that their model is more precise in prediction of roll force in continuous rod rolling where strain rates are in the range of $100-400 \, \mathrm{s}^{-1}$ and temperature is in the range of $900-1050 \, °C$.

Recently, some researchers have started exploring the possibility of utilizing soft computing techniques (particularly neural network and fuzzy set theory) for investigating the roll force and torque in the hot rolling process. As an improvement, Poliak, Shim, Kim, and Choo (1998) examined a proposed linear regression model to predict the roll force in hot plate rolling. Lee and Lee (2002) proposed a long-term learning method using neural network to improve the accuracy of rolling-force prediction in hot rolling mill. They combined neural network method with the conventional learning algorithm in the pre-calculation stage to reduce the thickness error at the head-end part of the strip. Neural networks were used by Yang, Linkens, Talamantes-Silva, and Howard (2003) to develop a rolling force and torque prediction model, without the requirement of a physically based or an empirical model. Lee and Choi (2004) adopted an online adaptable network for the rolling force setup and according to their field test results, they deduced that the prediction ability has been improved about 30% by the proposed method than the conventional method (mathematical models). Moussaoui, Selaimia, and Abbassi (2006) discussed the combination of an artificial neural network with analytical models to improve the performance of the prediction model of finishing rolling force in HSR process. Gudur and Dixit (2008) used a radial basis function neural network for prediction of velocity field and neutral point location in the cold flat rolling process that were further refined and post-processed in the FE code to calculate required parameters such as roll force, torque, and equivalent strain and thus provided highly accurate results with less computational time. The method of fuzzy control modification and Elman dynamic recursive network-combined prediction was used by Jia, Shan, and Niu (2008), to predict the rolling force for a tandem cold mill. A multiple radial basis function neural network model to predict rolling force based on wavelet analysis was proposed by Liu, Tong, and Lin (2012). In their model, the multi-resolution wavelet analysis method has been employed to separate the rolling force signal into several sub-signals corresponding to different factors.

In summary, many techniques have been utilized for predicting the roll separating force and torque in the hot rolling process. However, almost none of the above-mentioned efforts investigated the fluctuations of roll force and torque during the process and they have generally focused on the prediction of the mean roll force and torque in the steady-state region of each pass. The roll force and roll torque are the main parameters in controlling the flatness and geometrical accuracy, yield strength and the hardness of the product, the mill stretch and vibrations of the rolling mill, and roll flattening (Lenard, 2002). Therefore, it is worth to make a more detailed investigation on the variation of these parameters and consider their changes in the steady-state region of rolling process. The present study is aimed at a precision analysis of the roll force and torque during the rolling process and understanding the effect of process condition on the instabilities of these parameters. In doing so, the FE approach has been utilized to model HSR of aluminum alloys. The model is capable of considering temperature-dependent thermo-physical properties of rolling material and 3-D rolling conditions as it occurs during HSR. To verify the validity of the analysis, the predicted results are compared with the experimental data.

## 2. Mathematical model

### 2.1. Deformation in the roll gap

The thermo-viscoplastic FEM based on the flow formulation of the penalized form of the incompressibility was used to analyze the HSR process. According to the variational principle, the basic equation for the FE formulation is expressed as Equation (1) (Kobayashi, Oh, & Altan, 1989):

$$\int_V \bar{\sigma}\delta\dot{\bar{e}}\,dV + \int_V K\,\dot{\varepsilon}_v\delta\dot{\varepsilon}_v\,dV - \int_S F_i\delta u_i\,ds = 0 \qquad (1)$$

Here, $\bar{\sigma}$ is the effective stress which is a function of strain, strain rate and temperature, $\dot{\bar{\varepsilon}}$ the effective strain rate, $F_i$ the surface tractions, $u_i$ the velocity field, $V$ the volume, $S$ the boundary surface of subdomain, $K$ the penalty constant, and $\dot{\varepsilon}_v$ is the volumetric stain rate.

The metal starts flowing or deforming plastically when the applied stress reaches the value of yield stress. For investigating the metal plastic deformation behavior, the material yield stress under uniaxial conditions, as a function of strain, strain rate, and temperature, should also be considered as flow stress. This was done using a hyperbolic sine equation, shown in Equation (2), which relates the steady-state flow stress of the material to the strain rate and temperature under which it is deformed (Brown, Kim, & Anand, 1989):

$$\dot{\varepsilon} = A\exp\left(\frac{-Q_{\text{def}}}{R\theta}\right)\left[\sin h\left(\xi\frac{\tilde{\sigma}}{s}\right)\right]^{\frac{1}{m}} \qquad (2)$$

where $A$, $\xi$, and $m$ are material constants, $\dot{\varepsilon}$ is the mean equivalent plastic tensile strain rate, $\tilde{\sigma}$ the equivalent flow stress, $Q_{\text{def}}$ the activation energy for deformation, $R$ the universal gas constant, $\theta$ the absolute temperature, and $s$ is a scalar internal variable with dimensions of stress, called the deformation resistance. The coefficients of the hyperbolic sine equation for AA1100 deformation are summarized in Table 1 (Brown et al., 1989).

The flow behavior in the metal interior follows Equation (1). However, the effect of contact friction on metal flow should be considered when the interaction among metal surfaces exists. Contact friction between strip and roll is solved using Coulomb model, which is used mostly for forming simulations. Interfacial friction for the contact area is proportional to the normal force as shown in Equation (3):

$$\tau_{\text{crit}} = \mu p \qquad (3)$$

where $\tau_{\text{crit}}$ is the critical shear stress, $\mu$ the coefficient of friction, and $P$ is the contact pressure. The friction coefficient $\mu$ varies in wide ranges due to different rolling conditions and different options of the authors.

### 2.2. Heat transfer in the roll gap

Hot rolling is a thermo-mechanical process analyzed by solving the heat conduction equation coupled to a deformation analysis. During the rolling process, the temperature

Table 1. Summary of hyperbolic sine constants for AA1100 (Brown et al., 1989).

| Material | $A$ | $Q$ (kJ/mol) | $\xi$ | $m$ |
|---|---|---|---|---|
| AA1100 | 1.91E+07 | 175.3 | 7 | .23348 |

distribution in the strip can be calculated using the governing partial differential equation shown in Equation (4) (Wang, Yang, & He, 2008):

$$\frac{\partial}{\partial x}\left(k\frac{\partial T}{\partial x}\right) + \frac{\partial}{\partial y}\left(k\frac{\partial T}{\partial y}\right) + \frac{\partial}{\partial z}\left(k\frac{\partial T}{\partial z}\right) + \dot{Q} = \rho c\frac{\partial T}{\partial t} \qquad (4)$$

where $\rho$ is the density of the rolled strip, $c$ the specific heat of the rolled metal, $k$ the thermal conductivity of the rolled strip, and $\dot{Q}$ represents the volumetric rate of heat generation arising from the deformation. The heat generation term $\dot{Q}$ is calculated using Equation (5):

$$\dot{Q} = \eta\sigma\dot{\varepsilon} \qquad (5)$$

where $\eta$ is the fraction of plastic work converted into heat. A distributed surface flux, $q_{fric}$ is also generated from frictional sliding and rises rapidly near the entry and exit regions along the arc of contact inducing a dramatic change in the relative slip. Its overall contribution to the thermal balance in the hot rolling process is low, but if it is considered, its value is determined as follows (Shahani, Setayeshi, Nodamaie, Asadi, & Rezaie, 2009):

$$q_{fric} = |\tau\upsilon| \qquad (6)$$

where $\tau$ is the shear stress and $\upsilon$ is the sliding velocity. Since the heat lost from the strip is gained by the work roll in the roll gap, a simultaneous solution of the governing equations of both the strip and the work roll is required. To do so, the heat transfer in the work roll can be written as (Serajzadeh, Karimi Taheri, & Mucciardi, 2002):

$$\frac{1}{r}\frac{\partial}{\partial r}\left(k_r\frac{\partial T}{\partial r}\right) + \frac{\partial}{\partial z}\left(k_r\frac{\partial T}{\partial z}\right) = \rho_r c_r\frac{\partial T}{\partial t} \qquad (7)$$

where $\rho_r$, $c_r$, and $k_r$ are the density, the specific heat, and the thermal conductivity of the work roll, respectively. At the contact interface between the strip and the work roll, an interfacial heat transfer coefficient is assumed. The boundary conditions for heat transfer between the strip and work roll and surrounding temperature are applied as the equations employed by Kumar, Samarasekera, and Hawbolt (1992), and the interface heat transfer coefficient is determined based on the experimental work that has been conducted by Chun and Lenard (1997).

### 2.3.  FE modeling

As it is observed there is a complex interconnection between governing equations. In order to solve these equations, the simulation was carried out by a 3-D coupled thermo-mechanical FE analysis using commercial FE software package Abaqus/Explicit. Combining the merits of both Lagrangian and Eulerian formulations, ALE formulation was developed to handle mesh distortion, mesh entanglement, and special boundary condition changes in hot rolling process. Employing symmetry along the center line, a quarter of the strip was considered in the model. A sensitivity analysis was employed to determine the effect of changing the mesh density on the predicted model results. The work roll geometry is limited to a 90° section with a thickness of 5 mm. In both the strip and work roll, eight-node isoparametric brick elements were employed. The chosen elements were suitable for coupled thermal-stress analysis. The steel work roll was defined as an elastic material with a Young's modulus of 200 GPa. The large differences in elastic

modulus between the work roll and the strip causes the work roll to behave as a virtually rigid material. The thermo-physical properties of the strip and the steel work roll
are taken from Biswas and Mandal (2011) and Shahani et al. (2009), respectively.

## 3.  Results and discussion

### 3.1.  *Roll force and torque variations during the process*

In this study, the experimental rolling programs of Hum, Colquhoun, and Lenard (1996)
are used to investigate the rolling force and torque changes during HSR of commercial
pure aluminum. Typical contact pressure distribution in the roll gap is given in Figure 1, in
which an AA1100 strip with 6.28 mm thickness and 50 mm width is rolled at 500 °C with
a 250 mm diameter work roll. The amounts of reduction and rolling speed are 30.4% and
80 rpm, respectively, and only one-quarter of the strip is considered in the model (as
detailed in previous section). It can be seen from this figure that the distribution of the roll
pressure is quite uniform after rising from zero rapidly following which a moderate rise is
indicated until the peak at the neutral point, after which the pressure drops gradually. As
Jing (2001) describes, besides work-hardening and temperature drop, the additional constraint caused by the friction forces (so called resistance to flow in rolling) is responsible
for pressure build-up near the natural point. This type of pressure distribution is often
referred to as a 'friction hill' in which the more homogeneous deformation pattern is
expected (Dvorkin et al., 1997). According to the Li and Kobayashi (1982) findings, the
amount of reduction, work roll radius, and the strip entry thickness are the most important
parameters that shape the rolling pressure distribution.

Having the pressure distribution, the separating force as well as the acting torque
can be calculated by integrating over the area of contact. But it is difficult to find the
actual pressure distribution theoretically and working with approximate values as the
average pressure will not lead to a precise analysis of rolling force and torque. Accurate

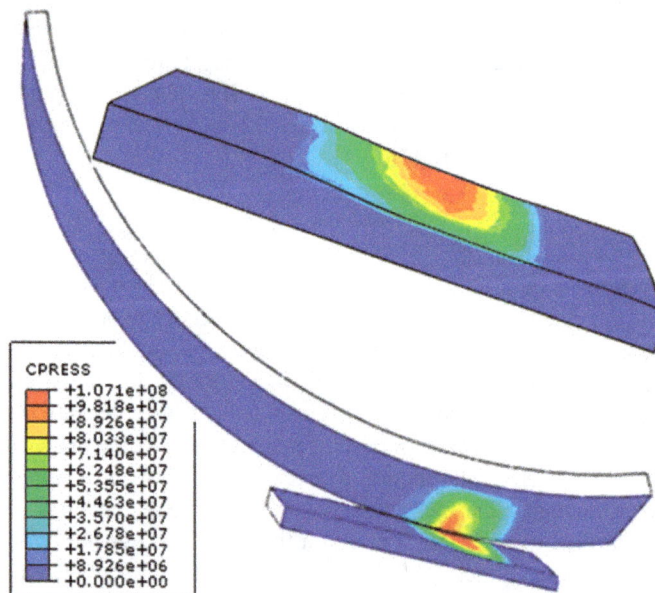

Figure 1.   Typical contact pressure distribution in the roll gap.

information about the rolling force is essential when designing a rolling pass schedule. The rolling force directly determines the thickness precision of rolled products. The rolling force is also the basis for computing the rolling torque. The accurate prediction of both of these parameters is imperative if mill breakdowns are to be avoided and to ensure maximum productivity in terms of geometric and property requirements. The predicted rolling force for the mentioned rolling program, as the material traverses the roll gap, is shown in Figure 2. As expected the rolling force increases gradually when the strip feeds into the mill gap and reaches a relatively steady value when the deformation is in the steady-state regime. Since only a quarter of strip is modeled, the actual rolling force is twice the average value in the steady-state deformation zone. As it shown in Figure 2, the rolling force fluctuates in the steady-state region and it varies between 21 and 27 kN.

Perhaps, the most important feature when designing the pass schedule is the calculation of energy requirement. An underpowered unit is the most obvious risk since this will lead to a reduction in productivity due to stalling. Thus, we must ensure that the motor is more than adequate for any immediate or future workloads. In the present

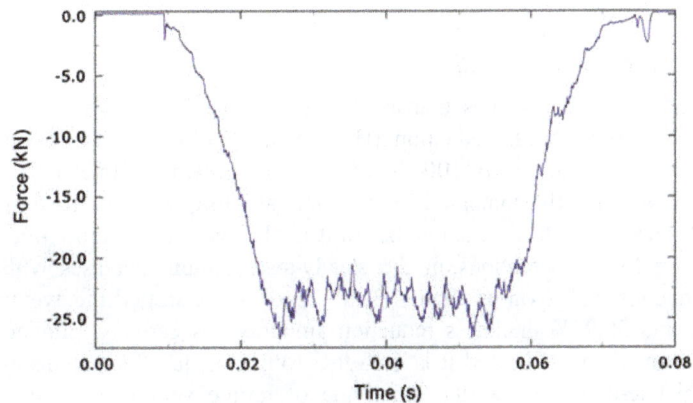

Figure 2.   Rolling force variations during the time.

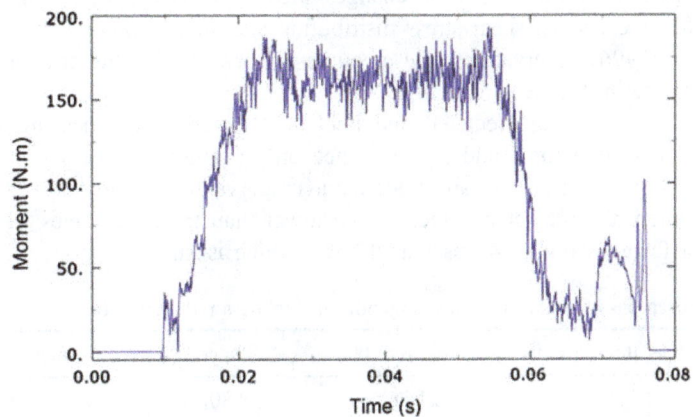

Figure 3.   Rolling torque variations during the time.

work, the rolling torque is investigated vs. different condition of process. The predicted rolling torque for described rolling condition is shown in Figure 3. As the rolling force, the rolling torque reaches a relatively steady value when the deformation is in the steady-state regime. But as it can be seen from Figure 3, the values of rolling torque fluctuate in the steady-state region with a high frequency.

The plots of Figures 2 and 3 indicate that the force and torque time history (in the steady-state region) show relatively large fluctuations about the average value. Such fluctuations of rolling force around the steady-state point cause variations in roll gap profile which may lead to strip flatness problems without a work roll bending system (Hu et al., 2006; Wang, Zhang, Xu, & Zhang, 2011). And the more fluctuations in roll torque leads to the shorter service life of rolling mill components (Pyun et al., 2013). Significant amplitude of the roll force and roll torque fluctuations (than the steady-state point value) can also cause mill vibrations and inhomogeneity in mechanical and micro-structural properties of the rolled material. In follows, we will try to consider the effects of main process parameters on the rolling force and torque predictions. It should be noted that in order for the results not to be mesh-dependent, the mesh of the strip was kept the same in all models.

### 3.2.  *Effects of thickness reduction*

Three real rolling pass schedules that are listed in Table 2 are chosen for considering the effects of strip thickness reduction (Hum et al., 1996). A 50 mm width strip of commercial pure aluminum (AA1100) is rolled with a 250 mm diameter work roll in all passes. Figure 4(a) and (b) compared rolling force and torque variations during the time for different thickness reduction amounts. As it is observed form Figure 4(a), the amplitude of rolling force fluctuations in the steady-state region increases with increasing thickness reduction and it varies about 7.3, 8.7, and 11.1% around the average value for 21.36, 30.4, and 39.05% thickness reduction amounts, respectively. Similar results can be deduced from Figure 4(b) and it is seen that rolling torque fluctuations are increased in higher thickness reductions (the amplitude of torque variations varies about 12.5, 13.7, and 14.2% about average value for 21.36, 30.4, and 39.05% reduction amounts, respectively). The most common effect of these phenomena is that there is more thick-ness variations during the rolled strip length in the higher thickness reduction amounts. It can be caused by surface condition changes such as interfacial friction in the contact area that affect the contact shear stress distribution and stick slip conditions. According to Hum et al. (1996) observations, the amounts of forward slip and friction coefficient rise with increase in thickness reduction.

Figure 5 compares the predicted and published experimental measurements (Hum et al., 1996) of rolling force and torque. Since only a quarter of strip is modeled, the actual rolling forces and torques are twice the average values in the steady-state regions. As it is observed, the predicted values are all lower than the actual ones. However, the relative error for all samples is less than 13.5%, which is known as a good accuracy in

Table 2.   Chosen pass schedules for investigation of thickness reduction effects.

| Pass no. | Inlet thick. (mm) | Thick. red. (%) | Roll speed (rpm) | Inlet strip temp (°C) |
|---|---|---|---|---|
| 1 | 6.32 | 21.36 | 80 | 500 |
| 2 | 6.28 | 30.4 | 80 | 500 |
| 3 | 6.3 | 39.05 | 80 | 500 |

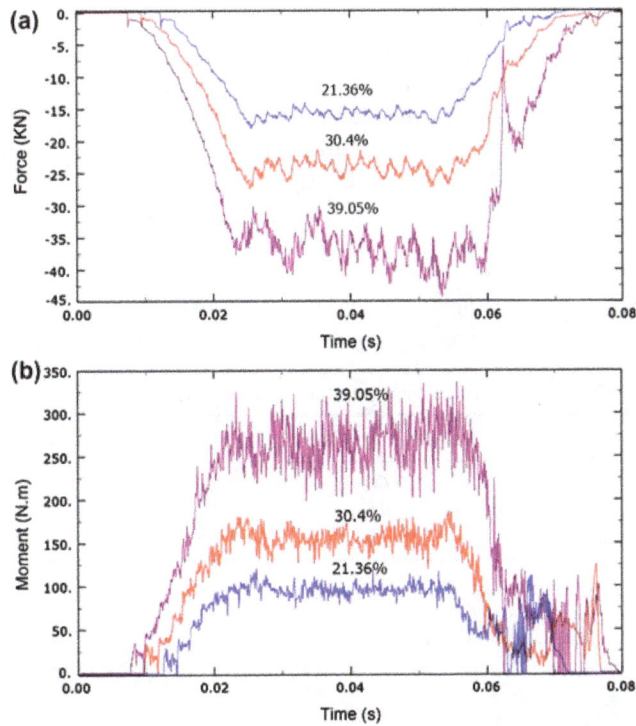

Figure 4.    The effect of thickness reduction amount on the variations of (a) roll force and (b) roll torque.

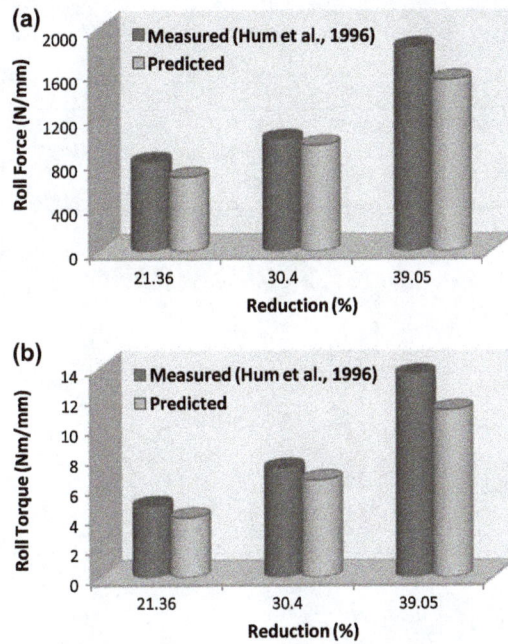

Figure 5.    Comparison of: (a) roll force and (b) roll torque predictions with experimental data at different reduction amounts.

metal forming theories. As expected, rolling forces and the torques clearly increase with increasing reduction. There are several mechanisms contributing to this finding, all of which are well known. Increasing the volume of plastic deformation, and hence more mechanical work is the dominant phenomena. The higher reduction also causes the more strain hardening and affects the continuously changing strip temperature distribution and hence, the metal flow behavior during the rolling.

### 3.3.  *Effects of rolling speed*

Rolling conditions that are summarized in Table 3 are chosen to consider the effects of rolling speed. Roll force and torque variations during the time are shown in Figure 6(a)

Table 3.   Chosen pass schedules for investigation of rolling speed effects.

| Pass no. | Inlet thick. (mm) | Thick. red. (%) | Roll speed (rpm) | Inlet strip temp (°C) |
|---|---|---|---|---|
| 4 | 6.29 | 30.68 | 20 | 500 |
| 5 | 6.28 | 30.41 | 60 | 500 |
| 6 | 6.29 | 31.48 | 100 | 500 |

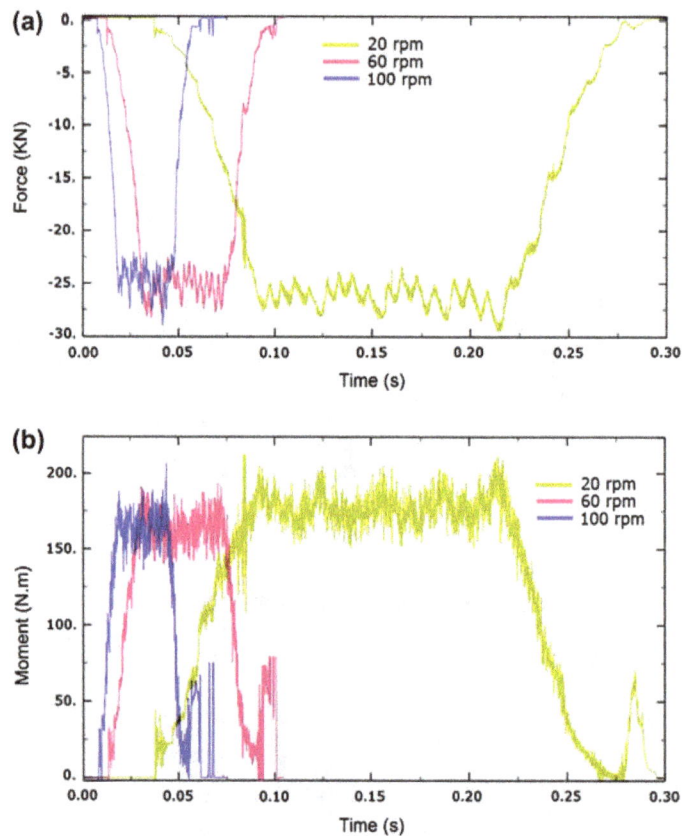

Figure 6.   The effect of rolling speed amount on the variations of (a) roll force and (b) roll torque.

and (b) for different rolling speeds. The magnitude of the rolling force fluctuations in the steady-state region increases as the rolling speed increases and it varies about 7, 7.9, and 9.7% around the average value for 20, 60, and 100 rpm rolling speed amounts, respectively (Figure 6(a)). For the rolling torque, amounts of variations are about 12.7, 13.1, and 14.6% around the average value for 20, 60, and 100 rpm rolling speeds (Figure 6(b)). Hence, more fluctuations of roll force and torque is expected at higher rolling speeds. The phenomena that contribute to the changes of fluctuation amplitude with rolling speed are the strip forward slip and the contact shear stress distribution which affect stick slip condition in different rolling speeds.

A comparison between the predicted and measured (Hum et al., 1996) rolling forces and torques vs. the rolling speed is shown in Figure 7. The prediction error has a mean of 13.2% for rolling force and 13.4% for rolling torque. As observed, the dependence of the rolling force and torque on the speed of rolling is not quite clear. The lack of speed effect on the forces and torques may be attributed to the balance between the strain rate dependence of the material's resistance to deformation which is increasing with the speed and the lower pass times (which decreases heat flow from the strip to the roll and environment) and higher internal heat generation (due to strain rate rises), both of which result in softening of the material and cause lower forces and torques. Therefore, effect of rolling speed on the rolling force and torque is completely related to the material flow behavior, heat transfer condition, and other process parameters such as initial strip thickness and the amount of reduction which affect the heat generation of plastic work and changing the strip material and process condition will affect the relation of rolling force and torque to the rolling speed.

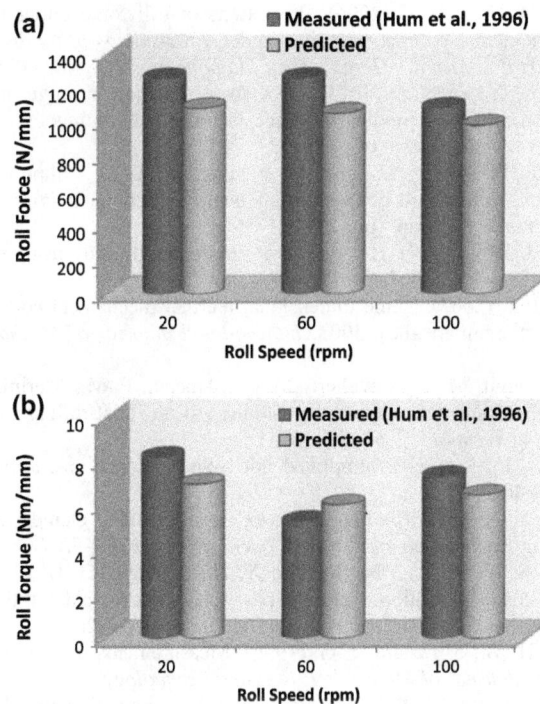

Figure 7.   Comparison of: (a) roll force and (b) roll torque predictions with experimental data at different rolling speeds.

## 4. Conclusions

A mathematical model based on the FE method has been developed to simulate hot rolling of commercial pure aluminum strips. The major innovation of this work is a precision analysis of the roll force and torque and their fluctuations during the rolling process. The effects of main process parameters such as rolling speed and strip reduction are considered. By comparing the model predictions with the experimental data, the performance of the model was proved with a reasonable accuracy.

The results show that the rolling force and torque values are not stable in the steady-state region and fluctuate around the average value. These fluctuations are increased in higher thickness reduction amounts that can be caused by contact condition changes. Furthermore, the average roll separating force and torque are clearly increased with increasing reduction. Predicted results indicate that rolling force and torque fluctuations rise as the rolling speed increases. However, the dependence of the average rolling force and torque on the speed of rolling is not quite clear.

## References

Bayoumil, S. (2007). A kinematic analytical approach to predict roll force, rolling torque and forward slip in thin hot strip continuous rolling. *Ironmaking & Steelmaking, 34*, 444–448.

Biswas, P., & Mandal, N. R. (2011). Effect of tool geometries on thermal history of FSW of AA1100. *Welding Journal, 90*, 129–135.

Biswas, S. (2003). *Simulation of thermo-mechanical deformation in high speed rolling of long steel products*. Worcester, MA: Worcester Polytechnic Institute.

Brown, S. B., Kim, K. H., & Anand, L. (1989). An internal variable constitutive model for hot working of metals. *International Journal of Plasticity, 5*, 95–130.

Byon, S. M., Kim, S. I., & Lee, Y. (2004). Predictions of roll force under heavy-reduction hot rolling using a large-deformation constitutive model. *Proceedings of the Institution of Mechanical Engineers, Part B: Journal of Engineering Manufacture, 218*, 483–494.

Byon, S. M., Na, D. H., & Lee, Y. S. (2013). Flow stress equation in range of intermediate strain rates and high temperatures to predict roll force in four-pass continuous rod rolling. *Transactions of Nonferrous Metals Society of China, 23*, 742–748.

Chen, S. Z., Zhang, D. H., Sun, J., Wang, J. S., & Song, J. (2012). Online calculation model of rolling force for cold rolling mill based on numerical integration. In *Proceeding of the control and decision conference*, Taiyuan, (pp. 3951–3955).

Chun, M. S., & Lenard, J. G. (1997). Hot rolling of an aluminum alloy using oil/water emulsions. *Journal of Materials Processing Technology, 72*, 283–292.

Duan, X., & Sheppard, T. (2002). Three dimensional thermal mechanical coupled simulation during hot rolling of aluminium alloy 3003. *International Journal of Mechanical Sciences, 44*, 2155–2172.

Dvorkin, E. N., Goldschmit, M. B., Cavaliere, M. A., Amenta, P. M., Marini, O., & Stroppiana, W. (1997). 2D finite element parametric studies of the flat-rolling process. *Journal of Materials Processing Technology, 68*, 99–107.

Ford, H., & Alexander, J. M. (1964). Simplified hot rolling calculations. *Journal of the Institute of Metals, 92*, 397–404.

Gudur, P. P., & Dixit, U. S. (2008). A neural network-assisted finite element analysis of cold flat rolling. *Engineering Applications of Artificial Intelligence, 21*, 43–52.

Hu, X. L., Zhang, Q. S., Zhao, Z., Tian, Y., Liu, X. H., & Wang, G. D. (2006). Application of approximation full-load distribution method to pass scheduling on plate mill with hydro-bending system. *Journal of Iron and Steel Research, International, 13*, 22–26.

Hum, B., Colquhoun, H. W., & Lenard, J. G. (1996). Measurements of friction during hot rolling of aluminum strips. *Journal of Materials Processing Technology, 60*, 331–338.

Jia, C. Y., Shan, X. Y., & Niu, Z. P. (2008). High precision prediction of rolling force based on fuzzy and nerve method for cold tandem mill. *Journal of Iron and Steel Research, International, 15*, 23–27.

Jing, L. (2001). *Rolling mill roll design*. Durham: Durham University.

Joun, M. S., & Hwang, S. M. (1992). An approximate analysis of hot-strip rolling – A new approach. *International Journal of Mechanical Sciences, 34*, 985–998.

Kim, N., Lee, S. M., Shin, W., & Shivpuri, R. (1992). Simulation of square-to-oval single pass rolling using a computationally effective finite and slab element method. *Transactions of ASME Journal of Engineering for Industry, 114*, 329–335.

Kim, Y. K., Kwak, W. J., Shin, T. J., & Hwang, S. M. (2010). A new model for the prediction of roll force and tension profiles in flat rolling. *ISIJ International, 50*, 1644–1652.

Kobayashi, S., Oh, S. I., & Altan, T. (1989). *Metal Forming and the finite-element method*. Oxford: Oxford University Press.

Kumar, A., Samarasekera, I. V., & Hawbolt, E. B. (1992). Roll-bite deformation during the hot rolling of steel strip. *Journal of Materials Processing Technology, 30*, 91–114.

Kwak, W. J., Lee, J. H., Hwang, S. M., & Kim, Y. H. (2002). A precision on-line model for the prediction of roll force and roll power in hot-strip rolling. *Metallurgical and Materials Transactions A, 33*, 3255–3272.

Lee, D. M., & Choi, S. G. (2004). Application of on-line adaptable neural network for the rolling force set-up of a plate mill. *Engineering Applications of Artificial Intelligence, 17*, 557–565.

Lee, D. M., & Lee, Y. (2002). Application of neural-network for improving accuracy of roll-force model in hot-rolling mill. *Control Engineering Practice, 10*, 473–478.

Lenard, J. G. (Ed.). (2002). *Metal forming science and practice*. Oxford: Elsevier.

Li, G. J., & Kobayashi, S. (1982). Rigid-plastic finite-element analysis of plane strain rolling. *Journal of Engineering for Industry, 104*, 55–63.

Liu, Y., Tong, C., & Lin, F. (2012). Rolling force prediction based on wavelet multiple-RBF neural network. In *Proceeding of the IEEE international conference on information and automation*, Shenyang, (pp. 941–944).

Moon, C. H., & Lee, Y. (2008). Approximate model for predicting roll force and torque in plate rolling with peening effect considered. *ISIJ International, 48*, 1409–1418.

Moussaoui, A., Selaimia, Y., & Abbassi, H. A. (2006). Hybrid hot strip rolling force prediction using a Bayesian trained artificial neural network and analytical models. *American Journal of Applied Sciences, 3*, 1885–1889.

Poliak, E. I., Shim, M. K., Kim, G. S., & Choo, W. Y. (1998). Application of linear regression analysis in accuracy assessment of rolling force calculations. *Metals and Materials, 4*, 1047–1056.

Pyun, Y. S., Cho, I. H., Suh, C. M., Park, J., Rogers, J., Kayumov, R., & Murakami, R. (2013). Reducing production loss by prolonging service life of rolling mill shear pin with ultrasonic nanocrystal surface modification technology. *International Journal of Precision Engineering and Manufacturing, 14*, 2027–2032.

Said, A., Lenard, J. G., Ragab, A. R., & Elkhier, M. A. (1999). The temperature, roll force and roll torque during hot bar rolling. *Journal of Materials Processing Technology, 88*, 147–153.

Serajzadeh, S., Karimi Taheri, A., & Mucciardi, F. (2002). Unsteady state work-roll temperature distribution during continuous hot slab rolling. *International Journal of Mechanical Sciences, 44*, 2447–2462.

Shahani, A. R., Setayeshi, S., Nodamaie, S. A., Asadi, M. A., & Rezaie, S. (2009). Prediction of influence parameters on the hot rolling process using finite element method and neural network. *Journal of Materials Processing Technology, 209*, 1920–1935.

Shangwu, X., Rodrigues, J. M. C., & Martins, P. A. F. (1999). Three-dimensional simulation of flat rolling through a combined finite element-boundary element approach. *Finite Elements in Analysis and Design, 32*, 221–233.

Sims, R. B. (1954). The calculation of roll force and torque in hot rolling mills. *ARCHIVE: Proceedings of the Institution of Mechanical Engineers 1847–1982 (vols 1–196), 168*, 191–200.

Wang, P. F., Zhang, D. H., Xu, L., & Zhang, W. X. (2011). Research and application of dynamic substitution control of actuators in flatness control of cold rolling mill. *Steel Research International, 82*, 379–387.

Wang, X., Peng, Y., Xu, L., & Liu, H. (2010). A 3-D differential method for solving rolling force of PC hot strip mill. *Journal of Iron and Steel Research, International, 17*, 36–39.

Wang, X., Yang, Q., & He, A. (2008). Calculation of thermal stress affecting strip flatness change during run-out table cooling in hot steel strip rolling. *Journal of Materials Processing Technology, 207*, 130–146.

Yanagimoto, J., Morimoto, T., Kurahashi, R., & Chikushi, I. (2002). Mathematical modelling for rolling force and microstructure evolution and microstructure controlling with heavy reduction in tandem hot strip rolling. *Steel Research, 73*, 56–62.

Yang, Y. Y., Linkens, D. A., Talamantes-Silva, J., & Howard, I. C. (2003). Roll force and torque prediction using neural network and finite element modelling. *ISIJ International, 43*, 1957–1966.

# 4

# A review of turning of hard steels used in bearing and automotive applications

Suha Karim Shihab*, Zahid A. Khan, Aas Mohammad and Arshad Noor Siddiquee

*Department of Mechanical Engineering, Jamia Millia Islamia (A Central University), MMA Jauhar Marg, New Delhi 110025, India*

Hard turning is a recent technology that involves machining of hard steels using modern machine tools. Hard machining presents challenges in terms of selection of tool insert with improved tool life and high-precision machining. Turning of hardened steels using single-point cutting tool has got considerable interest among manufacturers of ball bearings, automotive, gear, and die industry. Hard turning in the automotive industry and bearing applications typically has a number of potential benefits over traditional form grinding including lower equipment costs, shorter setup time, and fewer process steps which in turn provides high flexibility and ability to cut complex geometries. Moreover, the hard turning process is usually carried out without coolant/lubricant and thus, the problem of storage, handling, and disposal of cutting fluid is eliminated, and at the same time, it probably favors the health of operators. This paper presents an overview of the past research hard turning using hard turning tools such as PCBN, cubic boron nitride, Ceramics, Carbide, etc. Major hard turning cutting materials and effect of hard turning process parameters on cutting forces, heat generation during cutting, surface finish and surface integrity, and tool wear have been discussed in light of the findings of the past research.

Keywords: hard turning; hard tool materials; cutting parameters

## 1. Introduction

High-hardness materials include various hardened alloy steels, tool steels, case-hardened steels, super alloys, nitrided steels, hard-chrome coated steels, and heat-treated powder metallurgical parts. Finishing of hardened steel, (e.g. through hardened AISI 52100 steel for bearing applications, and case hardened steel 16MnCr5 for automotive gears and shafts) using hard turning using super hard cutting tools (PCBN, cubic boron nitride (CBN), Ceramics, and Carbide) was early recognized by the automotive industry as a means of manufacturing of precisely finished transmission components (Davim, 2011).

If hard turning is applied to the manufacture of complex parts, manufacturing costs can be reduced up to 30%, and US industries exploited the advantages of hard turning for an annual gain of up to $6 billion (Huang, Chou, & Liang, 2007). A qualitative comparison of the capabilities of hard turning and grinding processes in terms of work-piece quality, process flexibility, dimension and shape accuracy, etc. has been made by M'Saoubi, Outeiro, Chandrasekaran, Dillon, and Jawahir (2008).

---

*Corresponding author. Email: suhakarim_10@yahoo.com

In the past, many studies have been carried out to explore different facets of the hard turning of alloy steel. Thiele and Melkote (1999) investigated the effects of tool edge geometry of CBN tools and workpiece hardness (45, 52, and 60 HRC) on the surface roughness and cutting forces in the finish hard turning of AISI 52100 steel. Their study showed that large edge hones result in higher forces in the axial, radial, and tangential directions as compared to those tools with small edge hone. Further, the effect of workpiece hardness on the axial and radial components of force was found to be significant, particularly for large edge hones. Ramesh, Melkote, Allard, Riester, and Watkins (2005) examined the differences in structure and properties of white layers formed during machining of hardened AISI 52100 steel (62 HRC) at different cutting speeds. Their results indicated that the grain sizes of white layers formed were considerably smaller than the grain sizes of the bulk. They also observed that white layers generated at higher machining speeds are coarser than those generated at lower speeds. Diniz, Gomes, and Braghini (2005) studied the hardened steel turning (SAE 01 steel with $58 \pm 2$ HRC hardness) with continuous, interrupted, and semi-interrupted cutting using two different kinds of CBN cutting tools (high CBN content material and CBN content material). Their main conclusion pointed out that the low CBN content tool is more suitable for continuous and semi-interrupted surfaces, whereas the high CBN content tool presents a slightly longer life during interrupted cutting. Umbrello, Rizzuti, Outeiro, Shivpuri, and M'Saoubi (2008) developed hardness-based flow stress and fracture models for machining AISI H13 (50-60 HRC) tool steel, which can be applied for a wide range of work material hardness using FEM. They implemented these models in a viscoplastic non-isothermal finite element model of the orthogonal cutting process and validated the results of these models using the data available in the literature. Umbrello, Ambrogio, Filice, and Shivpuri (2008) determined residual stresses distribution and optimal cutting conditions during hard turning of AISI 52100 bearing steel using the hybrid model based on the artificial neural networks (ANNs) and finite element method (FEM). Adesta, Riza, Hazza, Agusman, and Rosehan (2009) investigated tool wear of cermet tools and also surface roughness under different rake angles (0, −3, −6, −9, and −12) and different cutting speeds (1000 and 300 m/min). Their experimental results showed that increase in negative rake angles causes higher wear, shorter duration of tool life, and poor surface finish. Further, high cutting speed turning gave shorter tool life, high wear rate but finer surface finish than conventional one. Gaitonde, Karnik, Figueira, and Davim (2009a) analyzed the effects of depth of cut and machining time on surface roughness and tool wear during turning of high chromium AISI D2 cold work tool steel (59/61 HRC) using ceramic inserts with TiN coating (CC650, CC650WG, and GC6050WH). From their analysis, it was observed that the CC650WG wiper insert performs better with reference to surface roughness and tool wear. They also observed that the surface roughness is minimal at lower values of depth of cut and machining time in case of CC650 and CC650WG inserts machining, while the minimum surface roughness occurs at 0.4 mm depth of cut for hard turning with GC6050WH insert. Finally, they found that the CC650 conventional insert is useful in reducing the cutting force. Fnides, Yallese, Mabrouki, and Rigal (2009) investigated the effect of three cutting parameters on surface roughness in turning of X38CrMoV5-1 hardened steel treated at 50 HRC using mixed ceramic tool. Their results revealed that the effect of feed rate on surface roughness is more significant than cutting speed, whereas the depth of cut is not significant. Moreover, the highest level of cutting speed, 180 m/min, the lowest level of feed rate, 0.08 mm/rev, and the highest level of depth of cut, 0.45 mm, resulted in larger amount of material removal and also a good surface finish. Thamizhmanii and Hasan

(2010) evaluated the performance of CBN and PCBN tools based on flank wear and cutting forces during machining of AISI 440 C hard martensitic stainless steel. They observed high cutting forces during machining and they suggested that this may be due to heat and flank wear combinations. In addition, they also found that lower cutting forces lead to low flank wear and provide good dimensional accuracy of the work material including low surface roughness. Meyer, Köhler, and Denkena (2012) studied the effect of geometrical contact conditions between the tool (WBN560 with a CBN content of 56%) and the workpiece (AISI 5115 steel) on the cutting forces and tool wear during hard turning. Their results showed that the geometrical contact conditions have a significant influence on the forces, tool, and workpiece load as well as on the tool wear behavior. Moreover, they suggested that by applying a load-specific tool and process design, the tool operation behavior and the productivity of the process can be enhanced in a beneficial way.

## 2. Major hard turning cutting tool materials

Material developments for the cutting tool, one of the most critical elements in metal cutting, have always been characterized by an increase in wear resistance to machine harder, tougher, or chemically reactive materials. For example, superhard materials such as ceramic and CBN were one of the main keys to enable the hard turning technology to be an alternative to grinding processes. Correlation between chemical, physical, and mechanical characteristics of cutting tool materials and their performances in cutting operations is therefore, a key issue for both tool manufacturers and users (Suresh, Basavarajappa, and Samuel, 2012).

Cutting tools must simultaneously withstand big mechanical loads and high temperatures. Temperature in the chip/tool interface reaches more than 700 °C in some cases such as turning of high hardness alloy steel (Davim, 2011). Additionally, the friction between tool and removed chip, on one hand, and tool against the new machined surface, on the other, is very severe (Davim, 2011). Bearing this in mind, the following main factors should be considered for a good tool design and post-manufacturing (Davim, 2011):

- Chemical and physical stability of cutting-tool substrate material must be maintained at high temperatures.
- Material hardness must be maintained to high temperatures suffered at the chip/tool interface.
- Tool material has to present high resistance for the abrasion and adhesion wear.
- Tool material must present enough toughness to avoid fracture, especially when operation to perform implies interrupted cutting.

The main tool materials that are used for hard turning include sintered carbides, ceramics (e.g. $Al_2O_3$ or $Si_3N_4$ etc.), and extra-hard materials (e.g. PCD, PCBN etc.). The selection of appropriate tool material is vital for process efficiency and depends on the accuracy and surface finish required. Within each type, hundreds of different grades are available from various tool material, cutting insert, and tool manufacturers. Therefore, the selection of appropriate tool-material grade is one of the most important tasks in hard turning in terms of obtaining efficient and stable machining process. The insert shape, tool holders, and optimal machining regime just add more complications to this multivariable optimization problem. Knowledge, as understanding the essence of hard machining, and experience are prerequisites for success (Davim, 2011). The studies carried out by the researchers using cutting tool of the above-mentioned materials are describe in the following sections:

## 2.1.  Sintered carbide (hardmetal)

Sintered carbide tools, also known as hardmetal tools or cemented carbide tools, are made by a mixture of tungsten carbide micrograins with cobalt at high temperature and pressure. Tantalum, titanium, or vanadium carbides can also be mixed in small proportions. This type of material in the straight grade or in the coated grades is mostly used today for hard machining and high-speed machining.

A sintered tungsten carbide also includes TiC (carbide with hardness 3200 HV) and in some cases TiCN, but they typically have a nickel–chrome binder. New grades with TaNbC and MoC increase the tool-edge strength against the cyclic impacts which finds typical application in milling. Tungsten carbide is very stable with respect to chemical and thermal aspects of machining, and is very hard as well. In most cases, cemented carbide degradation starts from the cobalt binder and the tungsten carbide–cobalt cohesion (Davim, 2011).

Aslan (2005) investigated the wear behavior of different cutting tools such as TiCN coated tungsten carbide, TiCN + TiAlN coated tungsten carbide, TiAlN coated cermet, mixed ceramic with $Al_2O_3 + TiCN$, and CBN, in end milling of X210 Cr12 cold-work tool steel hardened to 62 HRC. The results of his investigations indicated that the TiAlN coated carbide and cermet tools perform slightly better than TiCN coated carbide tool which can be attributed to better high-temperature properties of TiAlN compared to TiCN. Further, he also found that CBN tool exhibits the best cutting performance in terms of both flank wear and surface finish and the highest volume of metal removal is obtained with CBN tool. Arsecularatne, Zhang, and Montross (2006) studied wear and tool life for tungsten carbide WC cutting tool during machining of AISI 4142 steel and AISI 1045 steel, and also for PCBN cutting tool during machining of AISI 52100 with two different hardness values of 60HRC and 62HRC. It was concluded from their study that the most dominant tool wear mechanism for WC is diffusion and for PCBN is chemical wear. Lima, Avila, and Abrao (2007) performed a study which is concerned with continuous turning of AISI 4340 steel hardened from 250 up to 525 HV using coated carbide tools. They assessed machining forces, tool life, and wear mechanisms. The results of their study indicated that the machining force components increase with the work material hardness, however, the cutting force components decrease slightly as the work hardness increase from 250 up to 345 HV. Further, tool wear was lower when machining the workpiece with a hardness value of 345 HV as compared to machining of the workpiece possessing hardness of 250 HV steel. Finally, it was observed that the abrasion is the principal wear mechanism and catastrophic failure takes place when attempting to machine the 525 HV steel. Chowdhury and Dhar (2011) investigated the tool wear and surface roughness for varying cutting parameters under dry and Minimum Quantity Lubrication (MQL) environment while turning hardened medium carbon steel using coated carbide insert. The results of their study indicated that the application of MQL technique significantly helps to obtain better performance of coated carbide insert in comparison to dry condition. Hwang and Lee (2010) studied the effect of cutting parameters (nozzle diameter, cutting speed, feed rate, and depth of cut) on cutting forces and surface roughness of carbon steel AISI 1045 using coated carbide inserts under MQL and wet turning conditions. They found that cutting speed and depth of cut show opposite effects on the cutting force and therefore, they suggested that cutting conditions should be set under a clear standard because the optimal combination of cutting parameters could be different depending on machinability. Moreover, cutting speed and depth of cut showed opposite effects on surface roughness in MQL turning and it produced better surface roughness compared to wet turning. They also observed that the MQL turning provides more advantages than wet turning.

## 2.2. Ceramics

Ceramics are very hard and refractory materials, withstanding more than 1500 °C without chemical decomposition. These features recommend them to be used for the machining of metals at high cutting speeds and in dry machining conditions. Ceramic tools are based primarily on alumina ($Al_2O_3$), silicon nitride (Si3N4), and sialon (a combination of Si, Al, O, and N). Alumina tools can contain titanium, magnesium, chromium, or zirconium oxides distributed homogeneously into the alumina matrix to improve toughness (Davim, 2011)

Several researches have been done to study the effect of hard turning on different types of ceramics tools. Luo, Liao, and Tsai (1999) in their research revealed the wear behavior of ceramic and CBN tools in the turning of AISI 4340 hardened alloy steels. It was found from their study that the main wear mechanism for the CBN tools is the abrasion of the binder material by the hard carbide particles of the workpiece and for the ceramic tools it is adhesive wear and abrasive wear. They also found that there is a protective layer formed on the chip–tool interface which plays an important role in wear behavior of CBN and ceramics tools. Kumar, Durai, and Sornakumar (2003) studied the machinability of EN 24 steel (HRC 40 and HRC 45) using two types of ceramic cutting tool materials namely, Ti[C, N] mixed alumina ceramic cutting tool and zirconia toughened alumina ceramic cutting tool. It was found from their study that the performance of ceramic cutting tools is good in the machining of hardened steel and Ti[C, N] mixed alumina ceramic cutting tool produces the best surface finish. Gaitonde, Karnik, Figueira, and Davim (2009b) analyzed the effects of depth of cut and machining time on machinability aspects such as machining force, power, specific cutting force, surface roughness, and tool wear, with the use of second-order mathematical models during turning of high chromium AISI D2 cold work tool steel with CC650, CC650WG, and GC6050WH ceramic inserts. From their analysis, it was revealed that the CC650WG wiper insert performs better with reference to surface roughness and tool wear, while the CC650 conventional insert is useful in reducing the machining force, power, and specific cutting force. Elmunafi (2012) evaluated the performance of wiper coated ceramic tool when turning ASSAB DF-3 grade hardened steel with hardness 55 HRC. Their study showed that the effects of cutting speed and feed rate on the tool life are statistically significant. They also found that the effects of cutting speed and feed rate on the surface roughness are statistically significant, specifically, at high cutting speed and low feed rate. Accordingly, wiper inserts were able to produce better surface finish. Maňková, Kovac, Kundrak, and Beňo (2011) studied the influence of cutting parameters on accompanying phenomena during hard turning of hardened steel with hardness of HRC 55 using mixed oxide ceramic inserts (70% $Al_2O_3$ + 30% TiC). They concluded that hard turning has many advantages in comparison to other processes in machining of hardened steels. Fnides, Yallese, Mabrouki, and Rigal (2011) stated the effect of cutting parameters on cutting force components in hard turning of AISI H11 hot work tool steel (50 HRC) using a mixed ceramic tool. They found that the depth of cut is the dominant factor affecting cutting force components. Additionally, the feed rate influences tangential cutting force more than radial and axial forces. Finally, the cutting speed affects radial force more than tangential and axial forces. Zhao, Yuan, and Zhou (2010) investigated the cutting performance, failure mode, and mechanism of the $Al_2O_3$-based composite ceramic tool material reinforced with WC microparticles and TiC nanoparticles in both continuous and intermittent turning of hardened AISI1045 steel. Their study revealed that the tool life of the $Al_2O_3$/WC/TiC ceramic tool increases when the cutting speed increases to 170 m/min. They also observed that the longer tool life of the $Al_2O_3$/WC/TiC composite ceramic tool is attributed to its synergistic strengthening/toughening mechanisms induced by the WC microparticles and TiC nanoparticles.

### 2.3. Extra-hard materials

PCD and PCBN are extra-hard materials. There are several grades in the PCD and PCBN groups. PCD is suitable for tools focused on machining abrasive non-ferrous metals, plastics, and composites. PCBN finds applications in the machining of hardened tool steels and hard cast irons. PCD plates are obtained by a high temperature and pressure process where synthetic diamond grains are sintered with cobalt. CBN is a polymorph boron-nitride-based material. It possesses high mechanical properties due to its crystalline structure and its covalent link. It has been industrially produced since 1957, starting from hexagonal boron nitride put under high pressures (8 GPa) and temperatures (1500 °C). With a lower hardness (<4500 HV) than diamond (>9000 HV), CBN is the second-hardest synthetic material. The CBN grains are sintered together with a binder to form a composite, PCBN. The size, shape, and ratio of CBN/binder define the different PCBN grade (Davim, 2011).

The literature reveals that many researchers have suggested effect of hard turning on CBN and PCBN cutting tools (Chowdhury and Dhar 2011; Diniz et al., 2005; Elmunafi 2012; Lima et al., 2007). In addition, many studies have been conducted to investigate the performance of CBN and PCBN tool in machining of various hard materials. Oliveira, Diniz, and Ursolino (2009) investigated the performance of PCBN and alumina-based ceramic reinforced with silicon carbide tools during turning of AISI 4340 steel with 56 HRC under continuous and interrupted cutting. The results of their study indicated that the longest tool life is achieved using PCBN in continuous turning, but similar tool longevity is attained in interrupted turning using both PCBN and ceramic. Also, the roughness values were lower for both continuous and interrupted surfaces when PCBN tools were used. Sahin (2009) compared the tool life of mixed alumina ceramic with that of $Al_2O_3$ (70%):TiC (30%) matrix and CBN cutting tools during machining of AISI 52100 steel. They showed that the CBN cutting tool has the best performance than that of ceramic-based cutting tool. Poulachon, Moisan, and Jawahir (2001) studied various modes of wear and damage of PCBN cutting tool under different loading conditions during turning of AISI 52100 steel in order to establish a reliable wear model. Their results led to the conclusion that the main wear mechanism of the PCBN tools is abrasion and it depends not only on the chemical composition of the PCBN, and the nature of the binder phase, but also on the hardness value and above all on the microstructure of the machining work material. They also found that use of TiN coated PCBN improves the tool-wear and hence the tool-life by reducing the diffusion wear between workpiece and tool rake face. Diniz, Ferreira, and Filho (2003) investigated the influence of cutting speed under three cutting conditions: dry cutting, wet cutting, and minimum volume of oil i.e. MVO (oil flow of 10 ml/h) on CBN tool wear in turning of AISI 52100 hardened steel. Their results revealed that dry and MVO cuttings produce most of the time, similar values of flank wear which is always smaller than the values obtained during wet cutting. They also observed that highest cutting speed for three cutting conditions gives the smallest value of surface roughness. Poulachon, Bandyopadhyay, Jawahir, Pheulpin, and Seguin (2003) examined the influence of the microstructure of different hardened tool steels (AISI D2 cold work steel, AISI H11 hot work steel, 35NiCrMo16 hot work steel, and AISI 52100 bearing steel) on the wear of PCBN cutting tools under dry turning. They observed a large variation in tool wear in the machining of these steels. Further, they found that the flank grooves are correlated to the microstructure of these steels, namely the presence of various carbides. Benga and Abrao (2003) studied the machinability of hardened 100Cr6 bearing steel (62-64 HRC) under dry turning using mixed alumina, whisker reinforced alumina, and PCBN inserts. As far as tool life is concerned, they obtained best results with the PCBN compact, followed by the mixed alumina tool at low feed rates and by the whisker reinforced alumina when feed rate was increased. Poulachon, Bandyopadhyay, Jawahir, Pheulpin, and Seguin (2004) investigated the tool wear mechanisms of CBN cutting tools in finish turning of the hardened steels (AISI D2 cold work steel, AISI H11 hot work steel, 35NiCrMo16 hot work steel, and AISI 52100bearing steel), treated at 54 HRC. They observed a large variation in tool-wear rate in machining of these steels. They also found that the generated tool flank grooves are correlated with the hard carbide content of the workpieces. In addition, they also performed a crater wear study in which they found the appearance of an adhesive third body which could induce a chemical wear in the tool. Kurt and Seker (2005) investigated the effects of chamfer angle of PCBN cutting tools on the cutting forces and the cutting tool stresses, principal, and Von Mises stress, in finishing hard turning of AISI 52100 bearing steel. Their results revealed a great influence of the chamfer angle on the cutting forces and tool stress. Based on their work, they concluded that critical chamfer angle is 20° in finishing hard turning of AISI 52100 bearing steel. Galoppi, Filho, and Batalha (2006) tested different types of CBN inserts with wiper geometry, coated with TiAlN and TiN as well as with no coated ones, during hard turning of tempered DIN 100Cr6 steel. Their results revealed significant cracking on the coating surface layer of both TiAlN and TiN coated tools which resulted in the removal of the surface hard coating by cracking. Further, they also found that the cratering that arises on insert surfaces is similar to that found on the no coated inserts. Finally, they observed that the inserts with a wiper geometry presents longer tool life compared with no coated tools. Remadna and Rigal (2006) explored the parameters of hard steel turning (alloyed steel 52HRC) with a CBN tool. They analyzed the correlation between tool wear and the direction of cutting force. Their results showed that the cutting forces increase gradually with increasing of cutting distance and tool flank wear. They also revealed that the wear of CBN does not directly or appreciably affect the manufactured surface. More, Jiang, Brown, and Malshe (2006) measured tool wear and evaluated machining performance of CBN–TiN coated carbide inserts and PCBN compact inserts in turning of AISI 4340 hardened steel. They observed that the flank wear is mainly due to abrasive actions of the martensite which is present in the hardened AISI 4340 alloy. Further, they found that the crater wear of the CBN–TiN coated inserts is less than that of the PCBN inserts because of the lubricity of TiN capping layer on the CBN–TiN coating. Arsecularatne, Zhang, Montros, and Mathew (2006) investigated the machinability of AISI D2 steel of hardness 62 HRC using PCBN tools. They found that the most feasible feeds and speeds fall in the ranges 0.08–0.20 mm/rev and 70–120 m/min, respectively, and that most of the tested PCBN tools reach the end of life mainly due to flank wear. Moreover, the highest acceptable values of tool life and volume of material removal were obtained at the lowest speed (70 m/min), indicating that this speed is more suitable for machining the selected tool/ work material combination. They also found that the highest feed results in the highest volume of material removal and lower feeds results in higher tool life values. They suggested that the most appropriate feeds for AISI D2 steel are 0.14 mm/rev for finishing operations and 0.20 mm/rev for roughing operations. Lin, Liao, and Wei (2008) studied tool wear mechanisms in turning of AISI 4340 alloy steels by CBN tools which contained 50% and 45% of CBN, respectively, and also TiC based-binder, under varying cutting speeds. It was observed from their results that at low cutting speed, the binder of the hard particles of the cutting tool is removed from the substrate due to high cutting force and the tool wear is mainly due to abrasion. Their study revealed formation of a protective layer, resulting from the diffusion of the bond material of the cutting tool, on the chip–tool interface when the cutting speed was increased. The layer so formed works as a diffusion barrier and reduces tool wear rate which results in the prolonged life of the CBN tool. They also found that at further higher cutting speed, (i) cutting temperature becomes the dominant factor instead of the cutting force which causes inhomogeneous shear strain, and a transition from continuous chip to saw-tooth chip, (ii) the friction force increases because of the very irregular chip–tool contact and it results in the removal of the protective layer, and (iii) the bond between tool particles gets weakened due to serious diffusion between the work material and the cutting tool. Subsequently, hard particles of the tool get detached from it and the tool life is reduced. Özel (2009) investigated the influence of different edge micro-geometry of PCBN tools on forces, stresses, friction, and tool wear during turning of alloy steel AISI 4340. On the basis of the results of their study, they made the following conclusions: (i) variable micro-geometry insert edge design reduces the heat generation and stress concentration along the tool cutting edge significantly which improves the tool life, (ii) variable micro-geometry insert cutting edge induces less plastic strain on the machined workpiece which improves surface integrity, and (iii) variable micro-geometry insert edge design reduces tool wear depth and wear rate. Yallese, Chaoui, Zeghib, Boulanouar, and Rigal (2009) investigated the behavior of a CBN tool during hard turning of 100Cr6-tempered steel. They observed that the cutting speeds ranging from 90 to 220 m/min is the most interesting cutting conditions for the system CBN7020-100Cr6. However, they found that beyond 280 m/min the machining system becomes unstable and produces significant sparks and vibrations after only a few minutes of work. Further, they found optimal productivity at the speed of 120 m/min with an acceptable tool flank wear below 0.4 mm. Godoy and Diniz (2011) made a comparison between the performance of CBN and ceramic tools in continuous and interrupted cutting of AISI 4340 steel with 56 HRC of hardness. They concluded that in both continuous and interrupted cutting, the CBN tools exhibit a much better performance with respect to both tool life and workpiece surface roughness than the ceramic tools. Zawada-Tomkiewicz (2011) evaluated the surface finish in a continuous dry turning of EN 41Cr4 low chromium alloy steel with 58 HRC using PCBN coated and uncoated tools. He observed that the hard turning with PCBN tools can produce a very smooth and uniform surface. Also, in the case of the PCBN coated tool material, the wedge's wear was almost unnoticeable. Further, the random part of the generated surface was significantly greater for the PCBN coated cutting tool material. Deepakkumar and Sadaiah (2011) investigated the effects of cutting speed, feed rate, and different cutting environments (dry, wet, and MQL) on the surface roughness and tool wear during turning of AISI 4340 steel using CBN insert. They analyzed the effect of cutting parameters and determined its optimum condition by using ANOVA technique. They found that increase in the cutting speed and feed rate results in decrease in $R_a$ value. In addition, he observed the $R_a$ value for dry and wet turning as 1.2 and 1.1 µm, respectively, while in the case of turning in MQL condition, the $R_a$ value was 0.9 µm. Finally, the results of their study indicated that as cutting speed and feed rate increase, the tool wear value increases.

## 3.   Effect of process parameters in hard turning:

Machining parameters such as cutting speed, feed rate, and depth of cut do affect the production costs and product quality. Thus, it is important to use optimization technique to determine optimal levels of these parameters so as to reduce the production costs and to achieve the desired product quality simultaneously. One of the main objectives in the optimization of a turning process is minimization of the cost of production and maximization of the production rate while keeping the quality of machined parts as per design specifications. The cost of machining is strongly related to the material removal rate. The material removal rate for a turning operation is given by the product of cutting parameters (cutting speed ($Vc$), feed rate ($f$), and depth of cut ($d$)). Therefore, if an increase in productivity is desired then an increase in these three cutting parameters is required. But, there are limits to these cutting parameters since they also have an effect on the tool life, tool wear, surface quality, cutting forces, cutting temperature, etc. Keeping this in view, many researchers have investigated effect of these parameters pertaining to the hard turning. The following sections present the findings of some of the research studies involving these parameters with reference to hard turning.

### 3.1.   Studies on cutting forces

The forces acting on the tool are important aspect of machining. Knowledge of the cutting forces is needed for the estimation of power requirements, the adequately rigid design of machine tool elements, tool-holders, and fixtures, for vibration free operations.

Many force measurement devices like dynamometers have been developed which are capable of measuring tool forces with increasing accuracy. Power consumed in metal cutting is largely converted into heat near the cutting edge of tool, and many of the economic and technical problems of machining are caused directly or indirectly by this heating action (Sharma, Dhiman, Sehgal, & Sharma, 2008).

By measuring the cutting forces, one is able to understand the cutting mechanism such as the effects of cutting variables on the cutting force, the machinability of the workpiece, the process of chip formation, chatter, and tool wear. It has been observed that engineering calculations used for obtaining the force values give some errors when compared to experimental measurements of the forces. The cutting force in even unsteady state conditions is affected by many parameters and the variation of cutting force with time has a typical characteristic. Cutting forces can be resolved into three components i.e. the radial thrust force (Fx), feed force (Fy), and tangential cutting force (Fz).Usually, the tangential cutting force is the largest of the three components, though in finish turning the radial thrust force is often larger, while the feed force is minimal. The findings of some of the research studies pertaining to the effect of cutting parameters on cutting forces are presented below.

The cutting forces increase drastically when machining materials with hardness higher than about 45 HRC (Davim, 2011). Kurt and Seker (2005) investigated the effects of chamfer angle on the cutting forces and the stresses on the PCBN cutting tools in finishing hard turning of AISI 52100 bearing steel. They found that the chamfer angles (20° and 30°) have a great influence on the passive cutting forces and Von Mises tool stresses distribution. Lee (2011) developed a theoretical model to predict cutting forces for machining AISI 4140-hardened materials (45 HRC) that contain more than 0.58% carbon. His predicted values of cutting forces from the model were found to be in good agreement with those measured from an experiment of hard machining of AISI 4140 steel heat-treated. Panzera, Souza, Rubio, Abrao, and Mansur (2012) investigated the effect of the cutting parameters (cutting speed, feed rate, and depth of cut) on the cutting force components during dry turning AISI 4340 steel using coated carbide inserts. The results of their study indicated that the three components of the turning force decrease slightly as cutting speed increases and also they increase linearly with feed rate and depth of cut. Further, the results of analysis of variance (ANOVA) revealed that the three components of the force are not significantly affected by cutting speed but they are significantly affected by feed rate and depth of cut. Aouici, Yallese, Fnides, and Mabrouki (2010) studied the influence of the cutting parameters (cutting speed, feed rate, and depth of cut) on cutting force components, surface roughness, temperature in the cutting zone, and tool life during turning of AISI H11 steel treated at 50 HRC using CBN tool (57% CBN and 35% Ti(C, N)). Their results indicated that (i) tangential cutting force is very sensitive to the variation in cutting depth, (ii) thrust force is dominating compared to both others cutting forces, (iii) surface roughness is very sensitive to the variation of the feed rate, and (iv) the temperature is greatly influenced by the cutting speed.

### 3.2. Studies on heat generation and cutting temperature

Most of the energy in the cutting process is converted into heat. This heat is generated by plastic deformation and friction at the tool–chip and the tool–workpiece interfaces. The generation of the heat during machining increases the temperature in the cutting zone which affects the strength, hardness, wear resistance, and life of the cutting tool and causes difficulty in controlling the dimensional accuracy and surface integrity. The temperature also causes thermal damage to the workpiece and affects its properties and service life.

The temperature in the cutting zone is affected by the cutting parameters. In addition, it also depends on the properties of the workpiece material, as well as on the physical properties of the tool. Therefore, considerable attention has been paid to the measurement and prediction of the temperatures at the tool, chip, and workpiece in the metal cutting (Aouici et al., 2010; Özel, 2009). The cutting tools that are used for machining should possess adequate hot hardness to withstand elevated temperatures generated at high-speed condition. Under these conditions, most tool materials generally lose their hardness resulting in the weakening of the interparticle bond strength and consequently, tool wear gets accelerated as reported by Ezugwu, Bonney, and Yamane (2003). Amritkar, Prakash, and Kulkarni (2012) performed machining of SAE 8620 material at various cutting speed and feed rate using uncoated tungsten carbide tool. They designed and developed a simple and economical technique of temperature measurement i.e. tool-work thermocouple setup for the measurement of the cutting temperature for which they calibrated the setup in order to establish a relationship between obtained e.m.f. during machining and the cutting temperature. They used regression analysis for establishing the relationship between temperature and the generated voltage.

They evaluated the performance of the setup for the different material like EN19, EN31, mild steel, SS 304, and SAE 8620 using uncoated tungsten carbide tool. Their obtained results confirmed that the setup is having better accuracy and good repeatability. Further, they observed that the tool-work thermocouple technique is the best method for measuring the average chip–tool interface temperature during metal cutting. Sutter, Faure, Molinari, Ranc, and Pina (2003) studied the effects of the cutting speed and depth of cut on the temperature profile at the chip during an orthogonal machining of 42 CrMo 4 steel using standard carbide tools TiCN coated. They performed the machining with a gas gun. It was found from their results that the temperature at the chip increases with the increase in both cutting speed as well as the depth of cut. Ren, Yang, James, and Wang (2004) determined the cutting temperatures during hard turning of high chromium hardfacing materials using PCBN tools. They found that the average cutting temperatures ranged from 600 to 700 °C which increased with higher cutting speed and feed rate. Abukhshim, Mativenga, and Sheikh (2005) used FEA to estimate the amount of heat flowing into the cutting tool in high-speed turning of AISI 4140 high strength alloy steel using uncoated cemented carbide. Their results showed that the maximum temperature at the tool–chip contact increases with cutting speed but not linearly and this could be attributed to the trend of the heat fraction flowing into the tool. Sutter and Ranc (2007) measured the temperature during machining of two steels i.e. C15 and 42CrMo4 for a range of cutting speed around 15–65 m/s. Their results showed that the increase in cutting speed from 10 to 65 m/s maximizes temperature at the chip continuously. List, Sutter, and Bouthiche (2012) predicted the interface cutting temperature and its relation with the crater wear mechanism. Their work was focused on the domain of the high-speed machining above 20 m/s. They analyzed the mechanical and thermal parameters that influenced the temperature distribution at the tool rake face.

### 3.3. Studies on surface finish and surface integrity

The surface integrity of a machined surface is defined in terms of residual stresses, surface roughness, microhardness, etc. Surface roughness and dimensional accuracy play an important role in the performance of a machined component. High cutting forces and high localized temperatures may dramatically affect the surface integrity, often resulting in the development of high tensile residual stresses in the machined surfaces. Residual stress on the machined surface and the subsurface is known to influence the service quality of a component, such as fatigue life, tribological properties, and distortion. Therefore, it is essential to predict and control it for enhanced performance as suggested by several researchers (Adesta et al., 2009; Aouici et al., 2010; Benga & Abrao, 2003; Chowdhury & Dhar, 2011; Diniz et al., 2003; Elmunafi, 2012; Fnides et al., 2009; Gaitonde et al., 2009; Godoy & Diniz, 2011; Kumar et al., 2003; More et al., 2006; Poulachon et al., 2001; Ramesh et al., 2005; Sharma et al., 2008; Thamizhmanii & Hasan, 2010; Thiele & Melkote, 1999; Yallese et al., 2009; Zawada-Tomkiewicz, 2011). In addition to the research studies cited above, Hua et al. (2005) analyzed the effects of cutting edge geometry, workpiece hardness, and cutting parameters, such as cutting speed and feed rate on subsurface residual stress in hard turning of AISI 52100 bearing steel. It was revealed from their analysis that hone edge plus chamfer cutting edge and aggressive feed rate help to increase both compressive residual stress and penetration depth. Moreover, using medium hone radius (0.02–0.05 mm) plus chamfer was good for keeping tool temperature and cutting force low, while obtaining desired residual stress profile. Rech et al. (2008) provided a comprehensive characterization of residual stresses that were developed in dry turning of a hardened AISI 52100 bearing steel using PCBN tools. For better understanding of the experimental results, they investigated the generated residual stresses in hard turning or in 'hard turning + belt finishing' by two complementary ways: an experimental X-ray diffraction characterization after each step of the process, and a finite element model of the belt finishing operation. Additionally, they explored the sensitivity of some parameters such as the lubrication and the indentation force during belt finishing. They observed that the belt finishing process improves the surface integrity by the induction of strong compressive residual stresses in the external layer and by a great improvement of the surface roughness. In addition, they also found that among the process parameters of the belt finishing technique, the lubrication is a key parameter to get compressive stresses. Caruso, Umbrello, Outeiro, Filice, and Micari (2011) examined the effects of the tool cutting-edge geometry, workpiece hardness, cutting speed, and microstructural changes (white and dark layers) on the residual stresses in dry orthogonal hard machining of AISI 52100 steel using PCBN tool inserts. Their results showed that tool geometry, workpiece hardness, and cutting parameters significantly affect the surface residual stress, maximum compressive residual stress below the machined surface, and its location. Moreover, microstructural analysis showed that thermally-induced phase transformations have a significant impact on the magnitude and location of this maximum compressive residual stress peak.

On the other hand, a number of investigations have been carried out in the past to assess the surface roughness that could be achieved with hard turning in comparison to grinding. Asilturk and Akkus (2011) investigated the effects of cutting speed, feed rate, and depth of cut on surface roughness during dry turning of AISI 4140 (51 HRC) with coated carbide cutting tools. Their results indicated that the feed rate has the most significant effect on the surface roughness. In addition, the effects of two factor interactions of the feed rate-cutting speed and depth of cut-cutting speed appear to be important. Aouici, Yallese, Chaoui, and Mabrouki (2012) investigated the effects of cutting speed, feed rate, workpiece hardness, and depth of cut on surface roughness in the hard turning of AISI H11 steel (hardened to (40; 45; and 50) HRC) using CBN which is essentially made of 57% CBN and 35% TiCN. Their results showed that both the feed rate and workpiece hardness have statistically significant effect on the surface roughness. Further, the best surface roughness was achieved at the lower feed rate and the highest cutting speed. Umbrello et al. (2011) investigated the effects of cryogenic coolant on surface integrity in orthogonal machining of hardened AISI 52100 bearing steel using chamfered CBN tool inserts. The results of their study showed that the use of cryogenic coolant significantly affects the surface integrity and improves product's functional performance. Grzesik, Żak, Prażmowski, Storch, and Pałka (2012) explored the effect of cryogenic cooling on the surface integrity produced in hard turning of low alloy 41Cr4 steel with hardness of $57 \pm 2$ HRC using low content CBN tools containing about 60% CBN. They confirmed that hard machining produces surfaces with acceptable surface roughness and, in some cases, with attractive service properties. They also observed that cryogenic hard cutting operations can partly eliminate grinding operations in cases when white layer is not produced. Abhang and Hameedullah (2012) determined the optimum cutting parameters (cutting speed, feed rate, tool nose radius, and concentration of solid–liquid lubricants (MQL)) for the multiperformance characteristics (surface roughness and chip thickness) in the turning of EN-31 steel by using the grey relational analysis coupled with factorial design. They concluded that gray relational analysis coupled with factorial design can effectively be used to obtain the optimal combination of cutting parameters. Additionally, they observed that the surface roughness and the chip thickness in the turning process can be improved effectively through this approach. Finally, they found that the minimum surface roughness and smallest chip thickness are 9.83 μm and 0.32 mm, respectively, which are obtained at optimal conditions at: 1200 rpm cutting speed; 0.06 mm/rev feed rate; 0.8 mm nose radius; and concentration of solid–liquid lubricant (10% boric acid + SAE-40 base oil. Islam (2013) investigated the effect of cooling method, blank size, and work material on the dimensional accuracy characteristics (diameter error and circularity) and surface roughness of turned parts under three turning conditions (dry, flood, and MQL) for different materials (aluminum (AISI 6061), mild steel (AISI 1030), and alloy steel (AISI 4340)). He analyzed the results by applying three methods: traditional analysis, Pareto ANOVA, and Taguchi method. He noticed that cooling method, blank size, and work material have demonstrated considerable effects on the dimensional accuracy characteristics, whereas the effect of these parameters on surface roughness is relatively low. He also revealed that while work material has the greatest effect on diameter error and surface roughness, the major contributor to circularity is blank size. Finally, dimensional accuracy and surface roughness of different work materials were influenced differently by the cooling methods, and in most cases the best result was achieved by MQL.

### 3.4. *Studies on the tool wear*

During machining, the cutting tools are subjected to severe forces and temperature which may cause tool wear and therefore, it is necessary to study and predict the tool wear during machining for the effective design of cutting tools and determination of cutting conditions that will lead to the formulation of the tool change strategies. Intensive research studies pertaining to the tool wear have been carried out in the past century which has contributed greatly to the understanding of the factors responsible for the tool wear and also the tool wear mechanisms. Many researchers focused their studies on the prediction of the tool wear during hard turning (Adesta et al., 2009; Arsecularatne et al., 2006, Arsecularatne et al., 2006; Aslan, 2005; Chowdhury & Dhar, 2011; Diniz et al., 2003; Elmunafi, 2012; Fnides et al., 2009; Gaitonde et al., 2009; Galoppi et al., 2006; Godoy and Diniz, 2011; Huang et al., 2007; Kumar et al., 2003; Lima et al., 2007; Lin et al., 2008; Luo et al., 1999; Meyer et al., 2012; More et al., 2006; Oliveira et al., 2009; Özel, 2009; Poulachon et al., 2001, 2003, 2004; Remadna & Rigal, 2006; Sahin, 2009; Suresh et al., 2012; Thamizhmanii & Hasan, 2010; Yallese et al., 2009; Zawada-Tomkiewicz, 2011; Zhao et al., 2010). In addition to the references cited before for the tool wear study, Grzesik and Zalisz (2008) investigated the wear phenomenon of the mixed ceramic tips during dry hard turning of AISI 5140 steel (60 HRC). They performed finishing cuts under varying feed rate, constant cutting speed of 100 m/min and small depth of cut of 0.2 mm. It was observed from their results that depending on the mechanical and thermal conditions generated on the wear zones, the wear mechanisms involve abrasion, fracture, plastic flow, material transfer, and tribochemical. El Hakim, Abad, Abdelhameed, Shalaby, and Veldhuis (2011) studied the performance of different tool materials, PCBN (CBN + TiN), TiN coated PCBN (CBN + TiN), mixed alumina ceramic ($Al_2O_3$ + TiC), coated tungsten carbide (TiN coated over a multilayer coating (TiC/TiCN/$Al_2O_3$)) in the machining of medium hardened steel AISI T15 HSS. Their results indicated that the mixed alumina ceramic and coated carbide tool materials have longer tool life than PCBN tools when they machined the selected workpiece material.

Chinchanikar and Choudhury (2013) investigated the effect of workpiece hardness, cutting parameters, and type of coating for coated cemented carbide inserts on flank wear during turning of hardened AISI 4340 steel at different levels of hardness. The results of their study revealed that flank wear is dominant wear form for CVD coated tool and crater wear is dominant wear form for PVD coated tool. Further, they found that abrasion and adhesion are the main causes for wear of CVD coated tool and abrasion, adhesion and diffusion leads to the wear of PVD coated tool. Gaitonde et al. (2009) studied the relationships between the cutting parameters (cutting speed, feed rate, and machining time) on tool wear. They used RSM to analyze the effects of process parameters on machinability during turning of high chromium AISI D2 cold work tool steel using CC650WG wiper ceramic inserts. They found that the maximum tool wear occurs at a cutting speed of 150 m/min for all values of feed rate and the tool wear increases with the increase in machining time. Dogra, Sharma, Sachdeva, Suri, and Dureja (2011) made comparison between the performance of CBN inserts with coated carbide and cryogenically treated coated/uncoated carbide inserts in terms of flank wear during finish turning of hardened AISI H11 steel (48–49 HRC). They indicated that the flank wear of CBN is lower than that of other inserts.

## 4.  Summary

Hard machining including hard turning is widely used by several manufacturing industries such as ball bearings, automotive, gear, and die-making industries, since it offers numerous advantages when compared with traditional methodology based on finish-grinding operation after heat treatment of workpieces. This technology possesses immense potential to machine very hard materials to produce near net shape components and also to contribute to a great extent to the sustainable manufacturing. In the recent past, this technology has created interest among researchers and attracted their attention, and consequently, many research studies have been conducted and the results have been reported which are available in the literature. This paper has presented an overview focusing mainly on turning of hardened steels that are used by ball bearings, automotive, gear, and die-making industries. On the basis of the research findings reported in the available literature reviewed and presented in this paper, following conclusions can be drawn:

- Hard turning offers a number of potential benefits over traditional form grinding, including lower equipment costs, shorter setup time, fewer process steps, greater part geometry flexibility, and elimination of the use of cutting fluid.
- During hard turning, the cutting tool is subjected to heavy mechanical loads and also it is exposed to very high temperature due to excessive heat generation, and therefore, cutting tools made of superhard materials such as coated cemented carbide, ceramic, PCD, CBN, etc. must be used as they perform well under severe machining conditions.
- The cutting tool geometry, workpiece hardness, and cutting parameters significantly affect the cutting forces, surface residual stress, surface roughness, surface integrity, tool wear, and tool life.
- The information gained through study and prediction of the tool wear during machining can be used as a basis for the effective design of cutting tools and determination of cutting conditions that will lead to the formulation of the tool change strategies.

- The complex phenomena involved in hard turning can be studied through simulation and modeling using techniques such as FEM, ANN, etc. and the results of the models can be validated with experimental results.

Thus, it is concluded that the information gathered through extensive literature review has been presented in a modular way in this paper. The review has been organized in terms of role of machining parameters on machining of hard steel, cutting force, heat generation, and temperature evolution during machining, surface integrity, and tool wear during hard machining, etc. The information presented are immensely useful to the researchers in identifying solutions to the several machining problems to mention some: (i) to identify strategies with regard to tool edge geometry, cutting parameters, etc. for specific work material hardness so as to obtain better surface integrity and surface finish; and (ii) to identify prevalent wear mechanisms and appropriate tool material for specified machining situations.

## References

Abhang, L. B., & Hameedullah, M. (2012). Determination of optimum parameters for multi-performance characteristics in turning by using grey relational analysis. *The International Journal of Advanced Manufacturing Technology, 63*, 13–24.

Abukhshim, N. A., Mativenga, P. T., & Sheikh, M. A. (2005). Investigation of heat partition in high speed turning of high strength alloy steel. *International Journal of Machine Tools and Manufacture, 45*, 1687–1695.

Adesta, E. Y. T., Riza, M., Hazza, M., Agusman, D., & Rosehan. (2009). Tool wear and surface finish investigation in high speed turning using cermet insert by applying negative rake angles. *European Journal of Scientific Research, 38*, 180–188.

Amritkar, A., Prakash, C., & Kulkarni, A. P. (2012). Development of temperature measurement setup for machining. *World Journal of Science and Technology, 2*, 15–19.

Aouici, H., Yallese, M. A., Chaoui, K., & Mabrouki, T. (2012). Analysis of surface roughness and cutting force components in hard turning with CBN tool: Prediction model and cutting conditions optimization. *Measurement, 45*, 344–353.

Aouici, H., Yallese, M. A., Fnides, B., & Mabrouki, T. (2010). Machinability investigation in hard turning of AISI H11 hot work steel with CBN tool. *MECHANIKA, 86*, 71–77.

Arsecularatne, J. A., Zhanga, L. C., & Montross, C. (2006). Wear and tool life of tungsten carbide, PCBN and PCD cutting tools. *International Journal of Machine Tools & Manufacture, 46*, 482–491.

Arsecularatne, J. A., Zhang, L. C., Montross, C., & Mathew, P. (2006). On machining of hardened AISI D2 steel with PCBN tools. *Journal of Materials Processing Technology, 171*, 244–252.

Asilturk, I., & Akkus, H. (2011). Determining the effect of cutting parameters on surface roughness in hard turning using the Taguchi method. *Measurement, 44*, 1697–1704.

Aslan, E. (2005). Experimental investigation of cutting tool performance in high speed cutting of hardened X210 Cr12 cold-work tool steel (62 HRC). *Materials and Design, 26*, 21–27.

Benga, G. C., & Abrao, A. M. (2003). Turning of hardened 100Cr6 bearing steel with ceramic and PCBN cutting tools. *Journal of Materials Processing Technology, 143–144*, 237–241.

Caruso, S., Umbrello, D., Outeiro, J. C., Filice, L., & Micari, F. (2011). An experimental investigation of residual stresses in hard machining of AISI 52100 steel. *Procedia Engineering, 19*, 67–72.

Chinchanikar, S., & Choudhury, S. K. (2013). Investigations on machinability aspects of hardened AISI 4340 steel at different levels of hardness using coated carbide tools. *International Journal of Refractory Metals and Hard Materials, 38*, 124–133.

Chowdhury, N. T., & Dhar, N. R. (2011). Experimental analysis and modeling of tool wear and surface roughness in hard turning under minimum quantity lubricant environment. Proceedings of the International Conference on Industrial Engineering and Operations Management, Malaysia, 326–330.

Davim, J. P. (2011). *Machining of hard materials*. London: Springer-Verlag. ISBN 1849964505, 9781849964500. doi:10.1007/978-1-84996-450-0

Deepakkumar, H. P., & Sadaiah, M. (2011). Investigations on finish turning of AISI 4340 steel in different cutting environments by CBN insert. *International Journal of Engineering Science and Technology, 3*, 7690–7706.

Diniz, A. E., Ferreira, J. R., & Filho, F. T. (2003). Influence of refrigeration/lubrication condition on SAE 52100hardened steel turning at several cutting speeds. *International Journal of Machine Tools & Manufacture, 43*, 317–326.

Diniz, A. E., Gomes, D. M., & Braghini, A., Jr. (2005). Turning of hardened steel with interrupted and semi-interrupted cutting. *Journal of Materials Processing Technology, 159*, 240–248.

Dogra, M., Sharma, V. S., Sachdeva, A., Suri, N. M., & Dureja, J. S. (2011). Performance evaluation of CBN, coated carbide, cryogenically treated uncoated/coated carbide inserts in finish-turning of hardened steel. *The International Journal of Advanced Manufacturing Technology, 57*, 541–553.

El Hakim, M. A., Abad, M. D., Abdelhameed, M. M., Shalaby, M. A., & Veldhuis, S. C. (2011). Wear behavior of some cutting tool materials in hard turning of HSS. *Tribology International, 44*, 1174–1181.

Elmunafi, M. H. S. (2012). *Performance evaluation of wiper coated ceramic tool when hard turning tool steel* (Dissertation thesis). University Technology of Malaysia.

Ezugwu, E. O., Bonney, J., & Yamane, Y. (2003). An overview of the machinability of aeroengine alloys. *Journal of Materials Processing Technology, 134*, 233–253.

Fnides, B., Yallese, M. A., Mabrouki, T., & Rigal, J.-F. (2009). Surface roughness model in turning hardened hot work steel using mixed ceramic tool. *MECHANIKA, 77*, 68–73.

Fnides, B., Yallese, M. A., Mabrouki, T., & Rigal, J.-F. (2011). Application of response surface methodologyfor determining cutting force model in turning hardened AISI H11 hot work tool steel. *Sadhana, 36*, 109–123.

Gaitonde, V. N., Karnik, S. R., Figueira, L., & Davim, J. P. (2009a). Machinability investigations in hard turning of AISI D2 cold work tool steel with conventional and wiper ceramic inserts. *International Journal of Refractory Metals & Hard Materials, 27*, 754–763.

Gaitonde, V. N., Karnik, S. R., Figueira, L., & Davim, J. P. (2009b). Analysis of machinability during hard turning of cold work tool steel (Type: AISI D2). *Materials and Manufacturing Processes, 24*, 1373–1382.

Galoppi, G. de S., Filho, M. S., & Batalha, G. F. (2006). Hard turning of tempered DIN 100Cr6 steel with coated and no coated CBN inserts. *Journal of Materials Processing Technology, 179*, 146–153.

Godoy, V. A. A. de, & Diniz, A. E. (2011). Turning of interrupted and continuous hardened steel surfaces using ceramic and CBN cutting tools. *Journal of Materials Processing Technology, 211*, 1014–1025.

Grzesik, W., Żak, K., Prażmowski, M., Storch, B., & Pałka, T. (2012). Effects of cryogenic cooling on surface layer characteristics produced by hard turning. *Archives of Materials Science and Engineering, 54*, 5–12.

Grzesik, W., & Zalisz, Z. (2008). Wear phenomenon in the hard steel machining using ceramic tools. *Tribology International, 41*, 802–812.

Hua, J., Shivpuri, R., Cheng, X., Bedekar, V., Matsumotob, Y., Hashimoto, F., & Watkins, T. R. (2005). Effect of feed rate, workpiece hardness and cutting edge on subsurface residual stress in the hard turning of bearing steel using chamfer + hone cutting edge geometry. *Materials Science and Engineering: A, 394*, 238–248.

Huang, Y., Chou, Y. K., & Liang, S. Y. (2007). CBN tool wear in hard turning: A survey on research progresses. *The International Journal of Advanced Manufacturing Technology, 35*, 443–453.

Hwang, Y. K., & Lee, Ch. M. (2010). Surface roughness and cutting force prediction in MQL and wet turning process of AISI 1045 using design of experiments. *Journal of Mechanical Science and Technology, 24*, 1669–1677.

Islam, M. N. (2013). Effect of additional factors on dimensional accuracy and surface finish of turned parts. *Machining Science and Technology, 17*, 145–162.

Kumar, A. S., Durai, A. R., & Sornakumar, T. (2003). Machinability of hardened steel using alumina based ceramic. *International Journal of Refractory Metals & Hard Materials, 21*, 109–117.

Kurt, A., & Seker, U. (2005). The effect of chamfer angle of polycrystalline cubic boron nitride cutting tool on the cutting forces and the tool stresses in finishing hard turning of AISI 52100 steel. *Materials and Design, 26*, 351–356.

Lee, T. H. (2011). Development of a theoretical model to predict cutting forces for hard machining. *International Journal of Precision Engineering and Manufacturing, 12*, 775–782.

Lima, J. G. de, de Avila, R. F., & Abrao, A. M. (2007). Turning of hardened AISI 4340 steel using coated carbide inserts. *Proceedings of the Institution of Mechanical Engineers, Part B: Journal of Engineering Manufacture, 221*, 1359–1366.

Lin, H. M., Liao, Y. S., & Wei, C. C. (2008). Wear behavior in turning high hardness alloy steel by CBN tool. *Wear, 264*, 679–684.

List, G., Sutter, G., & Bouthiche, A. (2012). Cutting temperature prediction in high speed machining by numerical modelling of chip formation and its dependence with crater wear. *International Journal of Machine Tools and Manufacture, 54–55*, 1–9.

Luo, S. Y., Liao, Y. S., & Tsai, Y. Y. (1999). Wear characteristics in turning high hardness alloy steel by ceramic and CBN tools. *Journal of Materials Processing Technology, 88*, 114–121.

M'Saoubi, R., Outeiro, J. C., Chandrasekaran, H., Dillon, O. W., Jr, & Jawahir, I. S. (2008). A review of surface integrity in machining and its impact on functional performance and life of machined products. *International Journal of Sustainable Manufacturing, 1*, 203–236.

Maňková, I., Kovac, P., Kundrak, J., & Beňo, J. (2011). Finite element analysis of hardened steel cutting. *Journal of production engineering, 14*, 7–10.

Meyer, R., Köhler, J., & Denkena, B. (2012). Influence of the tool corner radius on the tool wear and process forces during hard turning. *The International Journal of Advanced Manufacturing Technology, 58*, 933–940.

More, A. S., Jiang, W., Brown, W. D., & Malshe, A. P. (2006). Tool wear and machining performance of CBN–TiN coated carbide inserts and PCBN compact inserts in turning AISI 4340 hardened steel. *Journal of Materials Processing Technology, 180*, 253–262.

Oliveira, A. J. de, Diniz, A. E., & Ursolino, D. J. (2009). Hard turning in continuous and interrupted cut with PCBN and whisker-reinforced cutting tools. *Journal of Materials Processing Technology, 209*, 5262–5270.

Özel, T. (2009). Computational modelling of 3D turning: Influence of edge micro-geometry on forces, stresses, friction and tool wear in PCBN tooling. *Journal of Materials Processing Technology, 209*, 5167–5177.

Panzera, T. H., Souza, P. R., Rubio, J. C. C., Abrao, A. M., & Mansur, T. R. (2012). Development of a three-component dynamometer to measure turning force. *The International Journal of Advanced Manufacturing Technology, 62*, 913–922.

Poulachon, G., Bandyopadhyay, B. P., Jawahir, I. S., Pheulpin, S., & Seguin, E. (2003). The influence of the microstructure of hardened tool steel workpiece on the wear of PCBN cutting tools. *International Journal of Machine Tools and Manufacture, 43*, 139–144.

Poulachon, G., Bandyopadhyay, B. P., Jawahir, I. S., Pheulpin, S., & Seguin, E. (2004). Wear behavior of CBN tools while turning various hardened steels. *Wear, 256*, 302–310.

Poulachon, G., Moisan, A., & Jawahir, I. S. (2001). Tool-wear mechanisms in hard turning with polycrystalline cubic boron nitride tools. *Wear, 250*, 576–586.

Ramesh, A., Melkote, S. N., Allard, L. F., Riester, L., & Watkins, T. R. (2005). Analysis of white layers formed in hard turning of AISI 52100 steel. *Materials Science and Engineering: A, 390*, 88–97.

Rech, J., Kermouche, G., Grzesik, W., Garcia-Rosales, C., Khellouki, A., & Garcia-Navas, V. (2008). Characterization and modelling of the residual stresses induced by belt finishing on a AISI52100hardened steel. *Journal of Materials Processing Technology, 208*, 187–195.

Remadna, M., & Rigal, J. F. (2006). Evolution during time of tool wear and cutting forces in thecase of hard turning with CBN inserts. *Journal of Materials Processing Technology, 178*, 67–75.

Ren, X. J., Yang, Q. X., James, R. D., & Wang, L. (2004). Cutting temperatures in hard turning chromium hardfacings with PCBN tooling. *Journal of Materials Processing Technology, 147*, 38–44.

Sahin, Y. (2009). Comparison of tool life between ceramic and cubic boron nitride (CBN) cutting tools when machining hardened steels. *Journal of materials processing technology, 209,* 3478–3489.

Sharma, V. S., Dhiman, S., Sehgal, R., & Sharma, S. K. (2008). Estimation of cutting forces and surface roughness for hard turning using neural networks. *Journal of Intelligent Manufacturing, 19,* 473–483.

Suresh, R., Basavarajappa, S., & Samuel, G. L. (2012). Some studies on hard turning of AISI 4340 steel using multilayer coated carbide tool. *Measurement, 45,* 1872–1884.

Sutter, G., Faure, L., Molinari, A., Ranc, N., & Pina, V. (2003). An experimental technique for the measurement of temperature fields for the orthogonal cutting in high speed machining. *International Journal of Machine Tools and Manufacture, 43,* 671–678.

Sutter, G., & Ranc, N. (2007). Temperature fields in a chip during high-speed orthogonal cutting – An experimental investigation. *International Journal of Machine Tools and Manufacture, 47,* 1507–1517.

Thamizhmanii, S., & Hasan, S. (2010). Relationship between flank wear and cutting force on the machining of hard Martensitic stainless steel by super hard tools. Proceedings of the World Congress on Engineering, London, U.K, 3, 2185–2190.

Thiele, J. D., & Melkote, S. N. (1999). Effect of cutting edge geometry and workpiece hardness on surface generation in the finish hard turning of AISI 52100 steel. *Journal of Materials Processing Technology, 94,* 216–226.

Umbrello, D., Rizzuti, S., Outeiro, J. C., Shivpuri, R., & M'Saoubi, R. (2008). Hardness-based flow stress for numerical simulation of hard machining AISI H13 tool steel. *Journal of Materials Processing Technology, 199,* 64–73.

Umbrello, D., Ambrogio, G., Filice, L., & Shivpuri, R. (2008). A hybrid finite element method–artificial neural network approach for predicting residual stresses and the optimal cutting conditions during hard turning of AISI 52100 bearing steel. *Materials and Design, 29,* 873–883.

Umbrello, D., Pu, Z., Carusoa, S., Outeiro, J. C., Jayal, A. D., Dillon, O. W., Jr, & Jawahir, I. S. (2011). The effects of cryogenic cooling on surface integrity in hard machining. *Procedia Engineering, 19,* 371–376.

Yallese, M. A., Chaoui, K., Zeghib, N., Boulanouar, L., & Rigal, J.-F. (2009). Hard machining of hardened bearing steel using cubic boron nitride tool. *Journal of Materials Processing Technology, 209,* 1092–1104.

Zawada-Tomkiewicz, A. (2011). Analysis of surface roughness parameters achieved by hard turning with the use of PCBN tools. *Estonian Journal of Engineering, 17,* 88–99.

Zhao, J., Yuan, X., & Zhou, Y. (2010). Cutting performance and failure mechanisms of an $Al_2O_3$/WC/TiCmicro- nano-composite ceramic tool. *International Journal of Refractory Metals and Hard Materials, 28,* 330–337.

**Appendix**

Major findings of the researches presented in this review are summarized in the following table.

| Author(s) | Tool material | Cutting parameters | Remarks |
|---|---|---|---|
| Thiele and Melkote (1999) | CBN tools | Tool edge geometries, workpiece hardness, and feed rate | Large edge hones of CBN tools result in higher forces in the axial, radial, and tangential directions as compared to the tools with small edge hone during turning AISI 52100 steel |
| Ramesh et al. (2005) | CBN | Cutting speed | The grains of white layers are smaller than the grains of the bulk material. Machining at higher speeds results in the formation of coarser grains |
| Diniz et al. (2005) | CBN | Two different kinds of CBN cutting tools (high CBN content material and CBN content material), cutting speed, three cutting operations continuous, semi-interrupted surfaces interrupted cutting) | Low CBN content tool is more suitable for continuous and semi-interrupted surfaces, whereas the high CBN content tool presents a slightly longer life during interrupted cutting |
| Adesta et al. (2009) | Cermet tools | Rake angles, cutting speed | Increase in negative rake angles causes higher wear, shorter duration of tool life, and poor surface finish. Further, high cutting speed turning gave shorter tool life, high wear rate but finer surface finish than conventional one |
| Gaitonde et al. (2009) | Ceramic inserts with TiN coating (CC650, CC650WG, and GC6050WH) | Depth of cut and machining time | CC650WG wiper insert performs better with reference to surface roughness and tool wear. The surface roughness is minimal at lower values of depth of cut and machining time in case of CC650 and CC650WG inserts machining, while the minimum surface roughness occurs at 0.4 mm depth of cut for hard turning with GC6050WH insert. Finally, the CC650 conventional insert is useful in reducing the cutting force |
| Fnides et al. (2009) | Mixed ceramic tool(insert CC650) | Cutting speed, feed rate, depth of cut | Statistical models of surface roughness criteria were developed |

| Reference | Tool material | Factors | Findings |
|---|---|---|---|
| Meyer et al. (2012) | CBN content of 56% | Geometrical contact conditions | It was suggested that by applying a load-specific tool and process design, the tool operation behavior and the productivity of the process can be enhanced in a beneficial way |
| Aslan (2005) | TiCN coated tungsten carbide, TiCN + TiAlN coated tungsten carbide, TiAlN coated cermet, mixed ceramic with $Al_2O_3$ + TiCN and CBN | Different cutting tool materials | TiAlN coated carbide and cermet tools perform slightly better than TiCN coated carbide tool which can be attributed to better high-temperature properties of TiAlN compared to TiCN. CBN tool exhibits the best cutting performance in terms of both flank wear and surface finish and the highest volume of metal removal is obtained with CBN |
| Lima et al. (2007) | Coated carbide | Workpiece hardness, cutting speed, feed rate, depth of cut, cutting time | The abrasion is the principal wear mechanism and catastrophic failure takes place when attempting to machine the AISI 4340 steel with 525 HV |
| Chowdhury and Dhar (2011) | Coated carbide insert | Tool geometry | The application of MQL technique significantly helps to obtain better performance of coated carbide insert in comparison to dry condition |
| Hwang and Lee (2010) | Coated carbide insert | Cutting speed, Feed rate, Depth of cut, Supplied air pressure, Nozzle diameter | Turning of AISI 1045 with coated carbide inserts under MQL results in better surface finish than wet turning |
| Luo et al. (1999) | CBN Ceramic Carbide P10 | Cutting speed, feed rate, depth of cut | A protective layer formed on the chip–tool interface plays an important role in wear behavior of CBN and ceramics tools |
| Kumar et al. (2003) | Ti[C,N] mixed alumina ceramic and zirconia toughened alumina ceramic | | Ti[C, N] mixed alumina ceramic cutting tool produces the best surface finish during turning of EN 24 steel (HRC 40 and HRC 45) |

(Continued)

**Appendix** (*Continued*).

| Author(s) | Tool material | Cutting parameters | Remarks |
|---|---|---|---|
| Gaitonde (2009) | CC650, CC650WG, and GC6050WH ceramic inserts | Cutting speed, feed rate, machining time | CC650WG wiper insert performs better with reference to surface roughness and tool wear, while the CC650 conventional insert is useful in reducing the machining force, power and specific cutting force during turning of high chromium AISI D2 cold work tool steel |
| Elmunafi (2012) | Wiper coated ceramic | Cutting speed, feed rate | Wiper inserts produce better surface finish when turning ASSAB DF-3 grade hardened steel with hardness 55 HRC |
| Maňková et al. (2011) | Mixed oxide ceramic inserts (70% $Al_2O_3$ + 30% TiC) | Cutting speed, feed rate | Ceramic cutting tools have an advantage in the machining of hard work piece materials at high speed |
| Fnides et al. (2011) | Mixed ceramic tool | Cutting speed, feed rate, depth of cut | Statistical models of cutting force components for turning of AISI H11 grade steel treated at 50 HRC, with a mixed ceramic tool (insert CC650) was developed |
| Zhao et al. (2010) | The $Al_2O_3$-based composite ceramic tool material reinforced with WC microparticles and TiC nanoparticles | Cutting speed, depth of cut | The tool life of the $Al_2O_3$/WC/TiC ceramic tool increases when the cutting speed increases to 170 m/min. The longer tool life of the $Al_2O_3$/WC/TiC composite ceramic tool is attributed to its synergistic strengthening/toughening mechanisms induced by the WC microparticles and TiC nanoparticles |
| Oliveira et al. (2009) | PCBN and alumina-based ceramic reinforced with silicon carbide tools | Continuous and interrupted surfaces | The longest tool life is achieved using PCBN in continuous turning, but similar tool longevity is attained in interrupted turning using both PCBN and ceramic. PCBN tools result in lower roughness for continuous and interrupted surfaces |
| Sahin (2009) | Mixed alumina ceramic with that of $Al_2O_3$ (70%):TiC (30%) matrix and CBN cutting tools | Cutting speed, feed rate, tool hardness | The CBN cutting tool has the best performance as compared to the ceramic based cutting tool |

| Reference | Tool | Parameters | Findings |
|---|---|---|---|
| Poulachon et al. (2001) | PCBN cutting tool | Workpiece hardness | TiN coated PCBN improves the tool-wear and hence the tool-life by reducing the diffusion wear between workpiece and tool rake face. |
| Diniz et al. (2003) | CBN | Cutting speed, and under three cutting conditions: dry cutting, wet cutting, and minimum volume of oil i.e. MVO (oil flow of 10 ml/h) | Surface roughness in all three cutting conditions i.e. dry cutting, wet cutting, and minimum volume of oil cutting is the lowest at the highest cutting speed |
| Poulachon et al. (2003) | PCBN | Steel workpiece with different hardness | A large variation in tool wear occurs in the machining of these steels |
| Benga and Abrao (2003) | Mixed alumina, whisker reinforced alumina, and PCBN inserts | Cutting speed, feed rate | At low feed rates PCBN compact tool gives best results in terms of the tool life followed by the mixed alumina tool. However, as feed rate is increased tool life of the whisker reinforced alumina tool is better |
| Poulachon et al. (2004) | CBN | Different steels (AISI D2 cold work steel, AISI H11 hot work steel, 35NiCrMo16 hot work steel and AISI 52100 bearing steel), cutting speed, and feed rate | The generated tool flank grooves are correlated with the hard carbide content of the workpieces |
| Kurt and Seker (2005) | PCBN | Chamfer angle | Chamfer angle significantly affects the cutting forces and the tool stress |
| Galoppi et al. (2006) | Different types of CBN inserts with wiper geometry, coated with TiAlN and TiN as well as with no coated ones | Cutting speed, and feed rate | CBN insert with a wiper geometry presents longer tool life compared to no coated tools |
| Remadna and Rigal (2006) | CBN | Cutting speed | The wear of CBN does not directly or appreciably affect the manufactured surface |
| More et al. (2006) | CBN–TiN coated carbide inserts and PCBN compact inserts | Cutting speed, and feed rate | The crater wear of the CBN–TiN coated inserts is less than that of the PCBN inserts because of the lubricity of TiN capping layer on the CBN–TiN coating |
| Arecularatne et al. (2006) | PCBN tools | Cutting speed, and feed rate | The highest acceptable values of tool life and volume of material removal is obtained at low speed |

(Continued)

**Appendix** (*Continued*).

| Author(s) | Tool material | Cutting parameters | Remarks |
|---|---|---|---|
| Lin et al. (2008) | CBN tools which contained 50% and 45% of CBN | Cutting speed | At high cutting speed a protective layer is formed on the chip–tool interface which prolongs life of the CBN tool. With further increase in the cutting speed (i) cutting temperature becomes the dominant factor instead of the cutting force which causes inhomogeneous shear strain, and a transition from continuous chip to saw-tooth chip, (ii) the friction force increases because of the very irregular chip–tool contact and it results in the removal of the protective layer, and (iii) the bond between tool particles gets weakened due to serious diffusion between the work material and the cutting tool. Subsequently, hard particles of the tool get detached from it and the tool life is reduced |
| Özel (2009) | PCBN | Different edge micro-geometry, cutting speed, and feed rate | Micro-geometry insert edge design improves surface integrity, the tool life, and reduces tool wear depth and wear rate |
| Yallese et al. (2009) | CBN | Cutting speed, feed rate, and depth of cut | At very high cutting speed the machining system becomes unstable and produces significant sparks and vibrations after only a few minutes of work and causes tool wear |
| Godoy and Diniz (2011) | CBN and ceramic tools | Continuous and interrupted cutting, cutting speed | The CBN tools exhibit a much better performance with respect to both tool life and workpiece surface roughness than the ceramic tools |
| Zawada-Tomkiewicz (2011) | PCBN coated and uncoated tools | | Turning with PCBN tools can produce a very smooth and uniform surface |
| | | | The wedge's wear is almost unnoticeable for PCBN coated tool material |

| | | | |
|---|---|---|---|
| Deepakkumar and Sadaiah (2011) | CBN | Cutting speed, feed rate, depth of cut, and different cutting environments (dry, wet, and MQL) | The tool wear increases with increase in the cutting speed and feed rate |
| Panzera et al. (2012) | Coated carbide inserts | Cutting speed, feed rate, and depth of cut | The three components of the turning force decrease slightly as cutting speed increases and also they increase linearly with feed rate and depth of cut |
| Aouici et al. (2010) | CBN tool (57% CBN and 35% Ti(C, N)) | Cutting speed, feed rate, and depth of cut | Tangential cutting force is very sensitive to the variation in cutting depth. Thrust force is dominating compared to both others cutting forces. Surface roughness is very sensitive to the variation of the feed rate. The temperature is greatly influenced by the cutting speed |
| Amritkar et al. (2012) | Uncoated tungsten carbide tool | Cutting speed and feed rate | The tool-work thermocouple technique is the best method for measuring the average chip–tool interface temperature during metal cutting |
| Sutter et al. (2003) | Carbide tools TiCN coated | Cutting speed and depth of cut | The temperature at the chip increases with the increase in both cutting speed as well as the depth of cut |
| Ren et al. (2004) | PCBN | | The average cutting temperature increases with increase in both the cutting speed and feed rate |
| Abukhshim et al. (2005) | Uncoated cemented carbide | Cutting speed | The maximum temperature at the tool–chip contact increases with cutting speed but not linearly and this could be attributed to the trend of the heat fraction flowing into the tool |

(Continued)

**Appendix** (*Continued*).

| Author(s) | Tool material | Cutting parameters | Remarks |
|---|---|---|---|
| Rech et al. (2008) | PCBN | Cutting speed, feed, depth of cut, tool geometry (all fixed values), and subsequent belt finishing | Belt finishing process improves the surface integrity by the induction of strong compressive residual stresses in the external layer and due to reasonable improvement in the surface roughness<br>Lubrication used in the belt finishing technique is a key parameter to get compressive stresses to improve surface roughness |
| Caruso et al. (2011) | PCBN | Tool geometry, workpiece hardness, cutting speed | Tool geometry, workpiece hardness, and cutting parameters significantly affect the surface residual stress |
| Asilturk and Akkus (2011) | Coated carbide | Cutting speed, feed rate, and depth of cut | The feed rate significantly affects the surface roughness |
| Aouici et al. (2012) | CBN which is essentially made of 57% CBN and 35% TiCN | Workpiece hardness, cutting speed, feed rate, and depth of cut | The feed rate and workpiece significantly affect the surface roughness |
| Umbrello et al. (2011) | CBN | Under dry and cryogenic conditions, cutting speed | Cryogenic coolant significantly affects the surface integrity and improves product's functional performance |
| Grzesik et al. (2012) | CBN tools containing about 60% CBN | Under dry and cryogenic conditions, feed rate | Hard machining produces surfaces with acceptable surface roughness and, in some cases, with attractive service properties. In addition, cryogenic hard cutting operations can partly eliminate grinding operations in cases when white layer is not produced |
| Abhang and Hameedullah (2012) | CNMA 120404, CNMA 120408, CNMA 120412, and diamond-shaped carbide | Cutting speed, feed rate, tool nose radius, and concentration of solid–liquid lubricants (MQL) | The surface roughness and the chip thickness in the turning process can be improved effectively through application of MQL |

| Author | Tool material / insert | Parameters | Findings |
|---|---|---|---|
| Islam (2013) | | Cooling method, blank size, and work material | The effect of these parameters on surface roughness is relatively low. Dimensional accuracy and surface roughness of different work materials get influenced differently by the cooling methods, and in most cases the best result is achieved by MQL |
| Grzesik and Zalisz (2008) | Mixed ceramic | Feed rate | The wear mechanisms involve abrasion, fracture, plastic flow, material transfer and tribochemical |
| El Hakim et al. (2011) | PCBN (CBN + TiN), TiN coated PCBN (CBN+TiN), mixed alumina ceramic ($Al_2O_3$ + TiC), coated tungsten carbide (TiN coated over a multilayer coating (TiC/TiCN/$Al_2O_3$)) | Tool materials | The mixed alumina ceramic and coated carbide tool materials have longer tool life than PCBN tools |
| Chinchanikar and Choudhury (2013) | Coated cemented carbide inserts | Workpiece hardness, cutting speed, feed rate, depth of cut, and type of coating for coated cemented carbide inserts | Flank wear is dominant wear form for CVD coated tool and crater wear is dominant wear form for PVD coated tool |
| Gaitonde et al. (2009) | CC650WG wiper ceramic inserts | Cutting speed, feed rate, and machining time | The maximum tool wear occurs at a cutting speed of 150 m/min for all values of feed rate and the tool wear increases with the increase in machining time |
| Dogra et al. (2011) | CBN with coated/uncoated carbide | Cutting speed under cryogenic cooling | The flank wear of CBN is lower than that of other inserts |

# 5

# Optimizing a bi-objective inventory model for a two-echelon supply chain management using a tuned meta-heuristic algorithm

Ehsan Afshar Bakeshlu[a], Javad Sadeghi[b]*, Tahereh Poorbagheri[b] and Mahziar Taghizadeh[b]

[a]*Department of Industrial Engineering, College of Engineering, University of Kharazmi, Alborz, Iran;* [b]*Faculty of Industrial and Mechanical Engineering, Qazvin Branch, Islamic Azad University, Qazvin, Iran*

Since vendor-managed inventory strategy plays an important role to reduce total inventory costs, the focus of this paper is to develop an economic order quantity model presented for a two-echelon supply chain management including one vendor and one retailer. In this bi-objective model, the vendor delivers several products to the retailer while shortages are allowed. The aim of this paper is to determine order sizes and maximum backorder levels for each product to simultaneously minimize total inventory costs and a storage space. Moreover, two main constraints, namely budget and the number of orders, are considered to simulate real-world operating conditions for the proposed model. Since the presented model belongs to integer non-linear programming problems, a meta-heuristic algorithm, particle swarm optimization (PSO), is employed to optimize it. In addition, because the quality of solutions depends on the values of parameters of meta-heuristics, the parameters of PSO are tuned using the Taguchi method. Then, the proposed algorithm is compared to branch and bound method.

**Keywords:** vendor managed inventory (VMI); shortages; particle swarm optimization (PSO); storage space; design of experiments (DOE)

## 1. Introduction

While great uncertainty in stock markets occurs by the global financial crisis, there is some evidence to suggest that those companies and organizations that have a flexible supply chain are able to cope with this crisis. The supply chain management (SCM) is an integrated approach to schedule and control materials and data. The supply chain consists of production and distribution operations, transportation, stores, and customers (Chopra & Meindl, 2007). In order to overcome fluctuant customer demands, SCM has to integrate their chains to make better decision in the inventory management, which leads to many studies. As the inventory plays an important role in the total cost, several approaches such as vendor-managed inventory (VMI) as a well-known strategy were provided to manage inventory levels for echelons in SCM. In VMI, a vendor satisfies retailer's orders so that the vendor controls the retailer's inventory levels while determining the retailer's order quantity and time (Disney & Towill, 2002). JcPenney and Wal-Mart, as successful retailers, took advantage of VMI policy (Dong & Xu, 2002) so

---

*Corresponding author. Email: Sadeqi@qiau.ac.ir

VMI is an obvious example of reduction in inventory costs (Achabal, Mcintyre, Smith, & Kalyanam, 2000).

There are several approaches to model VMI problems, and two policies are often used in inventory systems: economic order quantity (EOQ) and economic production quantity (EPQ). EOQ is one of the most popular strategies in SCM because of its simplicity (Axsäter, 2010). Regarding deterministic demands, the retailer's inventory system can be modeled using EOQ policy (Dong & Xu, 2002).

Elvander, Sarpola, and Mattsson (2007) presented a framework to implement VMI policy. Next, Yao, Evers, and Dresner (2007) considered an analytical model for a single-vendor and single-retailer supply chain, which results showed inventory costs. Then, Zammori, Braglia, and Frosolini (2009) presented a flexible framework to implement VMI in several industrial fields. Afterwards, Darwish and Odah (2010) extended a VMI model for several retailers such that the vendor faced the penalty if items exceeded definite bounds, and then they presented an algorithm that reduced the computational efforts significantly. Additionally, Borade and Bansod (2010) provided a case study to employ VMI in Indian industry. Finally, Pasandideh, Niaki, and Nia (2011) proposed a one-vendor one-retailer VMI model with two constraints, namely the number of orders and the warehouse space; then, they used a genetic algorithm (GA) for solving their proposed model.

Moreover, Yang, Chan, and Kumar (2012) developed a VMI problem considering single-warehouse, and then they proposed a GA to solve it. In other environments, to model VMI problems, cases involving no shortages are allowed; Sadeghi, Mousavi, Niaki, & Sadeghi (2013) investigated the multi-vendor multi-retailer single-warehouse case; in addition, Sadeghi, Sadeghi, & Niaki (2014) hybridized an inventory problem with a redundancy allocation optimization problem near real-world problems.

In brief, Yao et al. (2007) presented a simple VMI model, and then they took advantage of VMI policy, and reducing inventory costs were shown by numerical examples. Next, Pasandideh et al. (2011) extended the model presented by Yao et al. (2007) for the multiple-product case. Therefore, this paper extends the model presented by Pasandideh et al. (2011) for the case of multi-objective optimization problem. In this paper, considering backordering shortages, the vendor is responsible to supply products for a retailer with two restrictions, namely the number of orders and the available budget. The aim of this bi-objective mathematical programming model is to find a near-optimal solution, including the order quantities and maximum backorder levels for each product at a cycle time, to simultaneously minimize the storage space and total inventory costs. Since the proposed model belongs to the integer non-linear programming (INLP) problems, a meta-heuristic, particle swarm optimization (PSO), is employed to optimize it. Moreover, a Taguchi method, as a design of experiments analysis, calibrates parameters of PSO to achieve a higher performance of PSO. To validate model and evaluate PSO, a branch and bound (BB) method via Lingo software is utilized.

This paper is organized as follows. The assumptions and the proposed model are presented in the next section. Section 3 describes PSO. Tuning parameters and the numerical examples are provided in Sections 4 and 5, respectively. Finally, the conclusion and the future research are presented in Section 6.

## 2.  The proposed VMI model

This paper uses the following notations to formulate the proposed model.

$j$          An index for products, $j = 1, ..., n$

$D_j$         Retailer's demand rate of $j$th product

$A_{jS}$     Vendor's fixed ordering cost per unit of $j$th product

$A_{jR}$     Retailer's fixed ordering cost per unit of $j$th product

$h_{jR}$     Carrying cost of $j$th product per unit held in the retailer's store during a period

$c_j$     The price of $j$th product

$C$     The available budget

$M$     Total number of orders for all items

$f_j$     Space occupied per unit of the $j$th product

$\hat{\pi}$     Fixed backorder cost per unit in time

$\pi$     Fixed backorder cost per unit

$Q_j$     Order quantity of the $j$th product

$b_j$     Maximum backorder level of the $j$th product in a cycle of VMI chain

$TC_{VMI}$     Total costs of VMI supply chain

$F$     Total storage space for all items

The single-vendor single-retailer supply chain is studied for $n$ products in which the proposed model is confronted with two constraints: the budget and the number of the orders. In VMI system, the integrated supply chain, including one retailer and one vendor, determines order quantity, not the vendor and the retailer individually. Shortages are allowed and backordered. It seems reasonable to assume that the retailer sells all the vendor's products; thus, the annual demand of retailer and vendor is equal. It is also assumed that the payment of the backorders is received when they occur in a period. Finally, the assumptions used in this paper are as follows: (i) the vendor determines the amount of orders; (ii) the delivery time is zero; (iii) discounts are not allowed, and costs are fixed; and (iv) the vendor's order and the available capital are restricted.

Considering the above-mentioned assumptions, the total inventory cost consists of ordering, holding, and shortage costs. In other words, in the proposed VMI model, a vendor, as a supplier to replenish a retailer, includes the ordering cost while the retailer's costs consist of the holding and shortage costs. The vendor, as a supplier, is responsible for providing goods for the retailer; thus, the vendor's cost includes the retailer and the vendor costs that are as follows.

$$\text{TOC} = \sum_{j=1}^{n} \left( \frac{D_j}{Q_j} (A_{jS} + A_{jR}) \right) \tag{1}$$

Then, total holding cost can be given as follows.

$$\text{THC} = \sum_{j=1}^{n} \left( \frac{h_{jR}}{2Q_j} (Q_j - b_j)^2 \right) \tag{2}$$

Since, the shortages are backordered, total shortage cost is,

$$\text{TSC} = \sum_{j=1}^{n} \left( \frac{\hat{\pi} b_j^2}{2Q_j} + \frac{\pi b_j D_j}{Q_j} \right) \tag{3}$$

So, total system cost can be written as follows.

$$\text{Min } TC_{VMI} = \text{TOC} + \text{THC} + \text{TSC} \tag{4}$$

The second objective is to optimize storage space, we have,

$$\text{Min } F = \sum_{j=1}^{n} f_j (Q_j - b_j) \tag{5}$$

In addition, there are two constraints on the proposed model. First, the available budget is limited to $C$, so,

$$\sum_{j=1}^{n} c_j(Q_j - b_j) \leq C \tag{6}$$

Note that the maximum inventory must be used for available capital constraint in EPQ environment or when shortages as backordered (Axsäter, 2010; Tersine, 1993).

Next, the number of ordering is restricted to $M$, thus,

$$\sum_{j=1}^{n} \frac{D_j}{Q_j} \leq M \tag{7}$$

Finally, the bi-objective VMI model can be written as follows.

$$\text{Min TC}_{VMI} = \sum_{j=1}^{n} \left( \frac{D_j}{Q_j}(A_{jS} + A_{jR}) + \frac{h_{jR}}{2Q_j}(Q_j - b_j)^2 + \frac{\hat{\pi}b_j^2}{2Q_j} + \frac{\pi b_j D_j}{Q_j} \right) \tag{8}$$

$$\text{Min F} = \sum_{j=1}^{n} f_j(Q_j - b_j) \tag{9}$$

s.t.,

$$\sum_{j=1}^{n} c_j(Q_j - b_j) \leq C \tag{10}$$

$$\sum_{j=1}^{n} \frac{D_j}{Q_j} \leq M \tag{11}$$

$$b_j \leq Q_j \tag{12}$$

$$b_j, Q_j : \text{integer}, \ j = 1, ..., n. \tag{13}$$

Considering Equation (12), backorder levels cannot be bigger than order quantity.

This paper has several innovations as follows. First, VMI model is extended to consider the second objective minimizing storage space. In addition, there is a considerable improvement on modeling (Equation 12). Moreover, the variables ($Q_j$ and $b_j$) are considered as integer variables. The next section presents a meta-heuristic algorithm to optimize the obtained model and compares it to a BB method.

## 3. Solution algorithm: PSO

Since the obtained model is an INLP, we use a PSO in meta-heuristic optimization methods to solve it similar to these studies Nachiappan and Jawahar (2007), Pasandideh et al. (2011), and Yu and Huang (2010) that presented an INLP model, and then they utilized a meta-heuristic to solve it. At first, Kennedy and Eberhart (1995) presented PSO in 1995, which simulated social behavior process of fish school or bird flock. PSO is a population-based search algorithm like GAs. PSO has been used as a powerful tool to solve problems (Shi et al., 2007).

To start PSO, the particles are randomly generated with respect to the number of population ($N_{pop}$) similar to chromosome in GA (see Figure 1). Since all particles have a velocity ($v_i^{gen}$) and a position ($x_i^{gen}$) in the generation of particles (*gen*), the velocities of all particles are calculated regarding acceleration coefficients that have been set by $\varphi_1$ and $\varphi_2$ coefficients. Next, the new positions of particles are determined considering their velocities and previous locations. Afterwards, particles are evaluated by the fitness function (Equations 8 and 9) to determine the best particle. This process is iterated until the stop condition is satisfied. It is hoped that the swarm, a large number of particles, moves near-optimal solution (for more details of PSO please refer to (Shi & Eberhart, 1998)).

This paper uses the fixed iteration number (*It*) to stop PSO. Note that the parameters of PSO (i.e. *It*, $\varphi_1$, $\varphi_2$, and $N_{pop}$) are tuned by the Taguchi method in the next section. In addition, since there are several constraints on the bi-objective VMI model in this paper, the candidate solution in PSO algorithm may never satisfy the constraints. Thus, in PSO, the death penalty method is employed to make feasible solutions. This penalty is added to the objective function evaluation. In brief, the flowchart of PSO to solve the proposed problem of this research is given in Figure 2.

## 4. Tuning parameters: Taguchi method

Considering various factors and levels, Fisher presented the experiments called the factorial design of experiments (Roy, 1990). Since the complete experiments confront financial and time restrictions, Taguchi developed the fractional factorial experiments.

In the Taguchi method, factors are divided into two groups: controllable (called signal) and uncontrollable (called noise). The aim of the Taguchi method is to maximize the rate of signal to noise (*S/N*). There are two approaches for analyzing results of experiments, analysis of variance, and *S/N* according to Taguchi's emphasis (Roy, 1990). Since PSO runs several times to obtain better solution, we employ *S/N* method for analysis. The quality design should be used at the beginning of production and not during it (Roy, 1990); thus, the amount of the parameters, presented in Table 1, is initialized via trial and error procedure.

The design of $L^9$, which includes 9 designs, and which is selected for the design of experiments in the Taguchi method, consists of four parameters and three levels. Taguchi suggested Equation (14) obtaining *S/N* for the experiments that follow minimizing problems.

$$S/N = -10 \log\left(\frac{1}{n}\sum_{i=1}^{n} Y_i^2\right) \tag{14}$$

where *n* is the iteration of experiments (here it is 5), and $Y_i$ is the response of experiments or the fitness function that is gained by the proposed model. Considering the example with five items in Table 2, Table 3 shows *S/N*s that are calculated from $Y_i$s,

$$\begin{aligned} Q_i &\rightarrow \begin{bmatrix} 91 & 12 & 22 & 32 & 650 \\ 25 & 43 & 33 & 234 & 63 \end{bmatrix} \\ b_i &\rightarrow \end{aligned}$$

Figure 1.   A particle for a problem with five items.

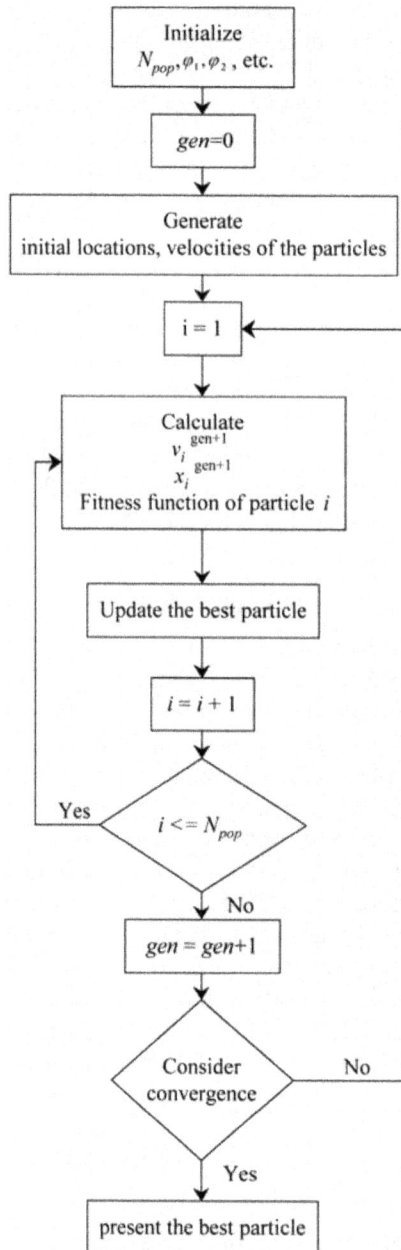

Figure 2.   Flowchart of PSO.

and Figure 3 shows the mean of $S/N$. Note that '1', '2', and '3' in Table 3 refer to the levels of parameters in Table 1.

In the Taguchi method, the design that has higher $S/N$ is the best design; therefore, the optimal values of the parameters are shown in Table 4. There is a marked tendency for those meta-heuristics to be tuned to have a better performance in the optimization of problems. Thus, in the next section PSO is compared to BB method.

Table 1.  The parameters of PSO.

| Variable | Value | | |
|---|---|---|---|
| | 1 | 2 | 3 |
| $It$ | 200 | 250 | 300 |
| $N_{pop}$ | 100 | 150 | 200 |
| $\Phi_1$ | 2 | 2.25 | 2.5 |
| $\Phi_2$ | 1.95 | 2 | 2.05 |

## 5. Numerical example

There are several methods to combine the two objective functions (Equations 8 and 9) into one scalar objective ($T$). This paper employs the weighted sum method, sometimes called linear scalarization, to solve the proposed model as follows (Yang, 2010).

$$\left.\begin{array}{c} \text{Min } TC_{VMI} \\ \text{Min F} \end{array}\right\} \rightarrow \text{Min } T = w_1 TC_{VMI} + w_2 F \qquad (15)$$

where $w_1$ and $w_2$ are the weights of the objectives ($w_1 + w_2 = 1$), so with different parameters for $w_1$ and $w_2$, different Pareto optimal solutions are produced. In this study, it is assumed that $w_1 = 0.8$ and $w_2 = 0.2$. Note that these values are assumed considering the Pareto principle and they are varied by a decision-maker.

This paper uses a meta-heuristic, PSO, to optimize the proposed model. There are two approaches to validate meta-heuristics used in optimization of INLP problems. First,

Table 2.  The example data (Pasandideh et al., 2011).

| Product | $D_j$ | $A_{jS}$ | $A_{jR}$ | $h_{jR}$ | $f_j$ |
|---|---|---|---|---|---|
| 1 | 420 | 1 | 3 | 4 | 3 |
| 2 | 360 | 2 | 2 | 9 | 2 |
| 3 | 540 | 3 | 1 | 7 | 3 |
| 4 | 390 | 5 | 4 | 2 | 1 |
| 5 | 480 | 2 | 2 | 4 | 4 |
| 6 | 510 | 4 | 2 | 6 | 3 |
| 7 | 530 | 1 | 3 | 5 | 2 |
| 8 | 380 | 2 | 1 | 3 | 1 |
| 9 | 430 | 3 | 4 | 2 | 3 |
| 10 | 580 | 4 | 2 | 8 | 4 |

Table 3.  PSO experimental results.

| $N_{pop}$ | $It$ | $\Phi_1$ | $\Phi_2$ | $Y_1$ | $Y_2$ | $Y_3$ | $Y_4$ | $Y_5$ | Mean | $S/N$ |
|---|---|---|---|---|---|---|---|---|---|---|
| 1 | 1 | 1 | 1 | 427.881 | 525.544 | 560.313 | 573.013 | 417.231 | 500.796 | −54.0676 |
| 1 | 2 | 2 | 2 | 469.173 | 462.047 | 472.186 | 515.404 | 562.416 | 496.245 | −53.9393 |
| 1 | 3 | 3 | 3 | 370.397 | 416.13 | 516.568 | 217.837 | 261.395 | 356.465 | −51.4175 |
| 2 | 1 | 2 | 3 | 510.491 | 266.19 | 364.644 | 560.698 | 414.495 | 423.304 | −52.7903 |
| 2 | 2 | 3 | 1 | 366.715 | 410.341 | 367.146 | 314.925 | 414.428 | 374.711 | −51.5142 |
| 2 | 3 | 1 | 2 | 419.875 | 514.106 | 465.968 | 317.932 | 260.539 | 395.684 | −52.1837 |
| 3 | 1 | 3 | 2 | 576.622 | 584.141 | 370.131 | 463.14 | 311.704 | 461.148 | −53.5117 |
| 3 | 2 | 1 | 3 | 318.79 | 370.798 | 270.052 | 293.722 | 610.675 | 372.808 | −51.8822 |
| 3 | 3 | 2 | 1 | 476.766 | 408.865 | 566.544 | 464.159 | 367.223 | 456.711 | −53.2868 |

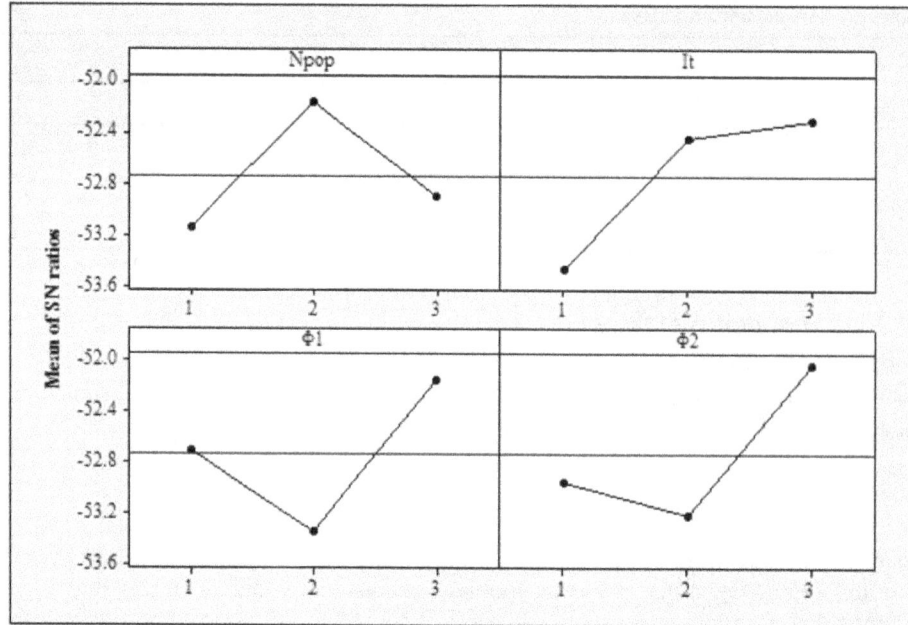

Figure 3.   The mean *S/N* plot for different levels of PSO parameters.

Table 4.   Tuned parameters for GA and PSO.

| Parameters PSO | $N_p$ | $It$ | $\varphi_1$ | $\varphi_2$ |
|---|---|---|---|---|
| Value | 150 | 300 | 2.5 | 2.05 |

another meta-heuristic in which this approach is utilized for the insolvable models with exact solvers is used to evaluate (see these studies Mousavi, Niaki, Mehdizadeh, & Tavarroth, 2012; Sadeghi et al., 2013). Second, optimizer solvers such as Lingo are used (see these studies (Esmaeili Aliabadi, Kaazemi, & Pourghannad, 2013; Nachiappan & Jawahar, 2007)). Indeed, there are many studies that have failed to perform comparing their proposed algorithms, so they are limited in their innovations such as these studies (Gupta, Bhunia, & Goyal, 2009; Pasandideh et al. 2011; Yang et al., 2012). Lingo, which is a systematic enumeration of all candidate solutions, presents the optimal or near-optimal solution for small-scale problems; therefore, this paper similar to this study (Nachiappan & Jawahar, 2007) employs Lingo optimization solver (ver. 11.0), BB method, to evaluate the performance of the proposed meta-heuristic to optimize the proposed model.

Table 5 shows the results of comparisons between PSO and BB method for three test problems. It can be concluded from Table 5 that PSO algorithm provides better solutions than BB method with respect to the cost functions (*TC*) and storage space (*F*) for the proposed model. Note that the third, the fourth, and the fifth rows in Table 5 present the near-optimal solution ($Q_j$ and $b_j$) and the objectives values (*TC* and *F*) for the test problems with one item, two items, and three items problems, respectively.

Table 5. The comparison PSO with BB method for three test problems (Pasandideh et al., 2011).

| | BB method Lingo | | | | | | | | PSO | | | | | | | |
|---|---|---|---|---|---|---|---|---|---|---|---|---|---|---|---|---|
| $j$ | $TC$ | $F$ | $Q_{01}$ | $Q_{02}$ | $Q_{03}$ | $b_{01}$ | $b_{02}$ | $b_{03}$ | $TC$ | $F$ | $Q_{01}$ | $Q_{02}$ | $Q_{03}$ | $b_{01}$ | $b_{02}$ | $b_{03}$ |
| 1 | 78.474 | 36 | 38 | | | 26 | | | 16.00 | 21 | 106 | | | 99 | | |
| 2 | 159.474 | 50 | 38 | 34 | | 26 | 27 | | 43.00 | 14 | 515 | 36 | | 511 | 35 | |
| 3 | 256.535 | 77 | 38 | 34 | 41 | 26 | 27 | 32 | 125.86 | 19 | 42 | 42 | 41 | 40 | 40 | 40 |

Table 6. The order quantity for 10 examples.

| $j$ | $TC$ | $F$ | $Time$ | $Q_1$ | $Q_2$ | $Q_3$ | $Q_4$ | $Q_5$ | $Q_6$ | $Q_7$ | $Q_8$ | $Q_9$ | $Q_{10}$ |
|---|---|---|---|---|---|---|---|---|---|---|---|---|---|
| 1 | 16 | 21 | 6 | 106 | | | | | | | | | |
| 2 | 43 | 14 | 7 | 515 | 36 | | | | | | | | |
| 3 | 125.86 | 19 | 8 | 42 | 42 | 41 | | | | | | | |
| 4 | 227 | 1005 | 8 | 564 | 721 | 582 | 618 | | | | | | |
| 5 | 506 | 948 | 9 | 924 | 678 | 737 | 210 | 119 | | | | | |
| 6 | 509 | 1418 | 9 | 413 | 370 | 676 | 338 | 399 | 269 | | | | |
| 7 | 778 | 2376 | 10 | 545 | 859 | 444 | 248 | 653 | 155 | 259 | | | |
| 8 | 903 | 2064 | 10 | 473 | 272 | 539 | 727 | 119 | 355 | 301 | 576 | | |
| 9 | 1451.8 | 3665 | 11 | 631 | 830 | 701 | 605 | 330 | 692 | 490 | 888 | 479 | |
| 10 | 2051.62 | 3492 | 11 | 155 | 913 | 479 | 276 | 826 | 431 | 269 | 412 | 432 | 972 |

Table 7. Maximum backorder level for 10 examples.

| $j$ | $TC$ | $F$ | $Time$ | $b_1$ | $b_2$ | $b_3$ | $b_4$ | $b_5$ | $b_6$ | $b_7$ | $b_8$ | $b_9$ | $b_{10}$ |
|---|---|---|---|---|---|---|---|---|---|---|---|---|---|
| 1 | 16 | 21 | 6 | 99 | | | | | | | | | |
| 2 | 43 | 14 | 7 | 511 | 35 | | | | | | | | |
| 3 | 125.86 | 19 | 8 | 40 | 40 | 40 | | | | | | | |
| 4 | 227 | 1005 | 8 | 380 | 709 | 496 | 447 | | | | | | |
| 5 | 506 | 948 | 9 | 899 | 436 | 692 | 120 | 78 | | | | | |
| 6 | 509 | 1418 | 9 | 198 | 330 | 674 | 156 | 350 | 166 | | | | |
| 7 | 778 | 2376 | 10 | 289 | 712 | 263 | 172 | 530 | 120 | 205 | | | |
| 8 | 903 | 2064 | 10 | 408 | 264 | 366 | 335 | 79 | 247 | 230 | 260 | | |
| 9 | 1451.8 | 3665 | 11 | 386 | 823 | 544 | 267 | 281 | 392 | 185 | 871 | 351 | |
| 10 | 2051.62 | 3492 | 11 | 105 | 871 | 401 | 257 | 764 | 21 | 53 | 185 | 248 | 893 |

In order to describe the sufficiency of the presented strategy regarding the tuned parameter by the Taguchi method, 10 test problems from this study (Pasandideh et al., 2011) are used to be solved by PSO; the solutions obtained for the order quantity ($Q_i$) and the maximum backorder ($b_i$) levels are shown in Tables 6 and 7.

Note that PSO algorithm is coded by MATLAB 2010b, where a PC with 2.2 GHz Intel Core 2 Duo CPU, and 4 GB of RAM memory via Windows 7 is used for all calculations. Moreover, Figure 3 is provided using Minitab 15.1.30.0.

## 6. Conclusion and recommendations for future research

To reduce bullwhip effects in SCM, the VMI is a common policy. According to the literature, the case of multi-item under shortage conditions with respect to minimizing warehouse space has yet to be investigated. Thus, this paper developed a hybrid VMI model in the SCM in which inventory costs and warehouse space were optimized simultaneously. While a vendor supplied several items to a retailer, shortages were allowed and backordered. The purpose of this paper was to find the order size and backorder levels to synchronously minimize total inventory cost and the warehouse space. Since the proposed model was an integer non-linear programming problem, a PSO was used to optimize it while the Taguchi method calibrated the parameters of PSO. Moreover, PSO was compared to the BB. Finally, this study can be developed as follows.

- Consider variable prices in modeling VMI problem such as discount and inflation.
- Consider the central warehouse in SCM.
- Extend VMI model for production environment such as the EPQ policy.
- Evaluate the Taguchi method with other tuning methods such as response surface methodology in the design of experiments.
- Investigate nondeterministic environments like fuzzy demand.
- Employ a multi-objective particle swarm optimization to present the Pareto solution.

## Acknowledgments

The authors are thankful for the constructive comments of the anonymous reviewers. Taking care of their comments certainly improved the presentation of the manuscript.

## References

Achabal, D. D., Mcintyre, S. H., Smith, S. A., & Kalyanam, K. (2000). A decision support system for vendor managed inventory. *Journal of Retailing, 76,* 430–454.

Axsäter, S. (2010). *Inventory control* (2nd ed.). New York, NY: Springer.

Borade, A. B., & Bansod, S. V. (2010). Study of vendor-managed inventory practices in Indian industries. *Journal of Manufacturing Technology Management, 21,* 1013–1038.

Chopra, S., & Meindl, P. (2007). *Supply chain management: Strategy, planning, and operation* (3rd ed.). Upper Saddle River, NJ: Pearson Prentice Hall.

Darwish, M. A., & Odah, O. M. (2010). Vendor managed inventory model for single-vendor multi-retailer supply chains. *European Journal of Operational Research, 204,* 473–484.

Disney, S. M., & Towill, D. R. (2002). A procedure for the optimization of the dynamic response of a vendor managed inventory system. *Computers & Industrial Engineering, 43,* 27–58.

Dong, Y., & Xu, K. (2002). A supply chain model of vendor managed inventory. *Transportation Research Part E: Logistics and Transportation Review, 38,* 75–95.

Elvander, M. S., Sarpola, S., & Mattsson, S.-A. (2007). Framework for characterizing the design of VMI systems. *International Journal of Physical Distribution & Logistics Management, 37,* 782–798.

Esmaeili Aliabadi, D., Kaazemi, A., & Pourghannad, B. (2013). A two-level GA to solve an integrated multi-item supplier selection model. *Applied Mathematics and Computation, 219,* 7600–7615.

Gupta, R. K., Bhunia, A. K., & Goyal, S. K. (2009). An application of genetic algorithm in solving an inventory model with advance payment and interval valued inventory costs. *Mathematical and Computer Modelling, 49,* 893–905.

Kennedy, J., & Eberhart, R. (1995). Particle swarm optimization. *Proceedings of IEEE International Conference on Neural Networks* (pp. 1942–1948). Perth, WA, Australia.

Mousavi, S. M., Niaki, S. T. A., Mehdizadeh, E., & Tavarroth, M. R. (2012). The capacitated multi-facility location–allocation problem with probabilistic customer location and demand: Two hybrid meta-heuristic algorithms. *International Journal of Systems Science, 44,* 1–16.

Nachiappan, S. P., & Jawahar, N. (2007). A genetic algorithm for optimal operating parameters of VMI system in a two-echelon supply chain. *European Journal of Operational Research, 182,* 1433–1452.

Pasandideh, S. H. R., Niaki, S. T. A., & Nia, A. R. (2011). A genetic algorithm for vendor managed inventory control system of multi-product multi-constraint economic order quantity model. *Expert Systems with Applications, 38,* 2708–2716.

Roy, R. (1990). *A primer on the Taguchi method: Society of manufacturing engineers.* New York, NY.

Sadeghi, J., Mousavi, S. M., Niaki, S. T. A., & Sadeghi, S. (2013). Optimizing a multi-vendor multi-retailer vendor managed inventory problem: Two tuned meta-heuristic algorithms. *Knowledge-based Systems, 50,* 159–170.

Sadeghi, J., Sadeghi, S., & Niaki, S. T. A. (2014). A hybrid vendor managed inventory and redundancy allocation optimization problem in supply chain management: An NSGA-II with tuned parameters. *Computers & Operations Research, 41,* 53–64.

Shi, Y., & Eberhart, R. (1998). A modified particle swarm optimizer. *Proceedings of IEEE International Conference on Evolutionary Computation* (pp. 69–73). Anchorage, Alaska.

Shi, X. H., Liang, Y. C., Lee, H. P., Lu, C., & Wang, Q. X. (2007). Particle swarm optimization-based algorithms for TSP and generalized TSP. *Information Processing Letters, 103,* 169–176.

Tersine, R. J. (1993). *Principles of inventory and materials management.* New Jersey: Prentice Hall.

Yang, X.-S. (2010). *Engineering optimization.* New Jersey: Wiley.

Yang, W., Chan, F. T. S., & Kumar, V. (2012). Optimizing replenishment polices using genetic algorithm for single-warehouse multi-retailer system. *Expert Systems with Applications, 39,* 3081–3086.

Yao, Y., Evers, P. T., & Dresner, M. E. (2007). Supply chain integration in vendor-managed inventory. *Decision Support Systems, 43,* 663–674.

Yu, Y., & Huang, G. Q. (2010). Nash game model for optimizing market strategies, configuration of platform products in a vendor managed inventory (VMI) supply chain for a product family. *European Journal of Operational Research, 206,* 361–373.

Zammori, F., Braglia, M., & Frosolini, M. (2009). A standard agreement for vendor managed inventory. *Strategic Outsourcing: An International Journal, 2,* 165–186.

# 6

# Numerical modeling and optimization of machining duplex stainless steels

Rastee D. Koyee[a]* ⓘ, Siegfried Schmauder[b], Uwe Heisel[a] and Rocco Eisseler[a]

[a]Institute for Machine Tools, University of Stuttgart, Holzgartenstr. 17, D-70174 Stuttgart, Germany; [b]IMWF, University of Stuttgart, Pfaffenwaldring 32, D-70569 Stuttgart, Germany

The shortcomings of the machining analytical and empirical models in combination with the industry demands have to be fulfilled. A three-dimensional finite element modeling (FEM) introduces an attractive alternative to bridge the gap between pure empirical and fundamental scientific quantities, and fulfill the industry needs. However, the challenging aspects which hinder the successful adoption of FEM in the machining sector of manufacturing industry have to be solved first. One of the greatest challenges is the identification of the correct set of machining simulation input parameters. This study presents a new methodology to inversely calculate the input parameters when simulating the machining of standard duplex EN 1.4462 and super duplex EN 1.4410 stainless steels. JMatPro software is first used to model elastic–viscoplastic and physical work material behavior. In order to effectively obtain an optimum set of inversely identified friction coefficients, thermal contact conductance, Cockcroft–Latham critical damage value, percentage reduction in flow stress, and Taylor–Quinney coefficient, Taguchi-VIKOR coupled with Firefly Algorithm Neural Network System is applied. The optimization procedure effectively minimizes the overall differences between the experimentally measured performances such as cutting forces, tool nose temperature and chip thickness, and the numerically obtained ones at any specified cutting condition. The optimum set of input parameter is verified and used for the next step of 3D-FEM application. In the next stage of the study, design of experiments, numerical simulations, and fuzzy rule modeling approaches are employed to optimize types of chip breaker, insert shapes, process conditions, cutting parameters, and tool orientation angles based on many important performances. Through this study, not only a new methodology in defining the optimal set of controllable parameters for turning simulations is introduced, but also the optimum set of process input variables for turning duplex stainless steels is defined.

**Keywords:** 3D-FEM; duplex stainless steel; JMatPro; VIKOR method; FANNS; machining performance index

## 1. Introduction

Empirical machining models are of limited use: They are generally only calibrated to be valid for a limited process range. Meanwhile, due to the simplification, analytical machining models are only partially suitable for describing complex processes such as can be described by finite element modeling (FEM). FEM can provide a comprehensive and in some cases complementary approach to empirical, mechanistic, or analytical approaches to study machining process. It makes possible to illustrate complex tools in

*Corresponding author. Email: rastee.ali@ifw.uni-stuttgart.de

2D or 3D configurations while simultaneously taking plasto-mechanical and thermal processes into consideration. However, it should be clear that in 2D-FEM of many machining operations: turning, drilling, milling, etc. (except broaching, sawing, etc.) the obtained final surface doesn't correspond to the final one that is obtained in 3D-FEM. Thus, some prediction issues cannot have any realistic meaning (Klocke, 2011; Quiza & Davim, 2011).

The success and reliability of any FEM depends upon accurate mechanical (elastic constants, flow stress, friction, fracture stress/strain, etc.) and thermo-physical (density, thermal conductivity, heat capacity, etc.) data. Characterization of material flow stress behavior is needed for the extreme conditions of machining which involve plastic strains of 5–20, strain rates up to $10^6$ s$^{-1}$, temperatures between 500 and 1400 °C, temperature gradient (1000 °C/mm), high heating rates close to $10^6$ °C s$^{-1}$, and high cutting pressures of 1–3 GPa (Arrazola, Özel, Umbrello, Davies, & Jawahir, 2013). For this purpose, researchers recently applied modified constitutive equations, a great variety of which are exist in the literature (He, Xie, Zhang, & Wang, 2014; Hou & Wang, 2010; Li, Li, Wang, Liu, & Wu, 2013; Liu et al., 2013; Samantaray, Mandal, & Bhaduri, 2009; Song, Ning, Mao, & Tang, 2013; Wang et al., 2013). Some of the researchers have utilized Java-based Materials Properties (JMatPro) software to model the material properties and behavior of steels and Ni-based superalloys (Guo, Saunders, Miodownik, & Schille, 2007; Olovsjö, Hammersberg, Avdovic, Ståhl, & Nyborg, 2012; Saunders, Guo, Li, Miodownik, & Schille, 2004). Inverse identification of the constitutive model parameters has been proposed as an alternative method of defining the model coefficients, so that the models are valid over large ranges of conditions during machining (Klocke, Lung, & Buchkremer, 2013; Pujana, Arrazola, M'Saoubi, & Chandrasekaran, 2007; Shrot & Bäker, 2012; Umbrello, M'Saoubi, & Outeiro, 2007). Recent researches proposed methods for characterization friction and heat partition coefficients at the tool-work material interface in metal cutting operations (Arrazola & Özel, 2010; Attanasio & Umbrello, 2009; Bonnet, Valiorgue, Rech, & Hamdi, 2008; Heisel, Storchak, & Krivoruchko, 2013; Iraola, Rech, Valiorgue, & Arrazola, 2012; Puls, Klocke, & Lung, 2012; Rech et al., 2013; Smolenicki, Boos, Kuster, Roelofs, & Wyen, 2014; Ulutan & Özel, 2013). Several contributions have improved the FEM of serrated chip formation, chip flow, and chip breaking (Aurich & Bil, 2006; Buchkremer et al., 2014; Calamaz, Coupard, & Girot, 2008; Chagas, Barbosa, Barbosa, & Machado, 2013; Ducobu, Rivière-Lorphèvre, & Filippi, 2014; Guo & Yen, 2004; Lorentzon & Järvstråt, 2008; Rhim & Oh, 2006; Sima & Özel, 2010; Vaziri, Salimi, & Mashayekhi, 2011). Other works have focused on FEM of machining stainless steel materials (Koné, Czarnota, Haddag, & Nouari, 2011, 2013; Maranhão & Paulo, 2010; Outeiro, Umbrello, & M'Saoubi, 2006). Finally, recent works have also investigated the machining of duplex stainless steels experimentally (De Oliveira Junior, Diniz, & Bertazzoli, 2014; Koyee, Schmauder, & Eisseler, 2013; Koyee, Heisel, Eisseler, & Schmauder, 2014; Krolczyk, Legutko, & Gajek, 2013; Nomani, Pramanik, Hilditch, & Littlefair, 2013; Paro, Hänninen, & Kauppinen, 2001).

A vividly testimony to the potential interest on duplex stainless steels as a key research topic by various researches in the world is obvious when a search with the keywords 'duplex stainless steel' in popular database such as Science Citation Index-Expand or Scorpus returns tens of recent publication. However, one can hardly find a research that addresses the FEM simulation of machining duplex grades. Furthermore, despite significant recent advances, FEM itself remains a 'plug and play' technique for predicting some process output, depending on the assumed boundary conditions, including the friction. For example, it has been reported that FEM simulation of the same processes can produce different results (Arrazola & Özel, 2009).

To address the FEM simulation of machining DSS correctly, the plug and play technique must be limited first. For this purpose, this study introduces a new method of inverse identification through converting the overall differences between simulation results and experimentation ones into a single measure using Taguchi-VIKOR method. FEM modeling under mixed Taguchi design $L_{18}(2^1 \times 3^7)$ is performed to tune input parameters. Thermal contact conductance $h_{tc}$, cutting speed $v_c$, feed rate $f_r$, cutting tool–workpiece interface hybrid Coulomb $\mu_c$ and shear $\mu_s$ friction coefficients, Taylor–Quinney coefficient $\kappa_t$, percentage reduction of original flow stress $\%p$, and Cockcroft–Latham critical damage criterion $D_{\text{crit.}}$ are considered as controllable input parameters. On the other hand, cutting experimentations are conducted and different performances are measured, recorded, and analyzed. The percentage difference between numerically and experimentally obtained performances such as thrust cutting force $\%E_t$, feed cutting force $\%E_f$, main cutting force $\%E_c$, chip thickness $E_h$, and tool nose temperature $\%E_T$ are considered as performance characteristics and are unified into a single index called VIKOR method. The derived indices are then optimized globally to determine the optimum set of input parameter using an effective neural network-based nature-inspired meta-heuristic algorithm known as Firefly Algorithm Neural Network System (FANNS). The optimum sets are next validated through experimentations. In the next stage of the research work, Taguchi optimization procedure is employed to numerically optimize the chip breaker types $CB$, insert geometries $Geo.$, cooling medium $CM$ such as still air, water-based and cryogenic coolants, cutting conditions such as cutting speed $v_c$ and feed rate $f_r$, and tool orientation angles such as normal rake $\alpha_n$ and inclination angle $\lambda$. Resultant cutting forces $(R)$, effective plastic stresses, chip-tool interface cutting temperatures, and tool wear rate are designated as performance characteristics. An expert system based on fuzzy rule modeling approach is adopted to derive a new index called numerical machining performance measure (NMPM). Finally, analysis of means (ANOM) is applied on the computed NMPMs to define the optimum levels of control factors. A schematic diagram summarizing the methodological framework developed in this study is shown in Figure 1.

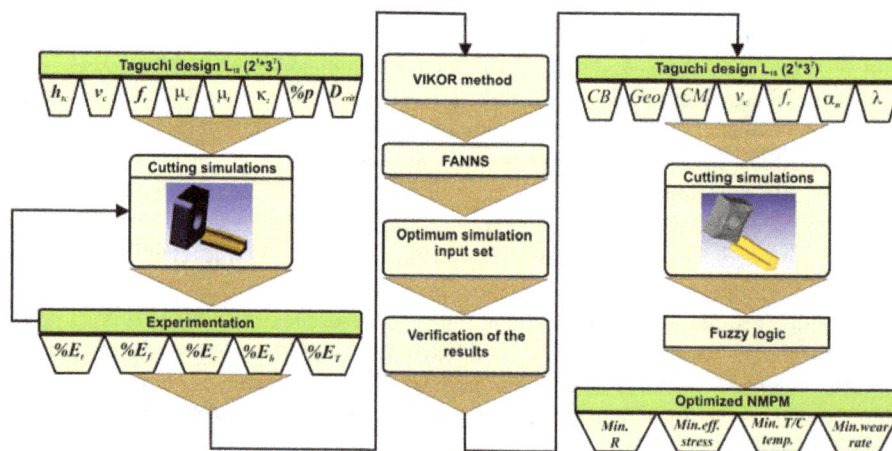

Figure 1. Framework of the research.

## 2. Experimentations

### 2.1. Longitudinal turning operation

EN 1.4462 standard duplex and EN 1.4410 super duplex rods of circular cross section (diameter 55 mm, length 200 mm) were machined on a CNC lathe (Gildemeister CTX 420 Linear V5) having maximum drive power of 25 kW and speed range of 35–7000 revolutions/min. The tool holder was a Sandvik Coromat referenced PCLNL 2525 M 12, and the inserts were uncoated rhombic cemented carbide of ISO designation; CNMA 120412-IC20. They had neither chip breaker nor coating. The reason behind the selection of these basic-shape inserts was mainly to facilitate the optimization of the chip breaker geometry in the final stage of the study. The chemical composition and mechanical properties of the involved work materials are tabulated in Table 1.

In the literature, depth of cut has shown very little influence on chip thickness, contact stress at the tool–chip interface, tool wear rate, and average temperature (Astakhov, 2006). Therefore, in this study, the cutting tests are conducted at constant cutting depth of 1.5 mm with the following cutting data: cutting speed $v_c = 80, 160$ and 240 m/min, and feed rate $f_r = 0.15, 0.225$ and 0.3 mm/rev. Finally, to avoid complications in cutting temperature measurements, the turning tests were carried out dry.

### 2.1.1. Cutting forces

The cutting force components $F_c$, $F_f$, and $F_t$ are measured using piezoelectric dynamometer from Kistler. To ensure a steady state force signal and reduce the tool wear effect on cutting forces and temperatures, a new cutting edge per each experimental trial was employed. The effect of feed rate and cutting speed on the components of cutting force is shown in Figure 2. For instance, increasing the cutting speed from $v_c = 80$–240 m/min at constant $f_r = 0.225$ mm/rev while turning EN 1.4462 had reduced $F_c$, $F_f$, and $F_t$ by 6.594, 40.613, and 19.8%, respectively. On the other hand, increasing feed rate from $f_r = 0.15$ to 0.3 mm/rev at constant $v_c = 160$ m/min while turning EN 1.4410 had increased $F_c$, $F_f$, and $F_t$ by 47.39, 22.076, and 52.82%, respectively.

Table 1. Properties of the work materials.

| Chemical composition (%Weight) | EN 1.4410 | EN 1.4462 |
|---|---|---|
| Carbon | .015 | .018 |
| Chromium | 24.92 | 22.42 |
| Nickel | 6.91 | 5.44 |
| Molybdenum | 4.06 | 3.12 |
| Manganese | .75 | .84 |
| Silicon | .25 | .37 |
| Nitrogen | .3 | .18 |
| Phosphor | .021 | .025 |
| Sulfur | .0007 | .003 |
| Copper | .1 | – |
| *Mechanical properties* | | |
| Yield strength (MPa) | 579 | 514 |
| Tensile strength (MPa) | 826 | 737 |
| Hardness (BHN) | 236 | 212 |
| Elongation (%) | 40 | 41 |

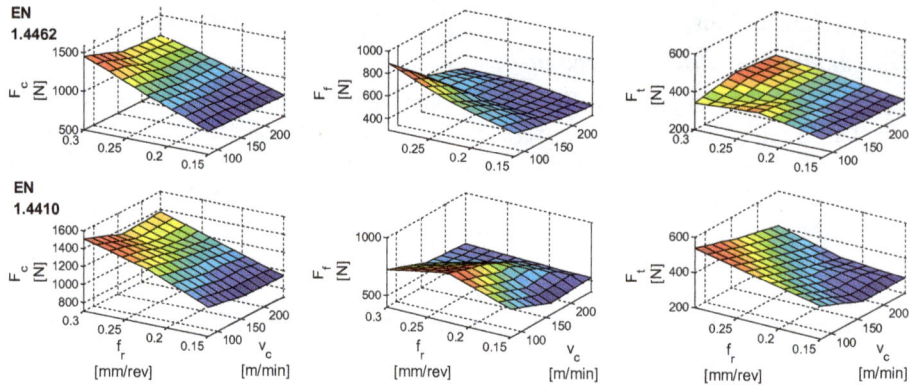

Figure 2.   Dependency of the components of cutting force on the cutting conditions.

### 2.1.2.   Cutting temperatures

Although infrared (IR) cameras are quite expensive, they remain one of the most promising solutions for temperature measurement since they allow for a non-contact, extensive measurement of temperature. Thus, the problem of perturbation of the heat flow in the tool and changes the results is avoided. The fast response of the IR camera lets high, cutting speeds to be used in machining experiments, since it can capture the transient changes and the stages of chip formation and entanglement around the workpiece, the tool holder, and the tool post. In this study, high-resolution thermo-graphic camera Image IR8300 by InfraTec was used to film the chip formation and measures the chip temperature. Each film made for an experimental trial is composed of 800 images. The IR camera is placed straight above the rake face of the tool. The lens of the camera is protected against possible impacts of the chips flying about. The images in the films are examined, and the mean of maximum temperature in the middle of the chips is recorded. Figure 3 shows the setup of IR camera.

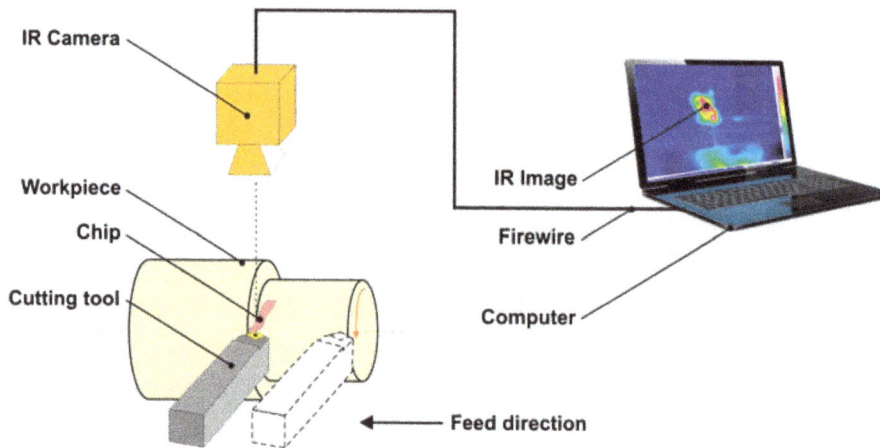

Figure 3.   Experimental setup of IR camera.

The dependency of maximum chip temperature on the cutting condition and the machined material can be seen in Figure 4. Examining the mean maximum chip temperature values, the average maximum chip temperature when machining EN 1.4410 was higher than EN 1.4462 by 3–5%. Experimental results have shown that high cutting speed and feed rates will not always lead to high chip temperatures. This seems in contradiction with the known proportional relations between cutting conditions and cutting temperature. The most likely explanation to this is that when the cutting speed and feed rate are increased, higher rate of removed volume is expected. In this case, the higher heat flux entering the chip resulting from the higher interface temperature is divided over a larger volume, hence lower intensity.

Due to the continuous entanglement of the DSS chips around the cutting tool and workpiece surface, the chip surface was not always available for direct observation by the IR camera. Furthermore, due to the concave surface of the scanned workpiece, highly reflective nature of DSS, surface and difficulties in accurately measuring surface emissivity, the erroneous temperature reading of tool–chip interface was inevitable (see Figure 5). During the early efforts to compare the maximum chip temperatures, similar

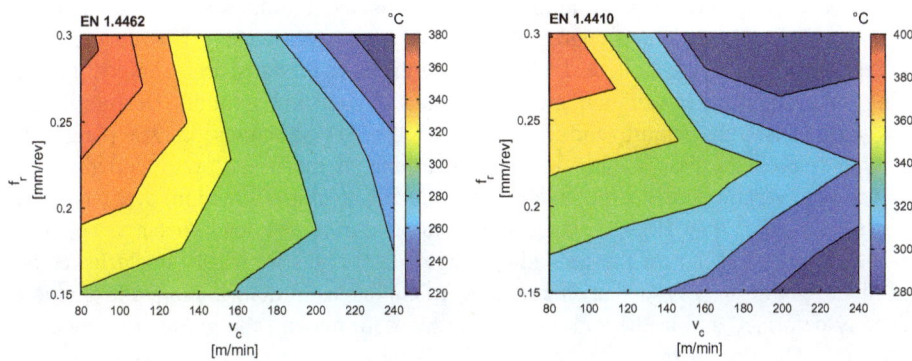

Figure 4.   Dependency of maximum chip face temperature on the cutting conditions.

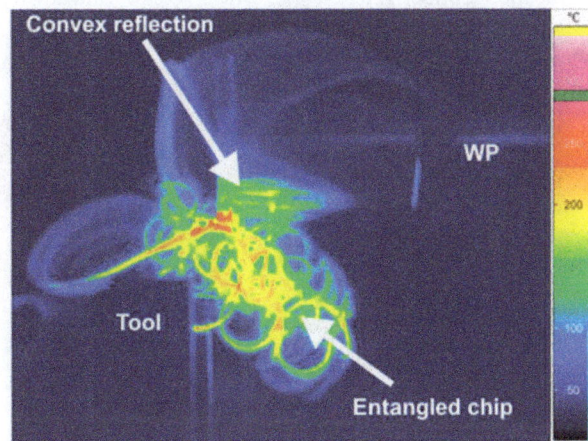

Figure 5.   Problems associated with IR camera temperature measurement during machining DSS.

problems of measuring maximum chip temperature due to the obstructed chip's free face have also been encountered. Therefore, the application of the obtained maximum chip temperature results is restricted only to the experimentation phase. Instead, the calibrated IR images of the black tool tip with emissivity value of 0.93 were obtained very shortly (0.25 sec) after the feed was halted and the temperature of the cutter at the end of steady cutting is recorded. The experimentally measured temperature data are then utilized to validate the results obtained in the simulation phase.

Figure 6 maps the experimentally measured maximum tool surface temperatures at location of 0.5 mm < 45° from the origin of the nose curvature. It can be seen that the maximum temperature on the tool face increases with cutting speed. With the increase in feed rate, the cross section of chip and tool–chip contact length increases and consequently friction rises. This is also involves the increase of temperature.

### 2.1.3. Micrographs of the chips

The serrated chip geometries, which were detected in some machining occasions, have to be correctly configured and measured. Thus, the chips are carefully collected and prepared for detailed analyses. Among many chip-related measurements, maximum and minimum chip thickness and degree of serration are recorded using two different techniques. In the first, the collected chips are embedded in transparent epoxy, polished using increasingly finer diamond grit, and etched in chemical solution. Despite the process time-consuming nature and reduction of the 3D structures to 2D planes, this method enabled the ease tracings of the streamline of material flow and paths of the grains. The variation of plastic deformation during a chip formation cycle is clearly shown in Figure 7(a). Higher plastic deformation during the loading stage and lower plastic deformation during the unloading stage of the chip formation cycle have influenced the grain shapes, so that in the zone of high plastic deformation, the grains are severely deformed and in the zone of low plastic deformation, the grains are moderately deformed. The tips of the cracks can be clearly observed at the boundary between the zones. The chip structure shown in Figure 7(b) is an example of saw-tooth continuous transitional chips obtained when DSSs are machined.

In the second measurement technique, the chips are photographed and measured using a high magnification optical microscope. The average values of five consecutive chip thickness ratios ($h_{max}/h_{min}$) are defined as the degree of serration:

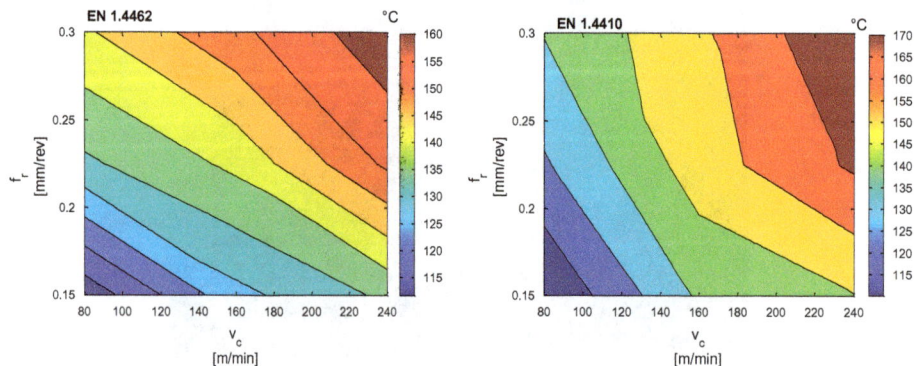

Figure 6.   Dependency of tool tip temperature on the cutting conditions.

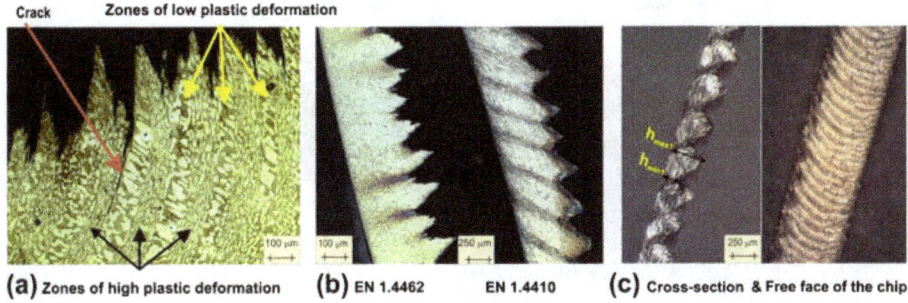

Figure 7.   Examples of formed chips when machining DSS. (a) EN 1.4462 at $v_c$ = 80 m/min and $f_r$ = 0.225 mm/rev, (b) Left, EN 1.4462 $v_c$ = 80 m/min and $f_r$ = 0.15 mm/rev, right, EN 1.4410 $v_c$ = 80 m/min and $f_r$ = 0.3 mm/rev, and (c) EN 1.4410 $v_c$ = 160 m/min & $f_r$ = 0.3 mm/rev.

$$H = \frac{1}{5}\sum_{i=1}^{5}\frac{h_{\max_i}}{h_{\min_i}} \tag{1}$$

where $h_{\max}$ and $h_{\min}$ are the perpendicular distances measured from the bottom of the chip to the peaks and valleys, respectively (see Figure 7(c)). The presence of fragments is also revealed by showing the image of free side of the chip. Only in the cases when $H \approx 1$, a smooth continuous chip is expected. Otherwise, the higher the ratio H, the more serrated the type of the chip is. Comparing the cutting of DSS, EN 1.4462 has generally led to produce friendlier to the machine, thinner and more segmented types of chips. This is possibly due to the higher content of chip breaking elements, such as phosphor and sulfur (see Table 1). In general, as cutting speed and feed rate are increased, there are increases in the degree of serration (H) as observed in the Figure 8. It is also observed that cutting speed, which adversely affected the $h_{\max}$, has contributed positively in the chip thinning process through reducing the consecutive distance between the peeks.

## 3.   Prerequisite conditions of machining numerical simulations

### 3.1.   Control factors

Since the success and reliability of machining modeling depends upon accurate mechanical and thermo-physical data, a brief description of the involved FEM control factors is considered important here, so that deeper understanding of the simulation of cutting DSS is gained.

### 3.1.1.   Damage

One of the reasons behind the poor machinability of standard and super DSS is their high tensile strength. Therefore, when machining of DSS, tensile stress is expected to play an important role. In this research, the effect of tensile strength is included in the material response employing Cockcroft–Latham's damage criterion, which is expressed mathematically as follows:

$$D = \int_0^{\varepsilon_f} \sigma_1 d\varepsilon \tag{2}$$

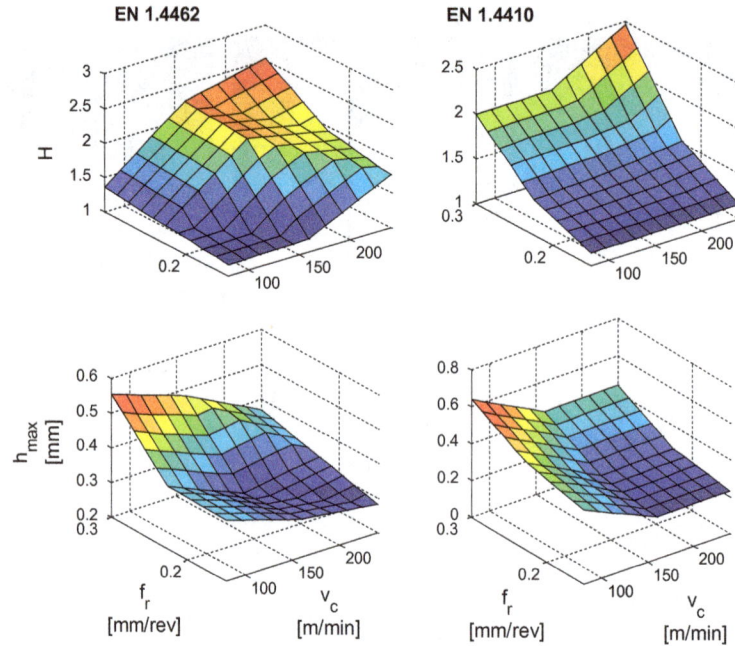

Figure 8.   Dependency of the chip characteristics on the cutting conditions.

where $\varepsilon_f$ is the effective strain, $\sigma_1$ is the principal stress, and $D$ is a material constant. Cockcroft and Latham's criterion states that when the integral of the largest tensile principal stress component over the plastic strain path in Equation (2).

Reaches the critical value $D_{\text{crit.}}$, usually called damage value, fracture occurs or chip segmentation starts, the flow stress is reduced to a lower value %$p$, which is expressed as percentage of the original flow stress (Umbrello et al., 2007).

### 3.1.2.   Friction

In this study, Zorev's friction model is adopted to overcome the shortcomings of the Coulomb and the shear friction law simultaneously. So that in the range of low normal pressure, the Coulomb law is applied while for high normal pressures, the shear friction law is used. This can be expressed by means of the following formulation:

$$\tau_f(x) = \begin{cases} \tau_p : \mu\sigma_n \geq \tau_p, & \left(l_p < x < l_c\right) \text{ Sticking region} \\ \mu\sigma_n : \mu\sigma_n < \tau_p, & \left(0 < x < l_p\right) \text{ Sliding region} \end{cases} \tag{3}$$

where the $\tau_f$, $\sigma_n$, and $\mu$ are the frictional stress, normal stress, and coefficient of friction, respectively (Zorev, 1963).

### 3.1.3.   Thermal contact conductance $h_{tc}$

The thermal contact conductance $h_{tc}$, also known as the heat transfer coefficient, describes the heat flux of two solids in contact and is defined as follows:

$$h_{tc} = \frac{q}{\Delta T} \tag{4}$$

in which $\Delta T$ is the temperature difference at the contacting surfaces and $q$, the heat flux, defined as:

$$h_{tc} = \frac{d}{dA}\left(\frac{dQ}{dt}\right) \tag{5}$$

where $A$ is the area of contact surface and $Q$ is the conducted heat in (J). It is recognized that thermal contact conductance is a function of several parameters, the dominant ones being the type of contacting materials, the macro- and micro-geometry of the contacting surfaces, the temperature, the interfacial pressure, the type of lubricant or contaminant, and its thickness (Rosochowska, Balendra, & Chodnikiewicz, 2003).

### 3.1.4. Taylor–Quinney coefficient

As the mechanical behavior is affected by temperature (softening effect), the plastic deformation is accompanied by heat generation which results in temperature rise. The heat generation due to this phenomenon is described by the following relationship:

$$\dot{q}_p = \kappa_t \bar{\sigma}\, \dot{\bar{\varepsilon}}_p \tag{6}$$

where $\dot{q}_p$ is the volumetric heat generation due to plastic work in (W/m$^3$), $\bar{\sigma}$ is the von Mises equivalent stress in (MPa), $\dot{\bar{\varepsilon}}_p$ is the von Mises equivalent plastic strain, and $\kappa_t$ is the Taylor–Quinney coefficient which represents the proportion of plastic works converted into heat. (Haddag & Nouari, 2013) have shown strong dependency of $\kappa$ on both strain and strain rate for various engineering materials. Within the confines of their constitutive framework, the assumption that $\kappa$ is constant is inconsistent with the rate independence of the stored energy of cold work, which is a fundamental consequence of thermodynamics. It would seem that the only justification for a priori assumptions on $\kappa$ is a lack of information on the stored energy of cold work. Therefore, this coefficient is considered as an unknown input value that has to be inversely determined in the next cutting simulations.

### 3.2. Prediction of chip serration

Serrated chips (also called segmented or non-homogeneous chips, see Figure 7(c)) are semi-continuous with large zones of low shear strain and small zones of high shear strain; hence, the latter zone is called shear localization. Metals with low thermal conductivity and strength that decreases sharply with temperature (thermal softening) exhibit this behavior (Kalpakjian & Schmit, 2006). Adiabatic shear bands have been observed in the serrated chip during high strain rate metal cutting process of titanium (Sima & Özel, 2010), carbon steel (Rhim & Oh, 2006), nickel-based alloys, and stainless steels (Lorentzon, Järvstråt, & Josefson, 2009). Adiabatic shear bands are narrow zones with thickness of the order of few micro-meters where shear deformation is highly localized. Each material has a different susceptibility to adiabatic shear because it depends on properties such as heat capacity, heat conductivity, strength level, microstructure, geometry, defects, and strain rates. It is also known that adiabatic shear banding precedes material failures at high strain rates. Adiabatic shear banding is usually accompanied by a loss in stress capacity owing to intense thermal softening in the shear bands and, in many cases, shear bands serve as sites for crack initiation and growth during subsequent dynamic fracture (Odeshi, Al-ameeri, & Bassim, 2005). Localized adiabatic shearing can be considered a unique consequence of severe plastic deformation at high strain rates. As both thermal and strain softening lead to rapid deformation localization, a shear band forms via a nearly adiabatic process. Also of note is that grain refinement can occur within shear bands and severe plastic strain (which can reach 5–20) can also appear within these shear bands (Xue, Liao, Zhu, & Gray, 2005).

In order to capture the formation of chip serrations numerically, a material model which accounts for thermal and strain softening should be employed. In last several years, numerous attempts have been made to predict serrated chip formation by finite element method. However, it was found to be difficult to predict the shear band and the serrated chip formation by the FEM technique, especially in three-dimensional configuration. In addition to the improper meshing size and strategy, the application of conventional flow stress models is one of the main reasons.

Unfortunately, a universal material model suitable for all cutting simulations remains one of the main unaccomplished tasks. Due to the typical machining high strain, strain rate temperature, and temperature gradient, it is not always easy to determine the flow stress curves experimentally. Additionally, experimental methods such as split Hopkinsonbartest are quite expensive and require a large number of experiments to conduct. On the other hand, empirical material laws which describe the flow stress as a function of strain, strain rate, and temperature contain specific material constants that have to be determined by regression analyses or by the least-squares method and verified experimentally. During the last three decades, many of such models are suggested. The most notably are the engineering-based and physically based models. Results of the first evaluative simulations trials carried out in this study have shown that chip morphologies predicted by engineering-based model such as Johnson–Cook (JC) model and physically based model such as Zerilli–Armstrong lacked the formation of the shear bands and strain localizations which are often denoted as characteristics of chips formed during machining materials that possess high ultimate and yield strength and low thermal conductivity such as duplex stainless steels. For example, the following conditions were considered in a 3D-FEM and extruded 2D-FEM modeling:

- Tool: normal rake angle = −5°, relief angle = 7°, inclination angle = 0°, radius of the cutting edge = 0.03 mm, and tool material is tungsten carbide (WC).
- Work material: EN 1.4462
- Work material models: (a) JMatPro and (b) JC model. The JC constitutive model is given by equation:

$$\sigma = (A + B\varepsilon^n)\left(1 + C\ln\left(\frac{\dot{\varepsilon}}{\dot{\varepsilon}_0}\right)\right)\left(1 - \left(\frac{T - T_r}{T_m - T_r}\right)^m\right) \tag{7}$$

where $\sigma$ is current von Mises flow stress; $A$ is initial yield strength; $B$ is strain hardening coefficient; $\varepsilon$ is equivalent plastic strain; $n$ is strain hardening exponent; $C$ is strain rate coefficient; $\dot{\varepsilon}$ is equivalent plastic strain rate; $\dot{\varepsilon}_0$ is reference plastic strain rate; $m$ is thermal softening exponent; $T$ is current temperature; $T_m$ is melting temperature; $T_r$ is reference temperature at which material constants are determined (Nasr, Ng, & Elbestawi, 2007).

The elastic-plastic, viscosity, and thermal softening terms are represented by the first, second, and third brackets of the above equation, respectively. The procedure of computing the constants of Johnson–Cook model from JMatPro stress flow curves involved the following steps:

- The value of $A$ is calculated from the yield stress of the metal given in Table 1, that is, $A = 514$ MPa.
- At $T = T_r$ and $\dot{\varepsilon} = \dot{\varepsilon}_0$, the curve $\ln(\sigma - A)$ vs. $\ln(\varepsilon)$ is plotted. The values of $B$ and $n$ are extracted from the intercept and slope of this plot, respectively.
- Substituting the values of $A$, $B$, and $n$ in Equation (7) and assuming $T = T_r$ which eliminates the thermal softening term, the strain rate coefficient ($C$) is obtained from the slope of the graph $\{\sigma/(A + B\varepsilon^n)\}$ vs. $\ln(\dot{\varepsilon})$.
  - At $\dot{\varepsilon} = \dot{\varepsilon}_0$, the viscosity term is eliminated. The exponent ($m$) is obtained from the slope of the graph $\ln[1 - \{\sigma/(A + B\varepsilon^n)\}]$ vs. $\ln\{(T - T_r)/(T_m - T_r)\}$. Finally, the overall correlation coefficient was found satisfactory at $R^2 = 0.85567$. The JC parameters of the EN 1.4462 DSSs are given in Table 2.
- Cutting regime: cutting speed $v_c = 100$ m/min and uncut chip thickness $f_r = 0.25$ mm.
- Damage criteria: $D_{\text{crit.}} = 50$ MPa and $\%p = 20$.
- Shear coefficient $\mu_t = 0.4$ and $h_{tc} = 45$ N/(mm s K).
- Mesh size = 20 μm and aspect ratio = 7.
- Stress flow curves of the work material are illustrated in Figure 9.

The left and right sides of the Figure 10 show the states of the deformation zones for JMatPro and JC models, respectively. For instance, Figure 10(b) shows the plastic strain distribution in the deformation zone, while Figure 10(c) presents the temperature distribution in the deformation zones. As clearly seen from the figures, the formation of transitional or saw-tooth chip types is hardly predicted in the chips correspond to the JC

Table 2. Parameters of the Johnson–Cook model for EN 1.4462.

| Parameter | $A$ (MPa) | $B$ (MPa) | $n$ | $C$ | $m$ | $\dot{\varepsilon}_0$ (1/s) |
|---|---|---|---|---|---|---|
| Value | 514 | 612.96 | .1801 | .01194 | .9765 | 1 |

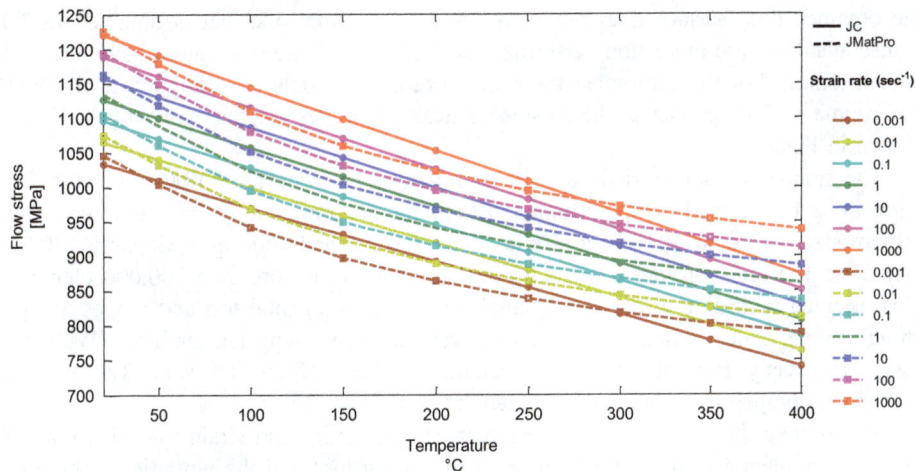

Figure 9.  EN 1.4462 JC and JMatPro flow stress curves at $\varepsilon = 1$.

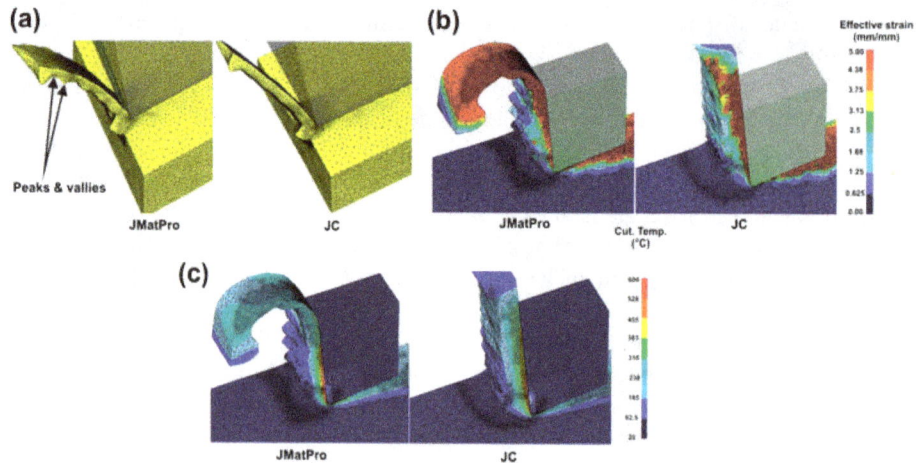

Figure 10.    JMatPro and JC-based cutting simulation results.

model. Furthermore, the morphologies of the chips when cutting stainless steel materials are often related with chip up-curling and side-curling are also hardly noticed in JC material model cases. On the other hand, FEM results employed JMatPro flow stress curves seem to perform better in terms of the prediction of chip curling and formation adiabatic shear bands phenomenon. Therefore, the option of implementing flow stress data generated by JMatPro software is adopted. A text file was first generated for each workpiece material with chemical composition and mechanical properties shown in Table 1 and copied into the keyword file of the actual simulation. The software is also used to model the elastic and thermo-physical behavior of the work material.

### 3.3.    *Finite element simulations*

The classical representation of cutting process by 2D models of chip formation suitable for orthogonal cutting has a limited assistance to tool and process designers. Moreover, the obtained final surface does not correspond to the final one that is obtained in 3D. Consequently, some prediction performances like residual stresses cannot have any realistic meaning. For these reasons, the oblique turning models are designed in 3D-FEM environment. The predicted thermo-mechanical properties of the work materials are shown in Figure 11.

The simulations are carried out with commercial software DEFORM-3D v10.1. The software is based on the implicit updated Lagrangian formulation. The workpiece, each 10 mm size, was considered as a plastic object and meshed with approximately 140,000 elements. The tool considered a rigid object meshed with more than 100,000 elements. It is oriented according to the cutting angles set in experimental test and moves along a straight path. The cutting tool material was uncoated tungsten carbide (WC) and assigned directly from the available material database of the software. Thermal and mechanical properties of WC tools are tabulated in Table 3.

To improve the gradients of temperature, stress, strain, and strain rate distributions, the mesh is refined in the vicinity of areas of cutting tool and the workpiece where the primary and secondary shear zone will be located. However, fixing the minimum element size at a very fine one could make the computation very expensive in terms of

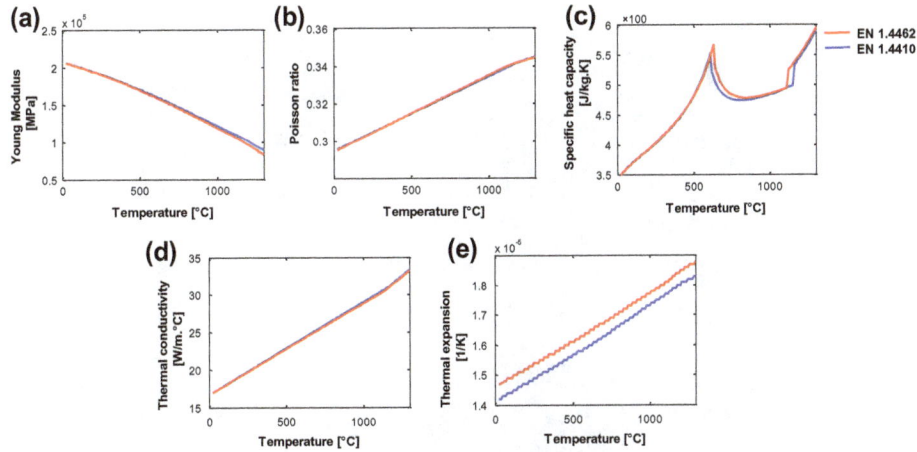

Figure 11.    Temperature-dependent mechanical and physical properties of the work materials.

Table 3.    Thermal and mechanical properties of WC.

| | |
| --- | --- |
| Elastic modulus | 650 GPa |
| Poisson's ratio | .25 |
| Thermal conductivity | 59 W/m K |
| Specific heat capacity | 15 J/kg K |
| Thermal expansion | $5 \times 10^{-6}$ 1/K |

running time and size of the final database. Therefore, the minimum mesh element size of the workpiece and aspect ratio was fixed at ¼ of the minimum uncut chip thickness and 7, respectively. The top sides of the workpiece as well as all sides of the cutting tool were allowed to exchange heat with the environment; the convection coefficient is considered constant at 0.02 N/(s mm K), which is the default value for dry cutting in DEFORM 3D. On the other hand, meshing of the cutting tool model expected to be a decisive factor for the simulation of the heat flux, temperature, and stresses in the material decohesion/deformation zone (Niesłony, Grzesik, Chudy, & Habrat, 2014). Therefore, the meshing density on the tool is increased in the potential chip-tool contact surface, while the rest of the tool is meshed with relatively coarser mesh. The mesh and boundary conditions for the finite element model are shown in Figure 12.

### 3.4.    Design of the FEM experiments

When the experimental design is a full factorial one, the final number of experiments is usually large, because all the possible combinations should be introduced in the design. In order to allow improvement in the methodology of simulation tests and study the entire control factors space with the smallest possible number of experiments, the concept of design of experiment (DOE) is applied. DOE is defined as process of planning of an experiment so that appropriate data will be collected, which are suitable for further statistical analyses resulting in valid and objective conclusions. One of the robust methods of experimental design is Taguchi method, which provides a simple, efficient, and systematic approach for the optimization of experimental designs for performance

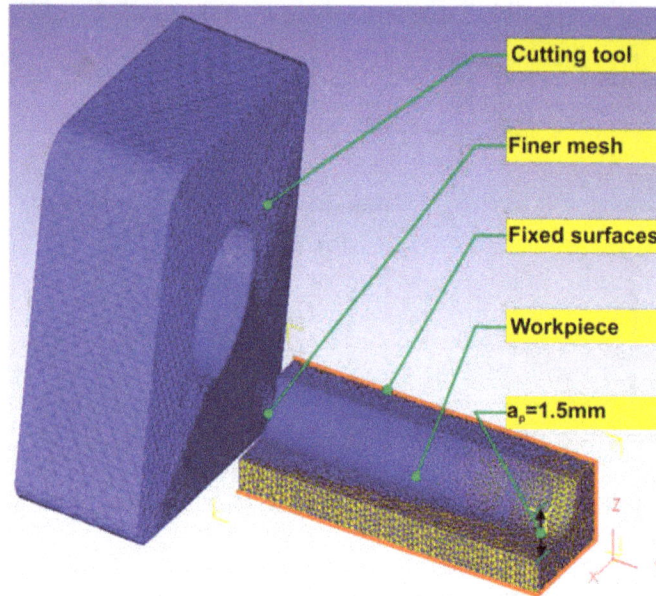

Figure 12.    Objects of the FEM model.

quality and cost. To evaluate the effect of each control factor or its interaction effects on the process outputs, Taguchi employs standard orthogonal arrays. After conducting the experiments and collecting the desired outputs according to the Taguchi arrays, the values of outputs have to be transferred into signal-to-noise (S/N) ratios. Usually, there are three S/N ratios available, ($\eta$), depending on the type of characteristic; the lower-the better (LB), the higher-the better (HB), and the nominal-the better (NB). Owing to the non-beneficiary nature of the considered performances (outputs) here, the LB is adopted as:

$$\eta_{ij} = -10 \log_{10} \left( \frac{1}{n} \sum_{i=1}^{n} Y_{ijk}^2 \right) \qquad (8)$$

where $\eta_{ij}$ is the S/N ratio for the response $j$ of experimental number $i$, and $Y_{ijk}$ is the experiment result for the response $j$ of the experiment $i$, in the $k$th replication; $n$ is the total number of replications, and $Y_0$ is the nominal response. Thereafter, analysis of means (ANOM) and analysis of variance are performed on the S/N ratio of orthogonal array to identify the optimal factor level combination and estimate the error variance for the effects and variance of the prediction error, respectively (Keleştemur, Arıcı, Yıldız, & Gökçer, 2014).

In this study, the eight considered control factors are as follows: (1) thermal contact conductance $h_{tc}$, (2) cutting speed $v_c$, (3) feed rate $f$, (4) Coulomb friction coefficient $\mu_c$, (5) shear friction coefficient $\mu_t$, (6) Taylor–Quinney coefficient $\kappa_t$, (7) percentage decrease in flow stress %$p$, and (8) Cockcroft–Latham critical damage value $D_{\mathrm{crit.}}$. The third and fourth control factors are referred to the cutting conditions and are included in the design to give the design more flexibility in dealing with other sets of cutting conditions that are planned in the second stage of the study. Taguchi mixed design $L_{18}$ ($2^1 \times 3^7$) is adopted so that the underestimation and overestimation problems are avoided by keeping the most parameter levels at three. Additional advantage of this factorial design is the reduction of experimentation trials from 18 to 9, because cutting speed and feed rate are the only input variables that can be designated as controlled factors during experimentation phase of the study. The control factors and their levels and the orthogonal array are shown in Tables 4 and 5, respectively.

## 4.  Results of the DOE

The first stage of performing numerical simulations has confirmed the strong influence of control factors on the cutting performances. Among the performances that have been strongly affected were the cutting temperatures, cutting forces, strain rates, and chip morphologies. Figure 13 depicts the impact of sets of control factors on the cutting temperature and chip morphology after cutting 8 mm of the workpiece length. It is difficult to draw clear conclusion points on the parametric effect of each control factor based on

Table 4.   Control factors and levels for the design of simulation experiments.

| Factors | Sym. | Unit | Levels | | |
|---|---|---|---|---|---|
| | | | 1 | 2 | 3 |
| Thermal contact conductance | $h_{tc}$ | N/(s mm K) | 100 | 1000 | |
| Cutting speed | $v_c$ | m/min | 80 | 160 | 240 |
| Feed rate | $f_r$ | mm/rev | .15 | .225 | .3 |
| Coulomb friction coefficient | $\mu_c$ | – | .5 | .75 | 1 |
| Shear friction coefficient | $\mu_t$ | – | .6 | .9 | 1.2 |
| Taylor–Quinney coefficient | $\kappa_t$ | – | .8 | .9 | 1 |
| Reduction in flow stress | %$p$ | – | 10 | 30 | 50 |
| Critical damage value | $D_{crit.}$ | MPa | 50 | 100 | 150 |

Table 5.   $L_{18}$ ($2^{1\times}3^7$) orthogonal array.

| Exp. No. | $h_{tc}$ (N/mm s K) | $v_c$ (m/min) | $f_r$ (mm/rev) | $\mu_c$ | $\mu_t$ | $\kappa_t$ | %$p$ | $D_{crit.}$ (MPa) |
|---|---|---|---|---|---|---|---|---|
| | | | Control factors | | | | | |
| 1 | 100 | 80 | .15 | .5 | .6 | .8 | 10 | 50 |
| 2 | 100 | 80 | .225 | .75 | .9 | .9 | 30 | 100 |
| 3 | 100 | 80 | .3 | 1 | 1.2 | 1 | 50 | 150 |
| 4 | 100 | 160 | .15 | .5 | .9 | .9 | 50 | 150 |
| 5 | 100 | 160 | .225 | .75 | 1.2 | 1 | 10 | 50 |
| 6 | 100 | 160 | .3 | 1 | .6 | .8 | 30 | 100 |
| 7 | 100 | 240 | .15 | .75 | .6 | 1 | 30 | 150 |
| 8 | 100 | 240 | .225 | 1 | .9 | .8 | 50 | 50 |
| 9 | 100 | 240 | .3 | .5 | 1.2 | .9 | 10 | 100 |
| 10 | 1000 | 80 | .15 | 1 | 1.2 | .9 | 30 | 50 |
| 11 | 1000 | 80 | .225 | .5 | .6 | 1 | 50 | 100 |
| 12 | 1000 | 80 | .3 | .75 | .9 | .8 | 10 | 150 |
| 13 | 1000 | 160 | .15 | .75 | 1.2 | .8 | 50 | 100 |
| 14 | 1000 | 160 | .225 | 1 | .6 | .9 | 10 | 150 |
| 15 | 1000 | 160 | .3 | .5 | .9 | 1 | 30 | 50 |
| 16 | 1000 | 240 | .15 | 1 | .9 | 1 | 10 | 100 |
| 17 | 1000 | 240 | .225 | .5 | 1.2 | .8 | 30 | 150 |
| 18 | 1000 | 240 | .3 | .75 | .6 | .9 | 50 | 50 |

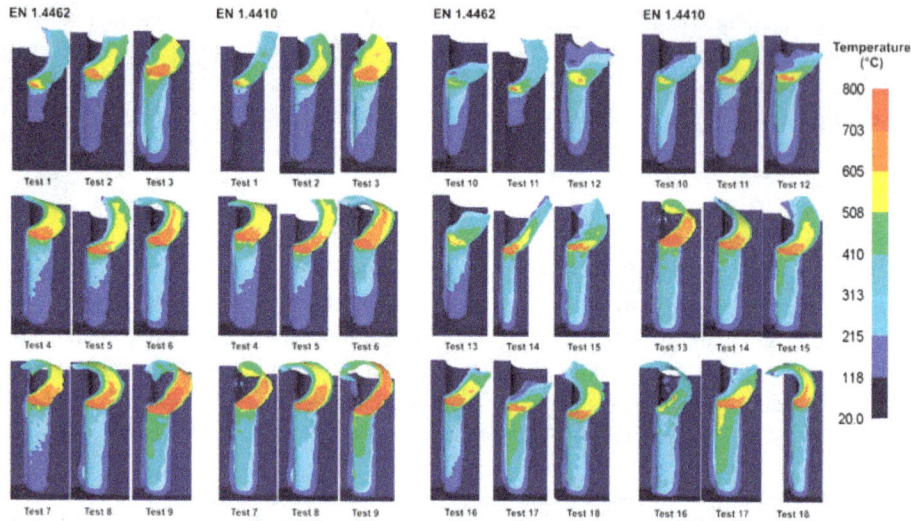

Figure 13.    Simulated workpiece temperature distributions.

the images shown, since at each cutting step, the values of performances change considerably. Therefore, statistical tools have to be employed to analyze the results and help emphasizing the effect of control factors separately.

Based on the Taguchi optimization procedure, the next step after recording the results of experimentation is to calculated the S/N ration of obtained performance characteristics such as cutting forces in feed ($F_f$), cutting ($F_c$), and thrust ($F_t$) directions, strain, strain rate, principal stress, mean tool–chip interface temperature, chip temperature, and chip thickness for each simulation. Owing to the non-beneficial nature of the studied performances, the lowest value is always desirable. Therefore, Equation (8) is employed in this case. The simulation of randomly specified trials was repeated and no differences between replications were noticed (i.e. $k = 1$). The overall mean of ($\eta_{ij}$) associated with $n$ trials is then determined using the principles of analysis of means (ANOM). The level which returns maximum S/N ratio is designated as the optimum, since it optimizes the response and minimizes the effect of the noise factors simultaneously.

Figure 14 shows the results of performing ANOM after normalization. The mean S/N ratios are normalized between the worst '0' and the best '1'. However, it should be noted here that explaining the main effect plots of each involved control factor on obtained performance characteristics is an enormous task and requires lots of paper space. For example, the optimum $factor^{level}$ combinations which minimize the main cutting force for machining EN 1.4462 and EN 1.4410 are $h_{tc}^1\ v_c^3\ f_r^2\ \mu_c^1\ \mu_t^1\ \kappa_t^3\ \%p^1\ D_{\text{crit.}}^1$ and $h_{tc}^1\ v_c^3\ f_r^1\ \mu_c^2\ \mu_t^1\ \kappa_t^3\ \%p^1\ D_{\text{crit.}}^1$, respectively. Another example, in the case of turning 25 EN 1.4410, simulation results showed that increasing the cutting speed value from $v_c = 80$ m/min to $v_c = 240$ m/min had reduced the main effects of $F_c$, $F_f$, $F_t$, strain, and chip thickness by 43.6%, 32.08%, 46.08%, 13.244%, and 39.44%, respectively. While the main effects of strain rate, normal stress, and chip/tool interface temperature had increased by 37.98%, 7.25%, and 19.045%, respectively.

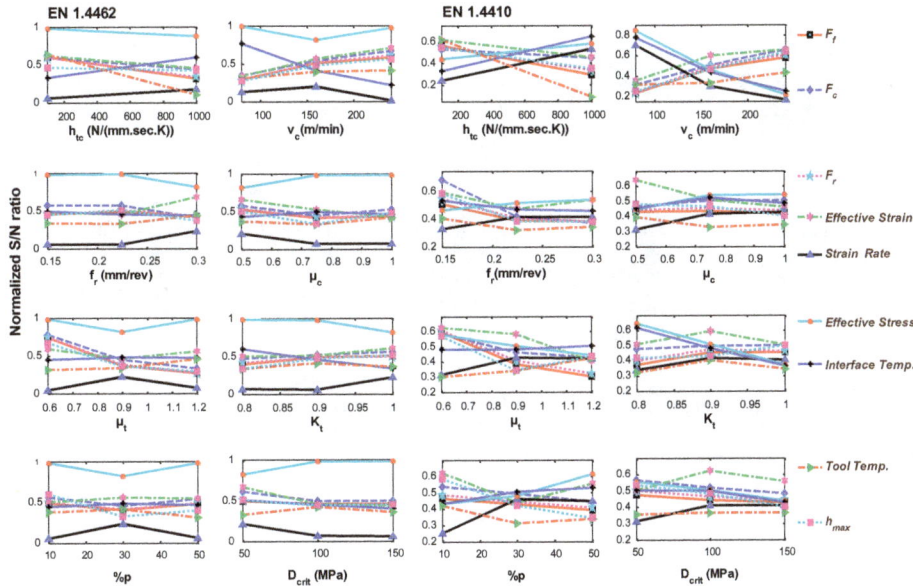

Figure 14. Main effect plots of the effect of control factors on the cutting performances during turning EN 1.4410 and EN 1.4462 DSSs.

## 5. Proposed methodology

The percentage of differences between the experimentally measured cutting forces, chip temperatures and thicknesses, and the FEM-predicted ones are computed using the percentage of difference expression (%E):

$$\%E = \frac{Y_{\mathrm{EXP}} - Y_{\mathrm{FEM}}}{Y_{\mathrm{FEM}}} \times 100 \qquad (9)$$

where $Y_{\mathrm{EXP}}$ represents the experimental performance values and $Y_{\mathrm{FEM}}$ the simulated performance values. To estimate the effect of the control factors and important interactions on the %E, percentage of contribution of each control factor on variance of the corresponding error percentage should be computed. The control factor that returns highest contribution percentage is designated as the prime factor. For the sake of more convenience, percentages are drawn in pie charts and shown in Figure 15. For instance, the control factors which are expected to play important roles in minimizing the percentage differences of feed force $\%E_f$, cutting force $\%E_c$, thrust force $\%E_t$, tool temperature $\%E_T$, and chip thickness $\%E_h$ during turning EN 1.4462 were $\mu_t$, $\kappa_t$, $\mu_t$ and $h_{tc} \times v_c$, $\mu_t$ and $h_{tc}$, and $\mu_t$, respectively.

Results of %E computations vs. the experimental order are illustrated in Figure 16. It can be seen that the predicted feed and thrust force components are generally underestimated except for some few cases at $D_{\mathrm{crit.}} = 150$ MPa. Fluctuations of $\%E_f$ and $\%E_t$ were generally in the range from −40 to 195%. On the other hand, $\%E_c$ has shown better agreements with the range of approximately −40 to 45%. Furthermore, comparing the $T_{\mathrm{tool}}$ and $h_{\mathrm{max}}$, the %E between experimental and FEM values also fluctuate between overestimation and underestimation, with very few cases where the absolute %E is less than 5%. These results show the strong impact of control factors on the outputs of the

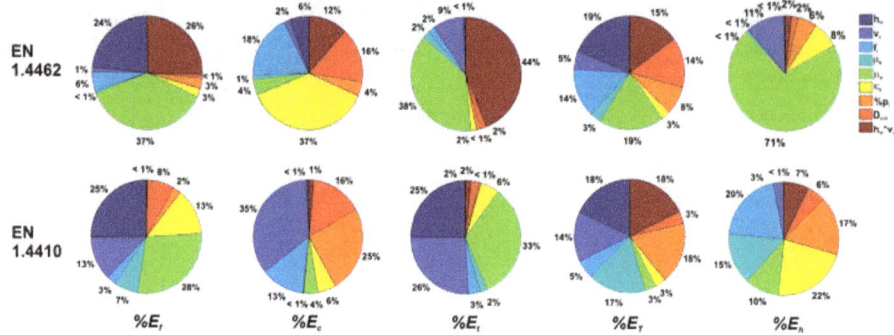

Figure 15.   The percentage contribution of FEM control factors of the percentage error.

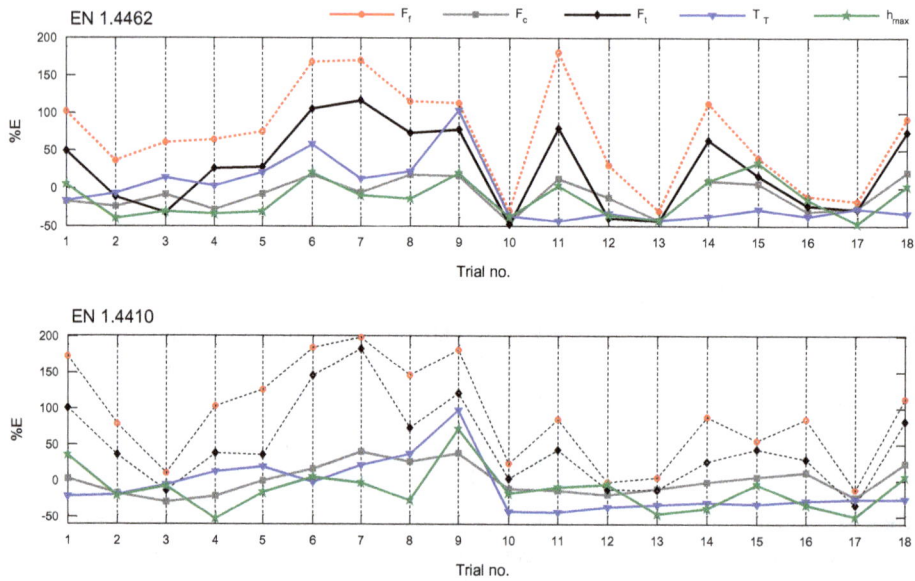

Figure 16. Percentage error between the experimental and predicted values for cutting performances.

FEM. The next sections describe the study approach intended to simultaneously minimize all the differences between measured and predicted performances.

## 5.1.  *VIKOR-based Taguchi*

The VIKOR method was developed for multi-criteria optimization of complex systems. It determines the compromise ranking-list, the compromise solution, and the weight stability intervals for preference stability of the compromise solution obtained with the initial (given) weights. This method focuses on ranking and selecting from a set of alternatives in the presence of conflicting criteria. It introduces the multi-criteria ranking index based on the particular measure of 'closeness' to the 'ideal' solution. On the other hand, VIKOR-based Taguchi provides a systematic and efficient methodology for determining the optimum combination of the control factors such that the output is effective and has high performance and also is robust to the noise factors. The multiple-percentage of error problem at this stage is converted into a single-response problem as follow:

**Step 1.** Construction of the decision matrix: The decision matrix in this case is comprised of the values of S/N ratios for computed percentages of errors, that is, $\eta_{ij}$ of $\%E_j$ where $i = 1, 2, ...,$ number of simulation experiments $(m = 18)$, $j = 1, 2, ...,$ number of performances percentage of errors $(n = 5)$.

$$D = \begin{bmatrix} \eta_{11} & \eta_{12} & \cdots & \eta_{1n} \\ \eta_{21} & \eta_{22} & \cdots & \eta_{2n} \\ . & . & \cdots & . \\ . & . & \cdots & . \\ \eta_{m1} & \eta_{m2} & \cdots & \eta_{mn} \end{bmatrix} \quad (10)$$

**Step 2.** Calculation of normalized ratings by the vector normalization:

$$r_{ij} = \left( \frac{\eta_{ij} - \min(\eta_j)}{\max(\eta_j) - \min(\eta_j)} \right) \quad (11)$$

**Step 3.** Calculate the relative weight of performances using the standard deviation method:

$$w_j = \frac{STDV_j}{\sum_{j=1}^{n} STDV_j} \quad (12)$$

where $STDV_j$ is the standard deviation of performance $j$. The summation of the individual weights should always yield one.

**Step 4.** Calculate weighted normalized ratings $(\vartheta_{ij})$:

$$\vartheta_{ij} = w_j r_{ij} \quad (13)$$

**Step 5.** Identification of positive ideal and negative ideal solutions $(A^*$ and $A^-)$: The positive ideal solution, $A^*$ $(A_i^*; i = 1, 2, ..., m)$, is made of all the best values (maximum S/N ratio) and the negative ideal solution, $A^-$ $(A_i^-; i = 1, 2, ..., m)$, is made of all the worst values(minimum S/N ratio) at the responses in the weighted normalized decision matrix. They are calculated using Equations (14) and (15). where $J$ is the set of cost type criteria and $J'$ is the set of benefit type criteria.

$$A^* = \left\{ \vartheta_1^*, ..., \vartheta_i^* \right\} = \left\{ \left( \max \vartheta_{ij} | j \in J \right), \left( \min \vartheta_{ij} | j \in J' \right) \right\} \quad (14)$$

$$A^- = \left\{ \vartheta_1^-, ..., \vartheta_i^- \right\} = \left\{ \left( \min \vartheta_{ij} | j \in J \right), \left( \max \vartheta_{ij} | j \in J' \right) \right\} \quad (15)$$

where $J = \{ j = 1, 2, ..., n | \vartheta_{ij},$ if desired response is large$\}$; $J' = \{ j = 1, 2, ..., n | \vartheta_{ij},$ if desired response is small$\}$.

**Step 6.** Compute the values of utility measure $S_j$ and regret measure $R_j$ by respective relations:

$$S_i = \sum_{j=1}^{n} w_j \frac{\left(\vartheta_j^* - \vartheta_{ij}\right)}{\left(\vartheta_j^* - \vartheta_j^-\right)} \qquad (16)$$

$$R_i = \max_j \left[ w_j \frac{\left(\vartheta_j^* - \vartheta_{ij}\right)}{\left(\vartheta_j^* - \vartheta_j^-\right)} \right] \qquad (17)$$

**Step 7.** Compute the VIKOR index ($Q_V$) values for attribute $j = 1, 2, ..., n$, by the relation

$$Q_V = \xi \frac{(S_i - S^*)}{(S^- - S^*)} + (1 - \xi) \frac{(R_i - R^*)}{(R^- - R^*)} \qquad (18)$$

where $S^* = \min_j S_j$, $S^- = \max_j S_j$; $R^* = \min_j R_j$, $R^- = \max_j R_j$.

$\xi$ is introduced as weight of the strategy of the maximum group utility, usually $\xi = 0.5$.

**Step 8.** Rank the alternatives, sorting by the values $S$, $R$, and $Q_V$. The results are three ranking lists.

**Step 9.** For given attribute weights, propose a compromise solution, alternative $A_{b1}$, which is the best ranked by $Q_V$, if the following two conditions are satisfied:

**Condition 1:** "Acceptable advantage": $Q_V(A_{b2}) - Q_V(A_{b1}) \geq (1/(m-1))$, where $A_{b1}$ and $A_{b2}$ are the first and second-best alternatives in the ranking order by $Q_V$.

**Condition 2:** "Acceptable stability in decision making": alternative $A_{b1}$ must also be the best ranked by $S$ and/or $R$. This compromise solution is stable within a decision-making process, which could be 'voting by majority rule' (when $\xi > 0.5$ is needed) or 'by consensus' (when $\xi \approx 0.5$) or 'with veto' (when $\xi < 0.5$). If one of the conditions is not satisfied, then a set of compromise solutions is proposed, which consists of:

(a)  Alternatives $A_{b1}$ and $A_{b2}$ if only condition 2 is not satisfied.
(b)  Alternatives $A_{b1}$, $A_{b2}$, ..., $A_{bi}$ if condition 1 is not satisfied; $A_p$ is determined by the relation $Q_V(A_{b2}) - Q_V(A_{b1}) \approx (1/(m-1))$ (Rao, 2013).

The procedure of calculating VIKOR indices for each work material case is started with the representation of decision matrix. S/N values are normalized using Equation (11). The criteria weights are then specified using the standard deviation method as per Equation (12). Thereafter, normalized decision matrices are calculated using Equation (13). The best and the worst values of all the criteria in weighted normalized decision matrices are identified using Equations (14) and (15). The values of utility ($S$) and regret ($R$) measures are calculated using Equations (16) and (17), respectively. The VIKOR index values ($Q_V$) are calculated using Equation (18). For both work material cases, the values of the first and second ranked alternatives are in acceptable advantage range, that is, $Q_V(A_{b2}) - Q_V(A_{b1})$ is always greater than $1/(18-1) = 0.05882$. For example, the $Q_V$ ($A_{b2}$) and $Q_V$ ($A_{b1}$) of EN 1.4462 are 0.106 and 0.012, respectively, which satisfies condition 1, that is, $0.106-0.012 = 0.094$ is greater than 0.05882. Both regret and utility measure values by consensus prove the stability in decision-making. Therefore, the values of $Q_V$ are directly used in the next analyses. Based on the outcomes of the

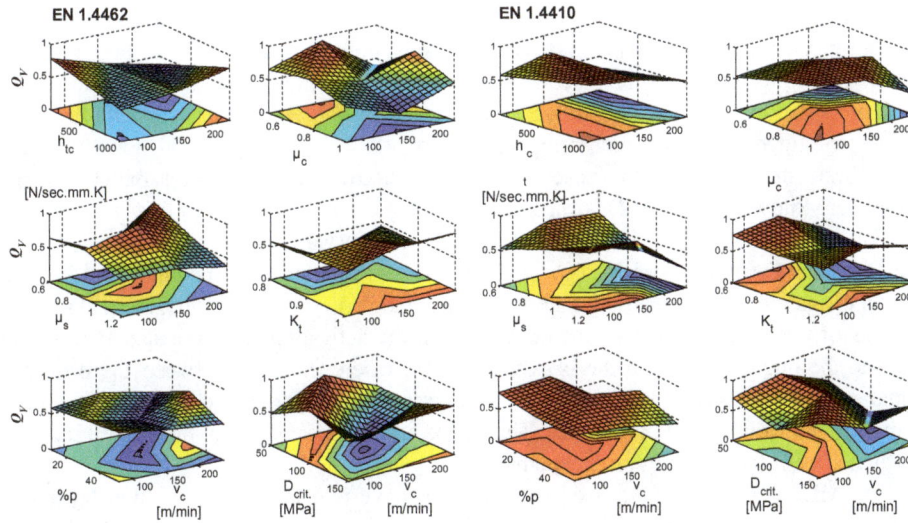

Figure 17.   Surface and contour plots of the VIKOR indices ($Q_V$).

VIKOR method, the best alternatives when cutting EN 1.4410 and EN 1.4462 were alternatives No. 15 and No. 6, respectively. Performing the ANOM, it can be proved that the optimum $factor^{level}$ sets when machining EN 1.4462 and EN 1.4410 are $h_{tc}^2\ v_c^1$ $f_r^1\ \mu_c^2\ \mu_t^2\ \kappa_t^3\ \%p^2\ D_{crit.}^3$ and $h_{tc}^2\ v_c^2\ f_r^3\ \mu_c^2\ \mu_t^3\ \kappa_t^3\ \%p^2 D_{crit.}^3$, respectively. The combination of these control factors should minimize the difference between experimental and numerical results. Figure 17 maps the effect of cutting speed interactions with the rest of control factors on the $Q_V$ values. The dark blue areas represent the favored regions where the difference between experimental and numerical performances is minimum.

## 5.2.   Firefly algorithm neural network system

Unfortunately, owing to the missing impact of interactions in Table 5, the choice of the selecting the optimal set is not always straightforward. Additionally, the above optimum sets are case specific to the given cutting conditions. To comprehend the $Q_V$ so that it covers all the domain of the control factors and their corresponding interactions, an effective modeling technique is required. Unfortunately, the traditional modeling techniques such as response surface methodology were seen not quite efficient to accurately predict all values of $Q_V$. Furthermore, to account for all interactions, adopting quadratic model for example, comprise 44 terms which makes the problem even more complex. Therefore, a multi-layer perceptron (MLP) artificial neural network (ANN) is proposed to accomplish the task. Once, this step is accomplished, a recently developed stochastic optimization algorithm has to be integrated with the proposed ANN to minimize $Q_V$ and determine the optimum set at any required cutting condition.

### 5.2.1.   Artificial neural network

Neural network is a logical structure with multi-processing elements, which are connected through interconnection weights. The knowledge is presented by the interconnection weights, which are adjusted during the training phase. The process of training a neural network can be broadly classified into two typical categories: supervised learning and unsupervised learning. The first requires using both the input and the target values for each sample in the training set. The most common algorithm in this group is the back propagation, used in the Multi-Layer Perceptron (MLP), but it also includes most of the training methods for recurrent neural networks, time delay neural networks, and thrust basis networks. Unsupervised learning is used when the target pattern is not completely known (Quiza & Davim, 2011).

To establish a useful relationship between independent variables ($h_{tc}$, $v_c$, $f_r$, $\mu_c$, $\mu_b$, $\kappa_b$, $\%p$, and $D_{crit.}$) and dependent variable ($Q_V$), supervised feed-forward MLP-neural networks with back propagation (BP) as learning algorithm were adopted. The back propagation algorithm consists of two phases: the forward phase, where the activations are propagated from the input to the output layer, and the backward phase, where the error between the observed actual and the requested nominal value in the output layer is propagated backwards in order to modify the weights and bias values. This procedure is repeated many times until the sum of squared error term reaches an acceptable level.

The MLP-neural networks used in this study have two layers: one hidden layer and one output layer (see Figure 18). The hidden layer uses a sigmoid-type transference function:

$$F(\chi) = \frac{1}{1 + \exp(-b - \sum \omega_i \chi_i)} \tag{19}$$

while the output layer uses a linear function:

$$F(\chi) = b + \sum \omega_i \chi_i \tag{20}$$

where $\omega$ and $b$ are the weights and biases of the network, respectively. To carry out the training process, all the inputs were normalized using the following equation:

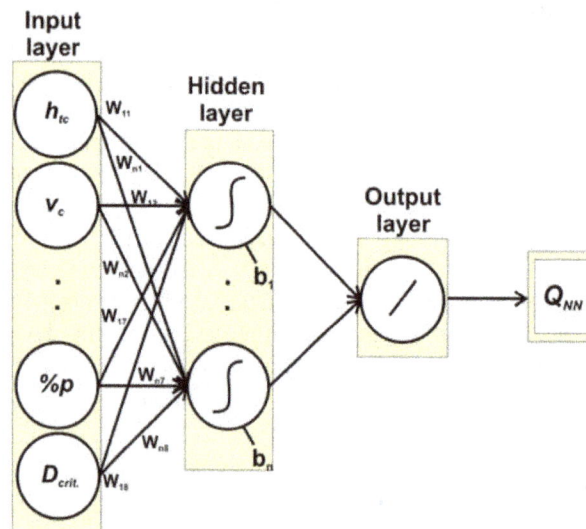

Figure 18.   The neural network architecture used in the FANNS.

$$\chi_i = \frac{2(\chi - \chi_{\min})}{(\chi_{\max} - \chi_{\min})} - 1 \qquad (21)$$

The final output ($Q_{\mathrm{NN}}$) of input vector (8*1)-number of hidden neurons (N)-output feed-forward neural network structure can be mathematically expressed as:

$$Q_{\mathrm{NN}} = \log \mathrm{sig}\left( \begin{bmatrix} \omega_{11} & . & . & \omega_{18} \\ . & . & . & . \\ . & . & . & . \\ . & . & . & . \\ \omega_{81} & . & . & \omega_{88} \end{bmatrix} \times \begin{bmatrix} h_{tc} \\ . \\ . \\ . \\ D_{\mathrm{crit.}} \end{bmatrix} + \begin{bmatrix} b_1 \\ . \\ . \\ b_N \end{bmatrix} \right) \times [\omega_1 \quad . \quad . \quad \omega_N] + b_{N+1}$$

$$(22)$$

In this study, the MLP uses sigmoid transfer functions in the hidden layers. These functions are often called 'squashing' functions, because they compress an infinite input range into a finite output range. Sigmoid functions are characterized by the fact that their slopes must approach zero as the input gets large. This causes a problem when you use steepest descent to train a multi-layer network with sigmoid functions, because the gradient can have a very small magnitude and, therefore, causes small changes in the weights and biases, even though the weights and biases are far from their optimal values. The purpose of the resilient back propagation training algorithm (RPROP) is to eliminate these harmful effects of the magnitudes of the partial derivatives. Only the sign of the derivative can determine the direction of the weight update; the magnitude of the derivative has no effect on the weight update. The size of the weight change is determined by a separate update value. The number of learning steps is significantly reduced in comparison to the original gradient-descent procedure as well as to other adaptive procedures, whereas the expense of computation of the RPROP adaptation process is held considerably small. Another important feature, especially relevant in practical application, is the robustness of the new algorithm against the choice of its initial parameter (Riedmiller & Braun, 1993).

The scaling or normalization ensures that the ANN will be trained effectively without any particular variable skewing the results significantly. The weights and biases of the network are initialized to small random values to avoid immediate saturation in the activation functions. The neural networks trained using the gradient descendent with adaptive velocity and momentum back propagation algorithm to model $Q_{V_5}$ where it is not easy to obtain analytical and good empirical relations. The optimum architecture was found out by varying network characteristics in MATLAB using trial and error technique. It was found that when the training function, RPROP; the number of hidden neurons, 13; maximum number of epochs to train, 100; the learning rate, 0.01; the increment to weight change, 1.2; decrement to weight change, 0.5; initial weight change, 0.07; and maximum weight change, 50, the root mean square of errors were the minimum.

Because neural networks are purely empirical models, validation is critical to operational success. Figure 19 shows the comparison of experimental results and modeling in verifying the network generalization capabilities. The results are almost identical.

### 5.2.2. Firefly algorithm (FA)

FA is one of the recent swarm intelligence methods developed by Yang (2010) in 2008 and is a kind of stochastic, nature-inspired, meta-heuristic algorithm that can be applied for solving the hardest optimization. The algorithm is inspired by the flashing lights of

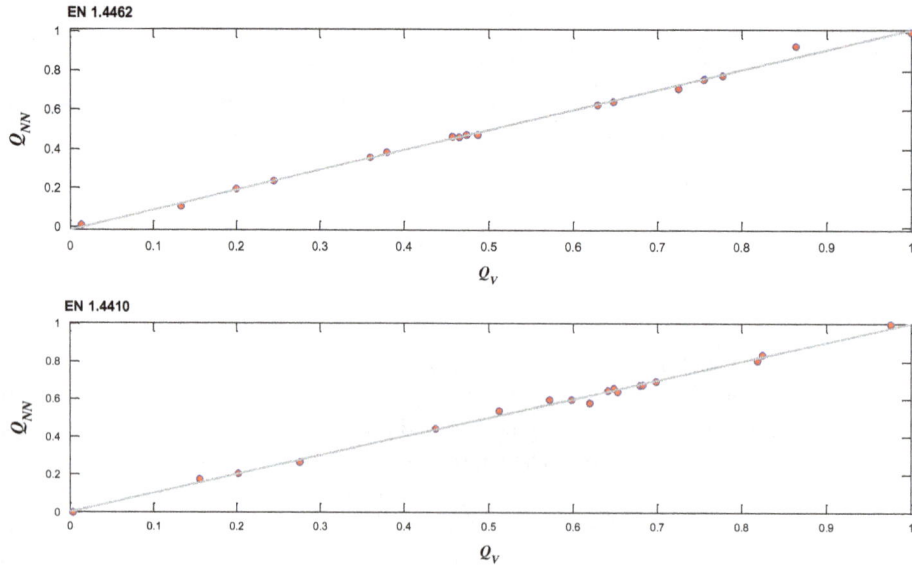

Figure 19.    Scatter plot of actual vs. ANN-predicted VIKOR indices.

fireflies in nature. Such flashing light may serve as the primary courtship signals for mating. Besides attracting mating partners, the flashing light may also serve to warn off potential predators. The algorithm has been formulated by assuming:

(1) All fireflies are unisex so that one firefly will be attracted to other fireflies regardless of their sex.
(2) Attractiveness is proportional to their light intensities. The less bright will be moving toward the brighter one. It will move randomly if there is no brighter one.
(3) The brightness of a firefly is affected or determined by the landscape of the objective function.

The light intensity and attractiveness are in some way synonymous. While the intensity is referred to as an absolute measure of emitted light by the firefly, the attractiveness is a relative measure of the light that should be seen in the eyes of the beholders and judged by other fireflies. The attractiveness $\beta(\delta)$ and light intensities $I(\delta)$ of each firefly are considered to decrease monotonically depending on the distance $\delta$ as:

$$\beta(s) = \beta_0 e^{-\gamma_a \delta_r^2} \tag{23}$$

$$I(s) = I_0 e^{-\gamma_a \delta_r^2} \tag{24}$$

where $\beta_0$ and $I_0$ denote the maximum attractiveness and light intensities, respectively (i.e. at $\delta_r = 0$), and $\gamma_a$ is the light absorption coefficient, which controls the decrease of the light intensity. The light intensity $I$ of a firefly representing the solution $s_i$ and is proportional to the value of fitness function $I(s_i) \propto f(s_i)$. The distance between any two firefliese $s_i$ and $s_j$ is expressed as the Euclidean distance by the base firefly algorithm, as follows:

$$\delta_{ij} = \left\|s_i - s_j\right\| = \sqrt{\sum_{k=1}^{k=d} \left(s_{ik} - s_{jk}\right)^2} \qquad (25)$$

where $d$ denotes the dimensionality of the problem. The movement of the $i$th firefly is attracted to another more attractive firefly $j$. In this manner, the following equation is applied:

$$s_i = s_i + \beta_0 e^{-\gamma_a s_{ij}^2}(s_j - s_i) + \alpha\psi_i \qquad (26)$$

where $\psi_i$ is a random number drawn from Gaussian distribution and $\alpha$ is the randomization parameter. In summary, FA is controlled by three parameters: the attractiveness $\beta$, the absorption coefficient $\gamma_a$, and the randomization parameter $\alpha$. Finally, the basic steps of the Firefly Algorithm (FA) can be summarized as the pseudo code and shown in Figure 20.

In order to obtain solutions that will provide useful information to the user during the phase of inverse identification of input parameters, neural network models should be integrated with the FA. The neural network models integrated with the FA optimizer were named FANNS and its architecture is shown in Figure 21.

The target of the optimization process in this study is to determine the optimal values of the input parameters that lead to the minimum value of ANN-predicted $Q_{NN}$ at any given cutting condition. The optimization problem can be defined as:

$$\text{Objective} = \text{Max}\ (1 - Q_{NN}) \qquad (27)$$

The decision variables are $h_{tc}$, $v_c$, $f_r$, $\mu_c$, $\mu_t$, $\kappa_t$, %$p$, and $D_{crit.}$ They are constrained within the range of simulation experiments (refer to Table 4):

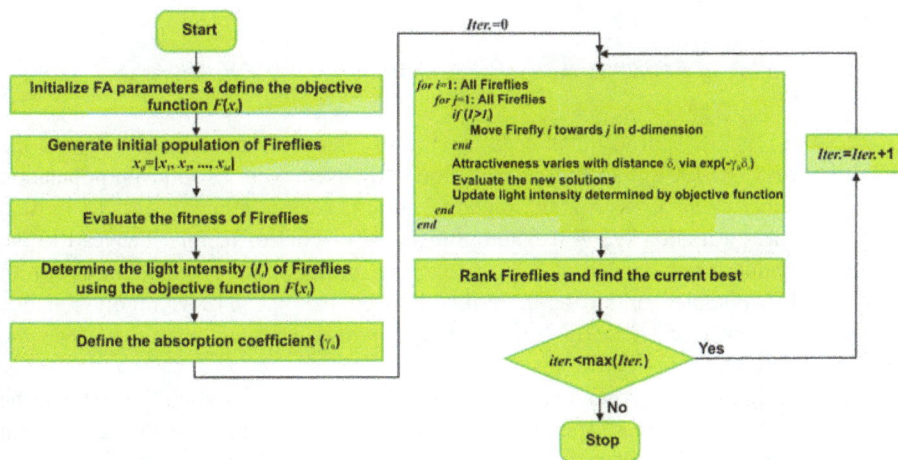

Figure 20.    Flow chart of the Firefly Algorithm (FA).

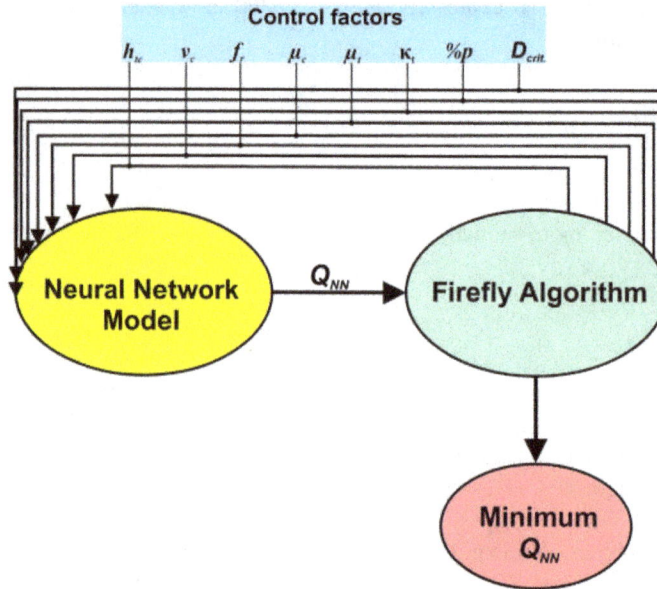

Figure 21.   Firefly algorithm neural network system (FANNS).

$$100 \leq h_{tc} \leq 1000 \text{ N}/(\text{s mm K})$$
$$80 \leq v_c \leq 240 \text{ m}/\text{min}$$
$$0.15 \leq f_r \leq 0.3 \text{ mm/rev}$$
$$0.5 \leq \mu_c \leq 1$$
$$0.6 \leq \mu_t \leq 1.2 \tag{28}$$
$$0.8 \leq \kappa_t \leq 1$$
$$10 \leq \%p \leq 50$$
$$50 \leq D_{crit.} \leq 150 \text{ MPa}$$

### 5.2.3.  *Application of FANNS to the global optimization of VIKOR indices and validation of the results*

The proposed FANNS approach has been applied to effectively model and optimize the VIKOR indices. The FA initializing optimization parameters were as follows: population size = 20, number of iterations = 2000, randomization ($\alpha$) = 0.5, attractiveness ($\beta_0$) = 0.2, and attractiveness variation ($\gamma_a$) = 1. The optimal sets of control factors that lead to the optimum $Q_V$ values are tabulated in Table 6, and the results of the FANNS are shown in Figure 22. The obtained optimization results showed that FANNS is highly reliable, converge consistently and quickly (the computation time was less than 3 min) to the optimum solution.

To validate the global FANNS results, the experimental cutting conditions should be identical to the numerical ones. Therefore, three longitudinal turning experiments per each work material were carried out at the exact optimum cutting conditions as listed in Table 6. The averages of the measured performances are computed and used to calculate the percentage difference between the numerical and experimental ones using Equation (9).

Table 6.   Optimum sets of control factors.

| Optimum control factor | Unit | EN 1.4462 | EN 1.4410 |
|---|---|---|---|
| Thermal contact conductance ($h_{tc}$) | N/(s mm K) | 652.306 | 647.945 |
| Cutting speed ($v_c$) | m/min | 181.175 | 130.601 |
| Feed rate ($f_r$) | mm/rev | .21321 | .27441 |
| Coulomb friction coefficient ($\mu_c$) | – | .63151 | .91255 |
| Shear friction coefficient ($\mu_t$) | – | .74952 | 1.01455 |
| Taylor–Quinney coefficient ($\kappa_t$) | – | .95352 | .96525 |
| Reduction in flow stress ($\%p$) | – | 19.3549 | 28.8121 |
| Critical damage value ($D_{crit.}$) | MPa | 69.2885 | 89.1997 |
| Maximum VIKOR index ($Q_{NN}$) | – | .92211 | .95785 |

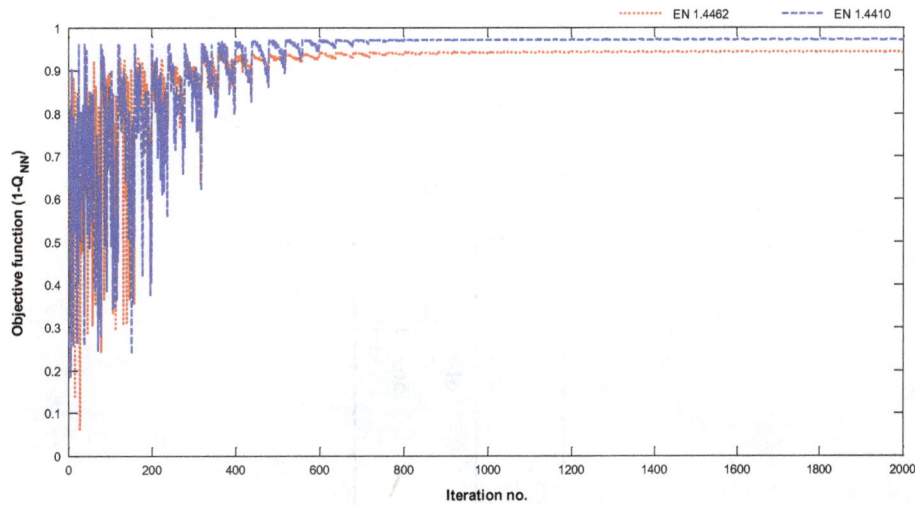

Figure 22.   Minimization of the VIKOR indices using FANNS.

The percentage difference between numerically obtained cutting forces, chip temperature, and maximum chip thicknesses, and the experimental ones for each work material are tabulated in Table 7. It can be seen that the calculated performances are in close agreement to the experimental results. The global optimum difference percentages are also compared with difference percentages of the rank No. 1. The lower percentages of difference confirm the advantage of FANNS application in identifying the input parameters while simulating the machining of duplex stainless steels.

### 5.3.   Extension of FANNS application: a case study

One of the aims of this study was to define a robust approach, so that the optimum set of control factors at any given cutting speed and feed rate is accurately determined. In this section, the adaptability of the described approach to the changing cutting speeds and feed rates is examined. Three arbitrary experimental cutting tests per each work material were conducted. The optimum sets of control factors at the exact cutting speeds

Table 7. Validation numerical and experimental results.

| | | $F_f$ (N) | | $F_c$ (N) | | $F_t$ (N) | | $T_{tool}$ (°C) | | $h_{max}$ (mm) | |
|---|---|---|---|---|---|---|---|---|---|---|---|
| | | Exp. | Num. | Exp. | Num. | Exp. | Num. | Exp. | Num. | Exp. | Num. |
| EN 1.4462 | | 385.167 | 369.941 | 935.654 | 9,720,614 | 298.646 | 290.139 | 134.635 | 140.519 | .3346 | .3516 |
| | %E | −4.116 | 3.745 | −2.139 | 4.187 | 5.125 | | | | | |
| EN 1.4410 | | 761.152 | 723.686 | 1485.63 | 1569.54 | 517.254 | 491.827 | 142.714 | 148.709 | .5535 | .5843 |
| | %E | −5.177 | 5.346 | −5.169 | 4.031 | 5.254 | | | | | |

Table 8. Optimum sets of control factors at fixed cutting speed and feed rates.

| | Parameter | Unit | A.1. | A.2. | A.3. |
|---|---|---|---|---|---|
| EN 1.4462 | $v_c$ | m/min | 75 | 150 | 225 |
| | $f_r$ | mm/rev | .325 | .175 | .2 |
| | $h_{tc}$ | N/(s mm K) | 665.558 | 771.698 | 820.049 |
| | $\mu_c$ | – | .53225 | .66125 | .69922 |
| | $\mu_t$ | – | .69525 | .70525 | .84552 |
| | $\kappa_t$ | – | .98308 | .90325 | .97825 |
| | %$p$ | – | 24.7349 | 21.1528 | 19.8571 |
| | $D_{crit.}$ | MPa | 65.7898 | 77.6984 | 71.2524 |
| EN 1.4410 | $h_{tc}$ | N/(s mm K) | 795.523 | 826.375 | 862.815 |
| | $\mu_c$ | – | .99125 | .84258 | .87134 |
| | $\mu_t$ | – | 1.04321 | 1.06925 | 1.08125 |
| | $\kappa_t$ | – | .98995 | .94582 | .92545 |
| | %$p$ | – | 30.7425 | 22.2481 | 26.4358 |
| | $D_{crit.}$ | MPa | 92.506 | 100.678 | 111.452 |

and feed rates were determined using the proposed FANNS (see Table 8); meanwhile numerical models, each at the exact optimum set of optimized control factors, were prepared and executed. Figure 23 shows the measured and predicted cutting performances for EN 1.4410 and EN 1.4462. The average of absolute percentage differences between the experimental and numerical results for each material case is 3.8652 and 4.9956%, respectively. Considering the wide range of applied cutting speeds and feed rates, the proposed methodology of inversely identifying the simulation input factors shows excellent results with respect to the examined performances.

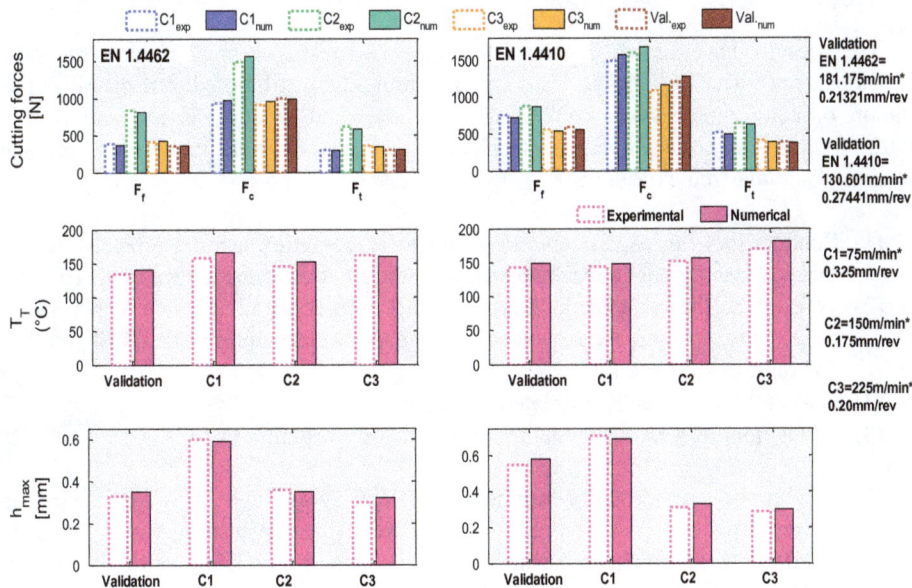

Figure 23. Experimental validation with respect to the performances for EN 1.4462 and EN 1.4410 DSS.

## 6.  Inverse identification of the Usui's wear law constants

Tool wear is a major consideration in all machining operations. It adversely affects tool life, quality of machining surface, and its dimensional accuracy, and consequently, the economics of cutting operations. Tool wear and the changes in tool geometry during cutting manifest themselves in different ways, generally classified as flank wear, crater wear, nose wear, notching, plastic deformation of the tool tip, chipping and gross fracture (Kalpakjian & Schmit, 2006). The allowable average wear land for various machining operations is usually determined by the width of flank wear (*VB*) which occurs on the relief (flank) face of the tool.

To characterize tool wear for a cutting operation, there exist two main approaches: empirical tool life models and tool wear rate models. In order to derive a reliable empirical tool life model, a large number of experimental tests are essential which is usually time-consuming, cost-intensive, restricted to the investigated tool–workpiece combinations and cannot predict the influence of work material or tool materials on the values of constants in the models. On the other hand, tool wear rate models involve process variables that are not directly measurable or very difficult to measure during a cutting operation, such as normal stress and temperature on the tool face, chip temperature, and chip sliding velocity along the tool rake face. However, in the last several years, the FEM has been successfully applied to estimate such variables (Attanasio, Ceretti, Fiorentino, Cappellini, & Giardini, 2010). Consequently, a better understanding of the fundamentals of cutting mechanics, engineering analyses of tool wear, and systematic approach for the process optimization is possible.

Considering the Show's equation of adhesive wear, tool wear rate model derived by Usui and co-workers involves variables such as temperature $T$, normal stress $\sigma_n$, sliding velocity $v_s$ at the contact surface, and two constants $A$ and $B$. It is expressed as:

$$\frac{dW}{dt} = A\sigma_n v_s \exp\left(-\frac{B}{T}\right) \tag{29}$$

Their results have shown that both the flank and crater wear rates have the same functional form. The variables in Usui's wear rate model can be predicted by FEM simulation of cutting process or combining analytical method and FEM (Yang & Liu, 2002). In this study, the procedure of calculating the Usui's wear model constants can be simply summarized as follows:

(1)  Conduct tool life experimentations under dry condition using a typically recommended insert grade to machine stainless steels to determine the wear rate (*dW/dt*).
(2)  To estimate the variables in Equation (29), prepare identical to the experimentations, fully coupled thermo-mechanical, and FANNS-optimized 3D-FE models, and run the simulations.
(3)  Collect the numerically estimated $\sigma_n$, $T$, and $v_s$.
(4)  Fit the nonlinear Usui's model to determine the constants A and B.

Seven arbitrary multi-pass dry turning tests of EN 1.4410 and EN 1.4462 DSS bars were performed on a variable spindle speed CNC lathe using rhombic 80° CNMA 120412-IC20 uncoated cemented carbide inserts. The solid bars had outside diameter of 55 mm. Constant material volumes of 29,405 mm$^3$ (i.e. 6 mm in thrust and 60 mm in axial directions) were removed per each experimental trial. An optical microscope with magnification 200 times was then used to measure the maximum wear on the flank surface after each cutting pass. Thereafter, logarithmic plots of tool wear vs. cutting time were drawn, and the slopes (*dW/dt*) were determined. Meanwhile, the 3D-FE cutting simulations were carried out based on the proposed procedure. The tool is considered rigid with no wearing possibility. Values of *T*, $\sigma_n$, and $v_s$ were directly extracted from simulation results. Matlab function nonlinear model fit was then utilized to fit the model and test its adequacy through ANOVA. The adjusted correlation factor ($R_{adj}$) was in very good agreement with predicted correlation factor ($R_{pre.}$) which supports the prediction power of the model and was generally above 0.94. The root mean square of errors (RMSE) for EN 1.4462 and EN 1.4410 Usui wear models were 0.000314 and 0.000469, respectively. The final forms of Usui's wear model for each workpiece material case are given as:

$$\left(\frac{dW}{dt}\right)_{EN\,1.4462} = 6.0478 \times 10^{-9}\sigma_n v_s \exp\left(-\frac{1172.8}{T}\right) \tag{30}$$

$$\left(\frac{dW}{dt}\right)_{EN\,1.4410} = 7.1704 \times 10^{-9}\sigma_n v_s \exp\left(-\frac{992.5203}{T}\right) \tag{31}$$

## 7. Optimization of Turning EN 1.4410 and EN 1.4462 DSS

In the second stage of this study, a hypothetical benchmark analysis based on the Taguchi optimization procedure is suggested. The objective was to select the best combination of criterions such as chip breakers type, tool geometry, process and cutting conditions, and cutting tool orientation angles for an effective machining of EN 1.4462 and EN 1.4410 DSS. For this purpose, Taguchi's mixed design $L_{18}(2^1 \times 3^7)$ is once again seen the most appropriate because of its attractive characteristics which combines the highest possible number of levels along with largest number of criterion and smallest number of experiments. Another attractive characteristic of the design is that the interaction between column 1 and 2 is orthogonal to all columns and hence can be estimated without sacrificing any column. The interaction can be estimated from the 2-way table of columns 1 and 2. The studied criterion and their corresponding levels are listed in Table 9. Columns 1 and 2 can be combined to form a 6-level column. Each level in column three represents an insert designation which is often commercially available. The next sub-sections briefly describe the studied criterion.

Table 9. Criterion and their corresponding levels.

| Factors | Symbol | Unit | 1 | 2 | 3 |
|---|---|---|---|---|---|
| Chip breaker type | *CB* | – | M3 M | PP | |
| Insert shape | *Geo.* | – | DNMG | CNMG | WNMG |
| Coling medium | *CM* | – | Still air | Water | Cryogenic |
| Cutting speed | $v_c$ | m/min | 75 | 150 | 225 |
| Feed rate | $f_r$ | mm/rev | .1 | .175 | .25 |
| Rake angle | $\alpha_n$ | degree | 0 | −6 | −12 |
| Inclination angle | $\lambda$ | degree | 0 | −6 | −12 |

## 7.1.  *Control factors*

### 7.1.1.  *Chip-breaker type*

In particular designed for machining stainless steels, chip breakers such as M3 M type are adapted with geometric features that improve the tool's life due to a reinforced cutting edge at the area where notch wear tends to occur when machining stainless steel, causing poor surface finish and risk of edge breakage (see Figure 24). On the other hand, PP-type chip breakers, which are also recommended for machining stainless steels, are characterized by having 3-step smart dot structures which provides smooth chip evacuation with a wide range of feed rates, and smooth taper cutting edge to reduce cutting forces.

### 7.1.2.  *Cutting insert shapes*

In metal cutting, the primary goal was to achieve the most efficient separation of chips from the workpiece. One of the main factors that contribute in optimizing chip morphology is the insert shape. For this reason, the selection of the right cutting tool geometry is critical. The three basic insert shapes which have to be investigated in this study were diamond 55°, rhombic 80°, and trigon 80°. Other geometric features, such as clearance angle, tolerance, size, thickness, nose radius, and cutting edge preparation, were identical.

### 7.1.3.  *Cooling medium*

Following the determination of the optimum thermal contact conductance of the tool–chip interface ($h_{tc}$) using FANNS approach, the influence of environmental heat convection in dry, wet, and cryogenic conditions on the cutting processes is investigated and optimized. A window for heat exchange was defined as shown in Figure 25. It was restricted to the secondary and tertiary deformation zones on the inserts using a cylindrical shape of 1.2 mm radius and 3 mm length and was not intended to change any other boundary conditions in the finite element model. Local convection coefficients in wet and cryogenic cutting were assigned in the areas covered by the window, while the rest of the cutting tool is still subjected to air convection. The convection coefficients of dry, wet, and cryogenic cutting mediums are taken from the literature as: 0.02, 10, and 5000 N/(s mm K), respectively (Pu et al., 2014; SFTC, 2010).

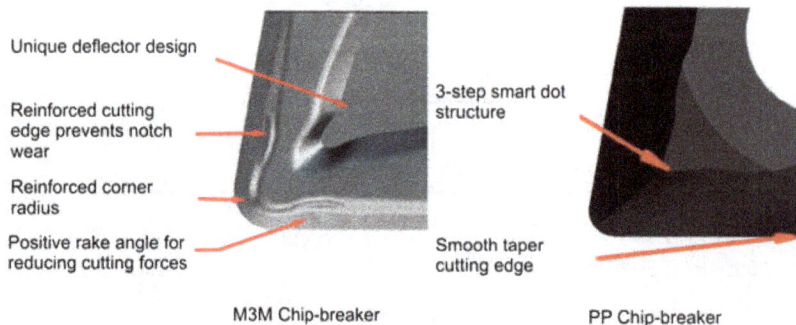

Unique deflector design

Reinforced cutting edge prevents notch wear

Reinforced corner radius

Positive rake angle for reducing cutting forces

M3M Chip-breaker

3-step smart dot structure

Smooth taper cutting edge

PP Chip-breaker

Figure 24.    Investigated chip breakers characteristics.

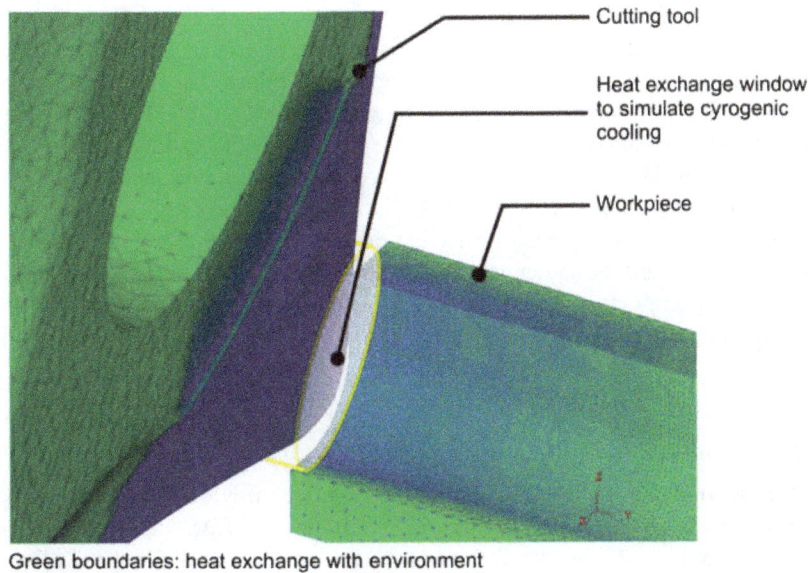

Green boundaries: heat exchange with environment

Figure 25.   An Illustrative thermal boundary condition of test no. 13.

### 7.1.4.   Cutting conditions

For all simulation trials, the finishing and semi-finishing set of cutting conditions are selected according to the maximum and minimum levels of cutting speed ($v_c$) and feed rate ($f$) as follows:

(a)  Finising:

$$150 \leq v_c \leq 225 \text{ m min}$$
$$0.1 \leq f_r \leq 0.175 \text{ mm/rev} \tag{32}$$

(b)  Semi-finishing:

$$75 \leq v_c < 150 \text{ m min}$$
$$0.175 < f_r \leq 0.25 \text{ mm/rev} \tag{33}$$

### 7.1.5.   Tool orientation angles

Tool inclination angles affect many performances such as cutting forces, chip thickness, tool–chip contact length, chip flow angle, etc. When machining DSS, the formed chips are tough, abrasive to the tooling, and tend to entangle around cutting tool, workpiece, and tool post. A proper cutting insert inclination and rake angle for example may cause the chips to flow to a position at which they are less likely to become entangled and rub and scratch the finished surface of the workpiece. Geometry of oblique cutting process with straight cutting edge is shown in Figure 26. The process is performed with tools having the tool cutting edge angle $K_r \neq 90°$ and tool inclination angle $\lambda \neq 0°$. The rake angle may be measured in more than one plane, and hence, more than one rake angles can be defined for a given tool and angle of obliquity. The different rake angles in oblique cutting are named as: normal ($\alpha_n$), velocity ($\alpha_v$), and effective ($\alpha_e$) rake angles. The flow of chip is at an angle to the normal to the cutting edge. The angle between normal to the cutting edge and chip velocity vector is called chip flow angle ($\Delta$). The shear angles could also be measured in different planes, such as a plane normal to the cutting edge ($\varphi_n$) and plane of effective rake angle ($\varphi_e$). Since chip flow and shear angles are often designated as dependent cutting variables, and the rake and inclination angles as independent cutting variables, therefore, when constructing the 3D-FE models, cutting tools are only positioned according to the combination of inclination and rake angles.

### 7.2.  3D-FE models

Due to the incorporation of chip breaker geometry in the array of control factors, a great deal of geometrical resolution is essential. Therefore, the 3D-CAD models of the cutting inserts are created in Solidworks first and exported as Standard for the Exchange of Product model data (STEP) files into Transmagic software to increase the resolution. The models are saved in Standard Tessellation Language (STL) format and exported to the Deform-3D environment. Both workpiece and cutting tools are meshed with grids of varying tetrahedral mesh densities so that the highest densities of meshes are assigned to the locations where temperature, strain, strain rate, …, etc. have the highest gradient. The total number of mesh elements in workpiece and cutting tool were 80,000 and 250,000, respectively. In addition to the described geometrical parameters, the cutting tools had identical nose radius of 0.8 mm, cutting edge radius of 30 µm, and clearance angle of 5°. The *WC* cutting tools are considered rigid and are worn out under Usui's

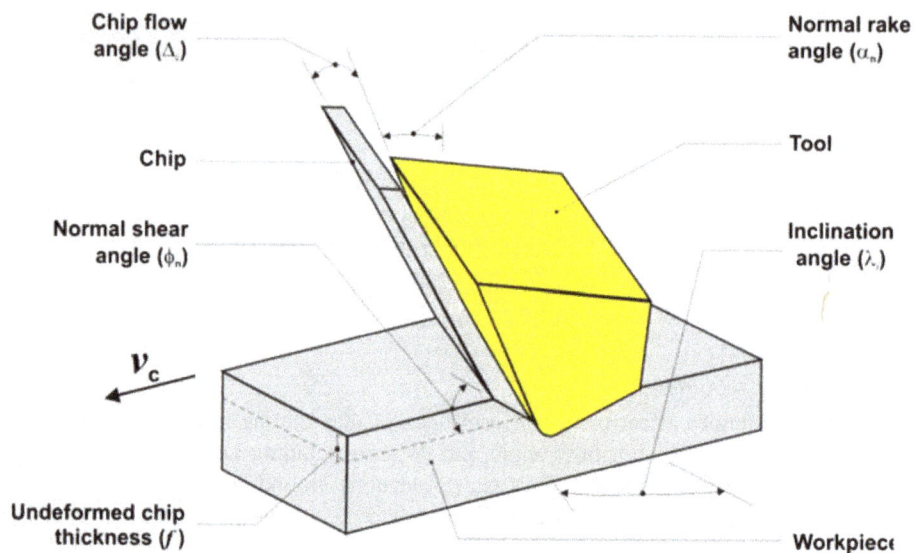

Figure 26.  Oblique cutting process.

wear model. Other FEM settings such as elasto-visco plastic and thermo-mechanical properties of the workpiece, the boundary conditions, and process conditions are covered under previous sections of this study.

### 7.3. *Simulation results and discussion*

Once the turning simulations has been finished, cutting performances such as cutting forces, temperatures, ..., etc. have been extracted. The S/N ratio of each performance is computed based on its beneficial or non-beneficial status. Almost all performances considered in this study were non-beneficial; therefore, *lower is better* concept applied and the values of S/N ratios are computed using Equation (8). To eliminate the complexity of problem, instead $F_f$, $F_c$, and $Ft$ components, resultant cutting force ($R$) has been considered:

$$R = \sqrt{F_f^{\,2} + F_c^{\,2} + F_r^2} \qquad (34)$$

The main effect of each control factor on each performance is determined by the ratio of sum of all S/N ratios corresponding to a factor at particular level to number of repetition of factor level. The optimum combinations corresponding to maximum average effect are directly depicted in the main effect plots as shown in Figure 27 below. Statistical analysis of variance (ANOVA) is performed to specify statistically significant control factors.

Figure 28 exhibits the percentage of contribution of each control factor on the variance of the corresponding cutting performance. In the following sub-sections, the effects of control factors on each single cutting performance are described.

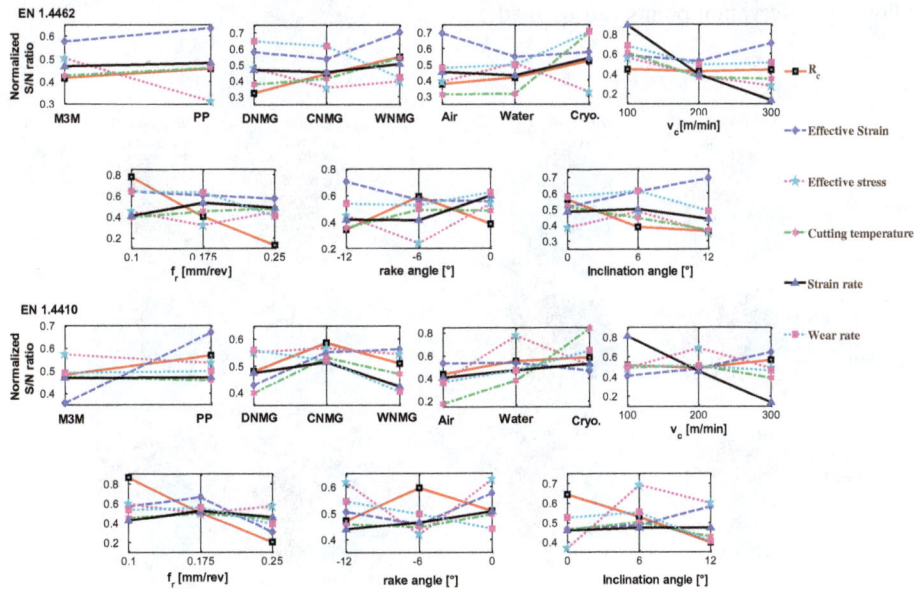

Figure 27.   Main effect plots of the effect of control factors on the cutting performances during turning EN 1.4410 and EN 1.4462 DSSs.

*7.3.1. Influence of control factors on the resultant cutting forces*

Based on a quick review of the results, the following conclusion points can be drawn:

(1) The optimum insert designations which minimize $R$ during turning EN 1.4462 and EN 1.4410 can be described in ISO norms as: WNMG 060408-PP and CNMG 120408-PP, respectively.
(2) In spite of the advantageous aspects of increasing cutting speed, applying coolant, and cryogenic conditions in increasing the productivity and reducing the overall $R$ values, results have shown a limited influence of these control factors on the mean $R$ values. This statement is supported by the slopes of the $R$ curves in Figure 27.
(3) Among the considered control factors, feed rate had shown the strongest impact on the resultant cutting forces (see Figure 28).
(4) The optimum values of cutting tool orientation angles expressed by ranges of investigated rake and inclination angles are −6° and 0°, respectively.

*7.3.2. Influence of control factors on the effective strain*

Without relating the hardening parameters in the loading function to the experimental uniaxial stress-strain curve, the work-hardening theory of plasticity cannot be applied in practical terms. In order to correlate the test results obtained by different load programs, the introduction of any strain and stress variables, that are functions of plastic strain and plastic stress, and can be plotted against each other is considered useful. These variables are often called effective strain and effective stress.

DSSs are a group of stainless steel alloys that are characterized by high work-hardening rate. They are machined at very high strain rates which increase the work-hardening rate and causing higher resistance to plastic deformation. Generally, the following observation points can be made:

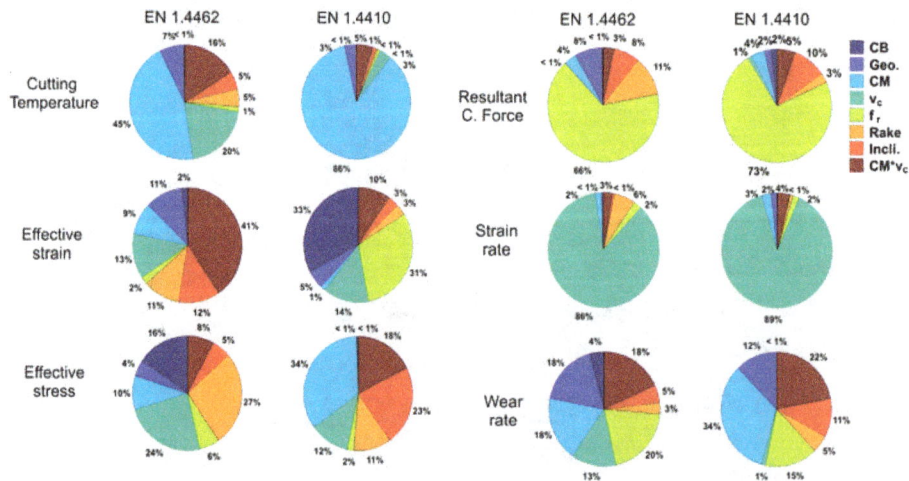

Figure 28.   The percentage contribution of FEM control factors in the cutting performances variances.

(1) The average effective plastic strain when machining EN 1.4462 was 13.2% lower than the corresponding EN 1.4410.

(2) Employing chip breaker type PP has generally caused lower effective strain values.

(3) Detrimental control factors vary considerably per work material family.

(4) Higher values of effective strains are observed when the cooling medium on the tool is cryogenic.

(5) Main effects of effective strain decrease with increasing cutting speed.

(6) Mean effective strains are found to be minimum at the feed rate range of 0.1–0.175 mm/rev and inclination angle range of 6°–12°.

(7) Minimum effective strain value when cutting EN 1.4462 and EN 1.4410 at $\alpha_n = 0°$ and $\alpha_n = -12°$, respectively.

### 7.3.3. Influence of control factors on the effective stress

Effective plastic stress is considered a relevant characteristic of the cutting process which characterizes the level of resistance to cutting. The correlation between the state of this performance imposed by the cutting tool in the layer being removed and the fracture strain of the work material could be used to estimate the physical efficiency of the cutting process. In summary to the results, the following conclusions can be drawn:

(1) The effective stress encountered in cutting EN 1.4410 12% higher than EN 1.4462.

(2) The optimum control $factor^{level}$ combinations of cutting EN 1.4462 and EN 1.4410 were $CB^1\ Geo.^1\ CM^2\ v_c^1\ f_r^1\ \alpha_n^3\ \lambda^2$ and $CB^1\ Geo.^2\ CM^2\ v_c^2\ f_r^1\ \alpha_n^3\ \lambda^2$, respectively.

(3) Increasing the rake angle in negative direction and inclination angle in positive direction generally increases the effective stress.

(4) The minimum effective plastic stress of 1261.48 MPa and 1492.01 MPa when cutting EN 1.4462 and EN 1.4410 was recorded at experiment No. 9, respectively.

(5) When cutting EN 1.4462, process conditions such as $v_c$, $f_r$, $CM$, and $v_c \times CM$ interaction have contributed to the effective stress values by 48%. On the other hand, when cutting EN 1.4410, they had contributed by as much as 66%.

### 7.3.4. Influence of control factors on the tool–chip interface temperature

Mean contact temperature at the tool–chip interface (also is referred to as the cutting temperature) is the basic tribological characteristics of the tool–chip interface. It plays a major role in the formation of crater on the tool face and leads to failure of tool by softening and thermal stresses. This temperature is the most suitable parameter to correlate the tribological conditions with tool wear. Control factors that affect the tool–chip interface temperature are workpiece and tool material, tool geometry, cutting conditions, and cutting medium. Fortunately, most of these factors are included in the original design of simulation experimentations. Figure 29 depicts the state of cutting temperature distribution in the primary and secondary deformation zones of the workpiece. The results of the numerical studies on temperatures in cutting DSS can be summarized as follows:

(1) Mean cutting temperatures at dry cutting when machining EN 1.4462 and EN 1.4410 were 36.3 and 53.5% higher than at corresponding cryogenic process conditions.

(2) The contribution of water-based coolant in lowering the cutting temperature has not exceeded 7.8% in both material cases.

(3) Minimum mean cutting temperature when cutting EN 1.4462 and EN 1.4410 has been recorded at experimental run 16 and 15, respectively.

(4) Cooling medium, cutting condition, and their interaction accounts for most of contributions in cutting temperature variations.

(5) The optimum control $factor^{level}$ combinations of cutting EN 1.4462 and EN 1.4410 were $CB^2\ Geo.^3\ CM^3\ v_c^1\ f^3\ \alpha_n^2\ \lambda^1$ and $CB^1\ Geo.^2\ CM^3\ v_c^1\ f^2\ \alpha_n^3\ \lambda^2$, respectively.

(6) The average cutting temperature when machining EN 1.4462 was 9.85% lower than the corresponding EN 1.4410.

### 7.3.5. Influence of control factors on the strain rate

The flow stress curves shown in Figure 9 exhibit sensitivity to temperature and strain rate. This sensitivity is directly related with time and temperature dependency of the mechanisms that govern the deformation and the evolution of the deformation in the material. The main mechanism by which plastic strain takes place is thermally activated

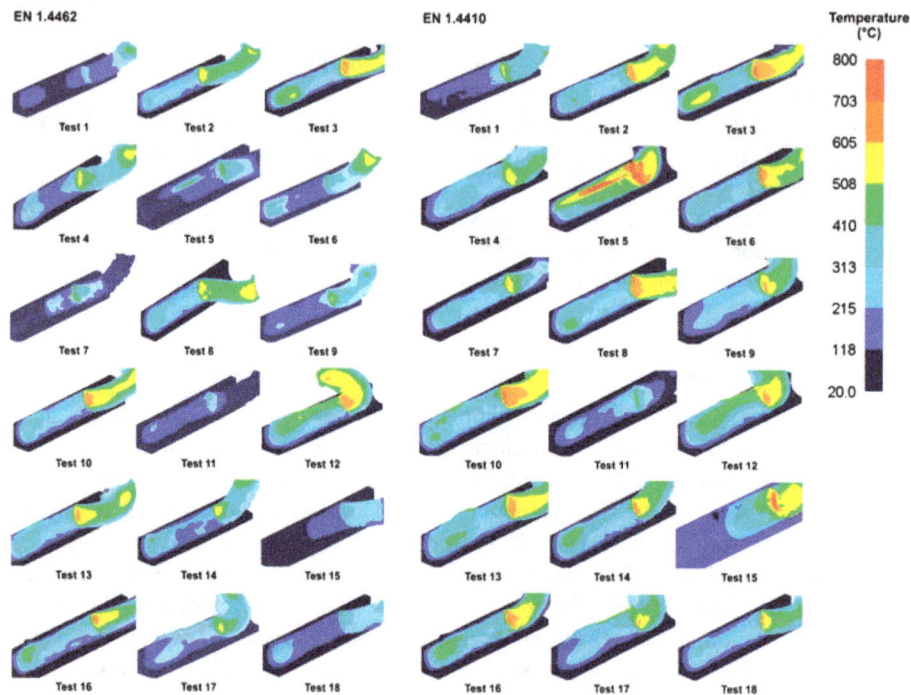

Figure 29.   Contour of temperature distribution in three-dimensional models of cutting DSS.

motion of dislocations past obstacles that exist within the lattice over a wide range of strain rates and cutting temperatures. The material response is significantly affected by the nature and density of the obstacles (which may change as the deformation takes place). When dealing with metals, experimental results show that the stress required for plastic strain often reduces with the increase of temperature and with the decrease of plastic strain rate (Gilat & Wu, 1997). It can then be said that temperature and plastic strain rate greatly influence the material response. In general, the stress decreases with the increasing of temperature and decreasing the plastic strain rate. Actually, temperature and strain rate effects are coupled, since one influences the other. Temperature affects the rate of deformation, which is controlled mainly by a thermally activated mechanism. On the other hand, plastic strain at high rate generates significant heating and cause an increase in temperature which leads to mechanical instability and the localization of deformation into narrow sheets of material (the adiabatic shear bands), which act as precursor for eventual material failure (Davim & Maranhão, 2009). As generalization to the given results in Figures 30 and 31, the following concluding points can be depicted:

(1) Average strain rate values under cutting EN 1.4462 and EN 1.4410 are of $2.067 \times 10^5$ 1/sec and $3.23 \times 10^5$ 1/sec, respectively.
(2) Increasing cutting speeds from 100 to 300 m/min at lower feed rates of 0.1 mm/rev had drastically increased the strain rate values.

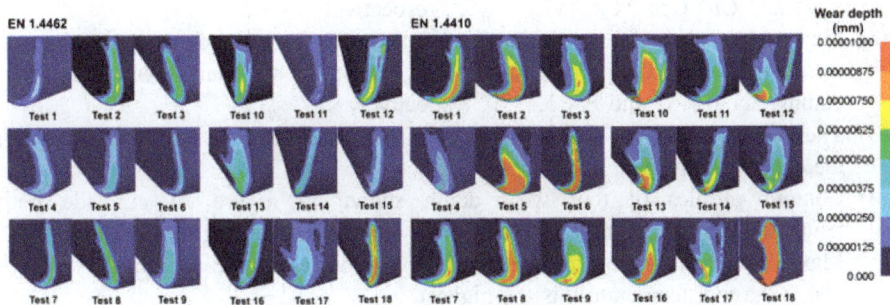

Figure 30.    Contours of tool wear depth in three-dimensional models of cutting tools.

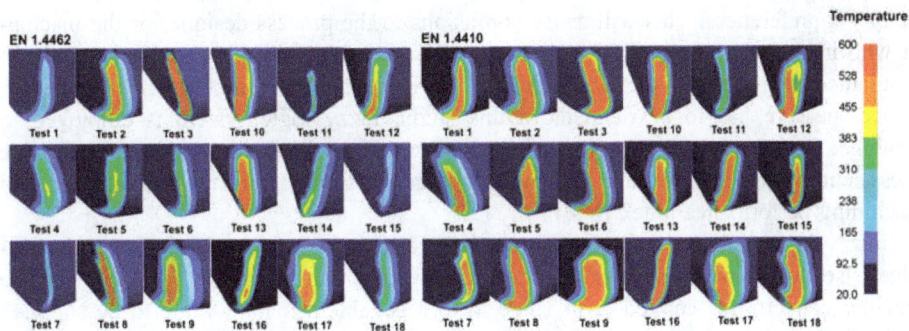

Figure 31.    Contours of tool temperature distributions in three-dimensional models of cutting tools.

(3) Mean strain rate values in cryogenic cutting conditions during cutting EN 1.4462 and EN 1.4410 were, respectively, 9.158 and 11.863% lower than the corresponding still air conditions.

(4) $CB^2\ Geo.^3\ CM^3\ v_c^1\ f^2\ \alpha_n^3\ \lambda^2$ and $CB^2\ Geo.^2\ CM^3\ v_c^1\ f^2\ \alpha_n^{33}\ \lambda^3$ were the optimum control $factor^{level}$ combinations which have minimized the strain rate when cutting EN 1.4462 and EN 1.4410, respectively.

### 7.3.6.  Influence of control factors on the tools wear

The Equation (34) has been used in conjunction with finite element simulation to model wear of WC tools in oblique cutting of DSS. The simulation results related to the tool wear were obtained in terms of tool wear rate, total wear depth, and tool temperature. Average values of tool wear rates are statistically analyzed and plotted as shown in Figures 27 and 28. Figure 30 maps the contours of tool wear depth in the nose area of the employed cutting tools after 7 mm longitudinal cutting. As it can be observed that the locations and intensity of tool wear depths are functions of work materials and control factor sets. Considering the effect of these factors, the summary of the findings is presented below.

(1) The optimum control $factor^{level}$ combinations which have minimized the tool wear rate during cutting EN 1.4462 and EN 1.4410 were $CB^2\ Geo.^1\ CM^3\ v_c^1\ f_r^1\ \alpha_n^3\ \lambda^2$ and $CB^2\ Geo.^1\ CM^3\ v_c^1\ f_r^1\ \alpha_n^1\ \lambda^2$, respectively.

(2) Major percentage of contributions of 69 and 72% in variance of tool wear rate were attributed to the process and cutting conditions and their interactions when cutting EN 1.4462 and EN 1.4410, respectively.

(3) The average tool wear rate when machining EN 1.4462 was 5.365% lower than the corresponding EN 1.4410.

(4) Contour profiles of tool wear depth shown in Figure 33 coincide with temperature distribution at rake and flank faces of the cutting tools in Figure 31. Maximum wear depths are occurred in positions at rake and flank surfaces where the tool temperature is the highest.

### 8.  Numerical machining performance measure (NMPM)

The previous analyses of machining performances have showed different control $factor^{level}$ preferences. This will cause confusions to the process designer or the machinist who interested in finding the best compromise combination of control factors which simultaneously optimize the machining of DSS. Therefore, a new machining performance measure has to be defined. In this study, fuzzy logic system is employed to combine the normalized performances, such as the resultant cutting forces, effective stress, cutting temperature, and tool wear rate into a single characteristics index called machining performance index (NMPM).

Fuzzy logic unit is composed of four main sub units namely fuzzification, knowledge base, fuzzy inference engine, and defuzzification unit as shown in Figure 32. The fuzzifier converts the entered crisp value in to a suitable semantic value to be supplied to the fuzzy set for the first part of fuzzy rule. In the knowledge base units, information about of fuzzy rule base and membership functions is given. The fuzzy rule base is a database mainly for saving fuzzy rule, which is composed of if–then expression to indicate a conditional descriptive sentence. 'If' is called the first part, and 'then' is called the conclusion part. They are used to describe the relationship between input and output. Fuzzy values are determined by the membership functions ($\mu$) that define the degree of membership of an object in a fuzzy set. So far there has been no standard method of choosing the proper shape of the membership functions for the fuzzy sets of the control variables; the present investigation adopts a triangular shape membership function (see Figure 33).

For a rule:

**Rule$_j$**: If $x_1$ is ($A_{i1}$) and $x_2$ is ($A_{i2}$), ..., $x_k$ is $A_{ik}$ then $y_i = B_i$, $j = 1, 2, ..., N$.

where $N$ is the total number of fuzzy rules, $x_i$ ($i = 1, 2, ..., k$) are the input variables, $y_i$ are the output variables, and $A_{ik}$ and $B_i$ are fuzzy sets modeled by membership functions $\mu_{Ai}(x_j)$ and $\mu_{Bi}(y_i)$, respectively. Based on the Mamdani implication method of inference, the membership function of the output of fuzzy reasoning can be expressed as:

$$\mu_{B_i}(y_i) = \max\left\{\min_i\left[\mu_{A_{i1}}(x_1), \mu_{A2}(x_2), ..., \mu_{A_{ik}}(x_k)\right]\right\} \tag{35}$$

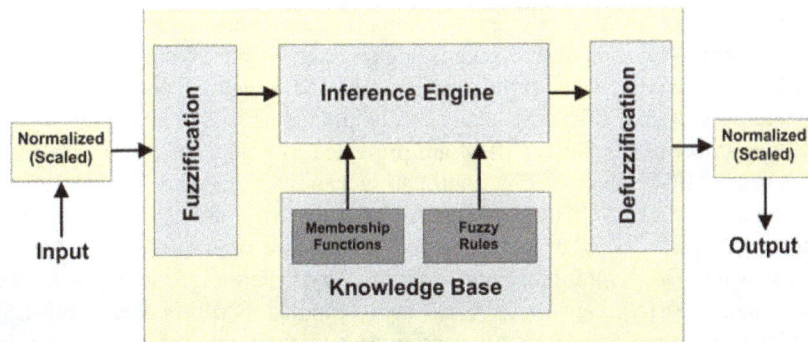

Figure 32.   Fuzzy Logic Unit (FLU).

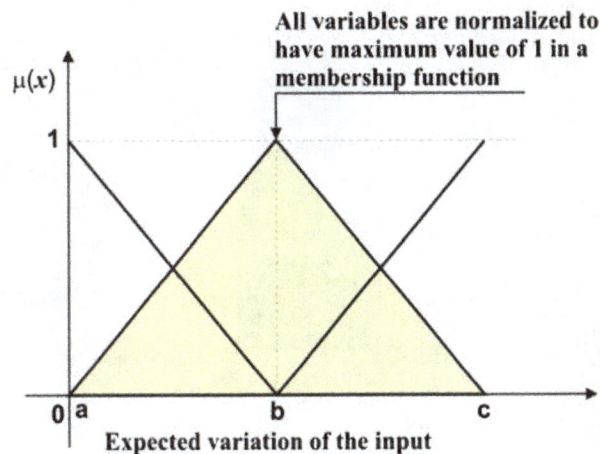

Figure 33.   Triangular membership function.

Finally, a defuzzification method, called the center-of-gravity method, is adopted to transform the fuzzy inference output $\mu_{Bi}(y_i)$ into a non-fuzzy value $y_0$:

$$y_0 = \frac{\int y_i \mu_{B_i}(y_i)dy}{\int \mu_{B_i}(y_i)dy} \tag{36}$$

The yielded value is the final crisp output value obtained from the input variables. It is worth mentioning here that in most cases, the membership functions are defined within a normalized interval (universe of discourse). Therefore, before the fuzzification can be started, the input values have to be scaled (normalized) so that they fit into the universe of discourse (Gupta, Singh, & Aggarwal, 2011).

Matlab software was used to construct the inference model of the NMPM. The S/N ratios of performance values were first adjusted to a notionally common scale between null and one, so that the digit 'one' represents the most desirable and 'null' is the least desirable alternative. The four input variables are assigned with the following fuzzy sets: Small (S), Medium (M), and Large (Lg). The output variable has the following nine levels: Extremely Low (EL), Very Low (VL), Low (L), Lower Medium (LM), Medium (M), Upper Medium (UM), High (H), Very High (VH), and Extremely High (EH). Mamdani implication method is employed for the fuzzy inference reasoning. The relationship between system input and output is expressed by an 'if–then' type. Totally $3^4$ fuzzy rules per material were formulated. Finally, the structural characteristics of the fuzzy inference system (FIS) in this study can be plotted as shown in Figure 34.

The main effect plots of the NMPM are presented in Figure 35. The figure indicates that the mean NMPM values of EN 1.4462 are generally higher than of EN 1.4410 and the final optimum $factor^{level}$ combinations for both materials are $CB^1$ $Geo.^3$ $CM^3$ $v_c^1$ $f_r^1$ $\alpha_n^3$ $\lambda^2$ and $CB^2$ $Geo.^2$ $CM^3$ $v_c^1$ $f_r^1$ $\alpha_n^2$ $\lambda^2$, respectively. Based on the NMPM values, it can be concluded that EN 1.4462 has better machinability in terms of the employed performances. Finally, performing ANOVA on the computed NMPMs has confirmed the major effects of process and cutting conditions and their interactions on the overall performance of cutting DSS (see Figure 36).

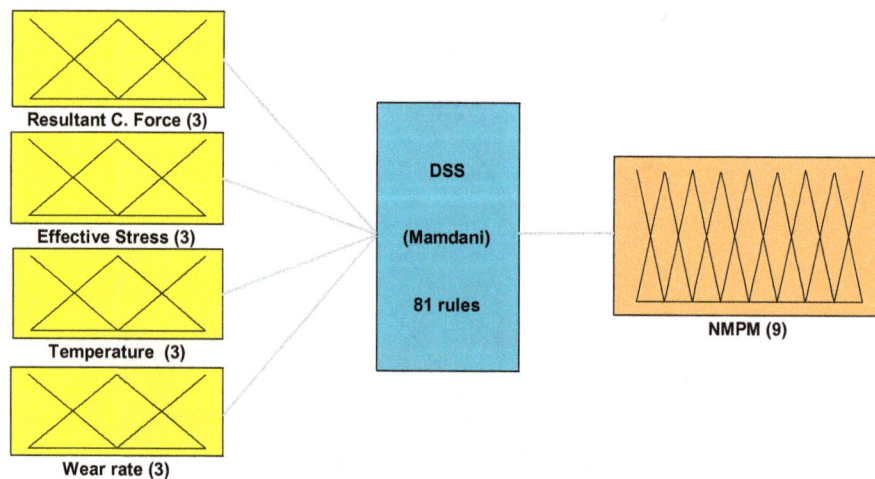

Figure 34.    FIS-NMPM: 4 inputs, 1 output, 81 rules.

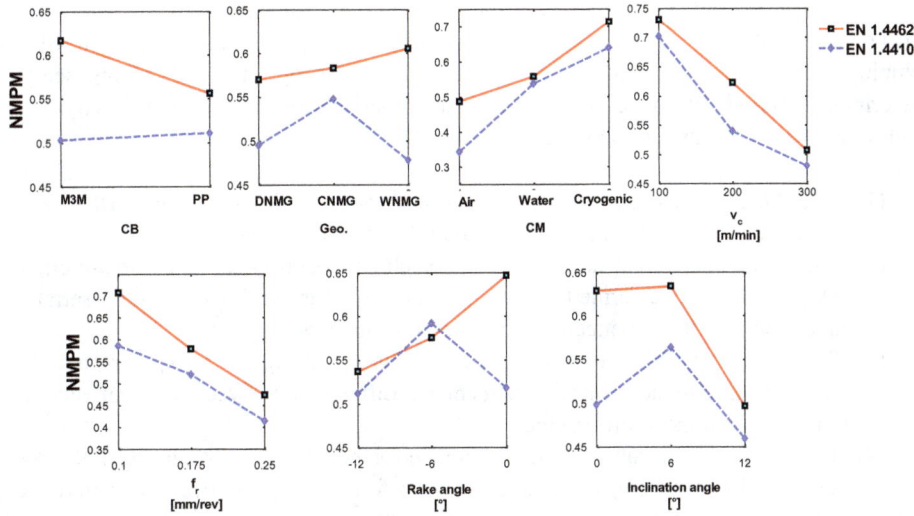

Figure 35. Main effect plots of the effect of control factors on the cutting performances during turning EN 1.4410 and EN 1.4462 DSS.

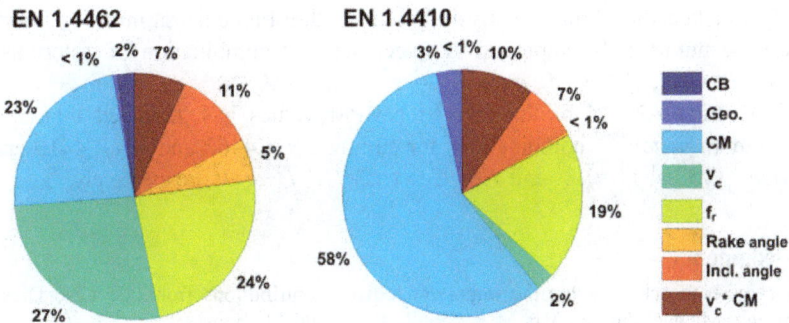

Figure 36. The percentage contribution of FEM control factors in NMPM variance.

## 9. Conclusions

In this study, three-dimensional numerical simulations of oblique cutting of standard EN 1.4462 and super EN 1.4410 DSS have been investigated. A new approach for an effective and accurate simulation of DSS using JMatSoft's-generated flow stress curves has been proposed. Taguchi design of experiments has been extensively used during the course of the study. Many FEM pre-processing variables are assigned as control factors for subsequent inverse identification process. The overall difference between the experimentally and numerically obtained results has been calculated and statistically analyzed based on the Taguchi optimization procedure. VIKOR method has been applied to combine the differences into one index. Thereafter, a new computational algorithm (FANNS) is suggested to effectively model and optimizes the indices in terms of the adopted control factors. Results of numerical simulation are validated and the approach prepared for the next stage of the investigation. In the second stage, a new methodology based on hypothetical numerical simulations is proposed to investigate the machining of DSS.

Parametric analysis of performances is undertaken, and many interim conclusion points are depicted. A new measure to evaluate machining performance of each of the employed DSS grades has also been suggested and the optimum machining setup is determined. Based on the experimental and numerical results and data analysis the following main conclusion points can be drawn:

(1) JMatPro-generated elasto-visco plastic and thermo-mechanical properties can be effectively used to numerically simulate the machining of DSS.
(2) Using time-dependent damage criteria such as Cockroft-Latham damage criteria, JMatPro has outperformed the JC material model in prediction of chip serrations under similar thermo-mechanical material properties.
(3) Pre-processing FEM control factors such as $h_{tc}$, $v_c$, $f_r$, $\mu_c$, $\mu_t$, $\kappa_t$, $\%p$, and $D_{\text{crit.}}$ have strong impacts on the percentage difference between experimental and numerical cutting performances.
(4) The lowest mean values of the overall error percentages for cutting EN 1.4462 and EN 1.4410 were, respectively, at the following pre-processing control $factor^{\text{level}}$ combinations; $h_t c^2\ v_c^1\ f_r^1\ \mu_c^2\ \mu_t^2\ \kappa_t^3\ \%p^2\ D_{\text{circ.}}^3$ and $h_t c^2\ v_c^2\ f_r^3\ \mu_c^2\ \mu_t^3\ \kappa_t^3\ \%p^2\ D_{\text{circ.}}^3$.
(5) Validations of numerical results through experimentations have revealed that the proposed VIKOR-FANNS approach can efficiently minimize the overall percentage differences at any desired cutting conditions.
(6) Hypothetical benchmark analysis based on the Taguchi optimization procedure can be numerically employed to select the best combination of criterions such as: $CB$, $Geo.$, $CM$, $v_c$, $f_r$, $\alpha_n$, and $\lambda$.
(7) Performing ANOM of the derived NMPM values has indicated that the final optimum $factor^{\text{level}}$ combinations for cutting EN 1.4462 and EN 1.4410 are $CB^1\ Geo.^3\ CM^3\ v_c^1\ f_r^1\ \alpha_n^3\ \lambda^2$ and $CB^2\ Geo.^2\ CM^3\ v_c^1\ f_r^1\ \alpha_n^2\ \lambda^2$, respectively.

## Acknowledgments

The authors wish to acknowledge the important logistic contributions from; Dr Uwe Diekmann, Herr. Dipl.-Ing. Michael Fiderer, Mrs M.A. Sibylle Krug, and Mr Bauer.

## Disclosure statement

No potential conflict of interest was reported by the authors.

## Funding

This work was supported by the Stuttgart University, Team 3, Hochschulschriften, Tausch, Geschenk grant number [01125435-9987].

## ORCID

*Rastee D. Koyee* ⓘ http://orcid.org/0000-0002-5215-3039

## References

Arrazola, P. J., & Özel, T. (2009). Finite element modeling of machining processes. In T. Özel & J. P. Davim (Eds.), *Intelligent machining: Modelling and optimization of the machining processes and systems* (pp. 125–163). London: ISTE and Wiley.

Arrazola, P. J., & Özel, T. (2010). Investigations on the effects of friction modeling in finite element simulation of machining. *International Journal of Mechanical Sciences, 52*, 31–42.

Arrazola, P. J., Özel, T., Umbrello, D., Davies, M., & Jawahir, I. S. (2013). Recent advances in modelling of metal machining processes. *CIRP Annals – Manufacturing Technology, 62*, 695–718.

Astakhov, V. P. (2006). *Tribology of metal cutting*. Oxford: Elsevier.

Attanasio, A., Ceretti, E., Fiorentino, A., Cappellini, C., & Giardini, C. (2010). Investigation and FEM-based simulation of tool wear in turning operations with uncoated carbide tools. *Wear, 269*, 344–350.

Attanasio, A., & Umbrello, D. (2009). Abrasive and diffusive tool wear FEM simulation. *International Journal of Material Forming* [Internet], *2*, 543–546. doi:10.1007/s12289-009-0475-z

Aurich, J. C. C., & Bil, H. (2006). 3D finite element modelling of segmented chip formation. *CIRP Annals – Manufacturing Technology, 55*, 47–50.

Bonnet, C., Valiorgue, F., Rech, J., & Hamdi, H. (2008). Improvement of the numerical modeling in orthogonal dry cutting of an AISI 316L stainless steel by the introduction of a new friction model. *CIRP Journal of Manufacturing Science and Technology, 1*, 114–118.

Buchkremer, S., Wu, B., Lung, D., Münstermann, S., Klocke, F., & Bleck, W. (2014). FE-simulation of machining processes with a new material model. *Journal of Materials Processing Technology, 214*, 599–611.

Calamaz, M., Coupard, D., & Girot, F. (2008). A new material model for 2D numerical simulation of serrated chip formation when machining titanium alloy Ti–6Al–4V. *International Journal of Machine Tools and Manufacture, 48*, 275–288.

Chagas, G. M. P., Barbosa, P. A., Barbosa, C. A., & Machado, I. F. (2013). Thermal analysis of the chip formation in austenitic stainless steel. *Procedia CIRP, 8*, 293–298.

Davim, J. P., & Maranhão, C. (2009). A study of plastic strain and plastic strain rate in machining of steel AISI 1045 using FEM analysis. *Materials and Design, 30*, 160–165.

De Oliveira Junior, C., Diniz, A., & Bertazzoli, R. (2014, October). Correlating tool wear, surface roughness and corrosion resistance in the turning process of super duplex stainless steel. *Brazilian Society of Mechanical Sciences and Engineering, 36*, 775–785.

Ducobu, F., Rivière-Lorphèvre, E., & Filippi, E. (2014). Numerical contribution to the comprehension of saw-toothed Ti6Al4V chip formation in orthogonal cutting. *International Journal of Mechanical Sciences, 81*, 77–87. doi:10.1016/j.ijmecsci.2014.02.017

Gilat, A., & Wu, X. (1997). Plastic deformation of 1020 steel over a wide range of strain rates and temperatures. *International Journal of Plasticity, 13*, 611–632.

Guo, Y. B., & Yen, D. W. (2004). A FEM study on mechanisms of discontinuous chip formation in hard machining. *Journal of Materials Processing Technology, 155–156*, 1350–1356.

Guo, Z. L., Saunders, N., Miodownik, A. P., & Schille, J. P. (2007). Quantification of high temperature strength of nickel-based superalloys. *Materials Science Forum, 546–549*, 1319–1326.

Gupta, A., Singh, H., & Aggarwal, A. (2011). Taguchi-fuzzy multi output optimization (MOO) in high speed CNC turning of AISI P-20 tool steel. *Expert Systems with Applications, 38*, 6822–6828.

Haddag, B., & Nouari, M. (2013). Tool wear and heat transfer analyses in dry machining based on multi-steps numerical modelling and experimental validation. *Wear, 302*, 1158–1170.

He, A., Xie, G., Zhang, H., & Wang, X. (2014). A modified Zerilli-Armstrong constitutive model to predict hot deformation behavior of 20CrMo alloy steel. *Materials and Design, 56*, 122–127.

Heisel, U., Storchak, M., & Krivoruchko, D. (2013). Thermal effects in orthogonal cutting. *Production Engineering* [Internet], *7*, 203–211. doi:10.1007/s11740-013-0451-9

Hou, Q. Y., & Wang, J. T. (2010). A modified Johnson–Cook constitutive model for Mg-Gd-Y alloy extended to a wide range of temperatures. *Computational Materials Science, 50*, 147–152.

Iraola, J., Rech, J., Valiorgue, F., & Arrazola, P. J. (2012). Characterization of friction coefficient and heat partition coefficient between an austenitic steel AISI304L and a tin -coated carbide cutting tool. *Machining Science and Technology, 16*, 189–204.

Kalpakjian, S., & Schmit, S. R. (2006). *Manufacturing engineering and technology* (5th ed.). Upper Saddle River, NJ: Pearson Education.

Keleştemur, O., Arıcı, E., Yıldız, S., & Gökçer, B. (2014). Performance evaluation of cement mortars containing marble dust and glass fiber exposed to high temperature by using Taguchi method. *Construction and Building Materials, 60*, 17–24.

Klocke, F. (2011). *Manufacturing processes 1: Cutting.* Heidelberg: Springer.

Klocke, F., Lung, D., & Buchkremer, S. (2013). Inverse identification of the constitutive equation of inconel 718 and AISI 1045 from FE machining simulations. In *Procedia CIRP* (Vol. 8, pp. 212–217), Turin, Italy.

Koné, F., Czarnota, C., Haddag, B., & Nouari, M. (2011). Finite element modelling of the thermo-mechanical behavior of coatings under extreme contact loading in dry machining. *Surface and Coatings Technology, 205,* 3559–3566.

Koné, F., Czarnota, C., Haddag, B., & Nouari, M. (2013). Modeling of velocity-dependent chip flow angle and experimental analysis when machining 304L austenitic stainless steel with groove coated-carbide tools. *Journal of Materials Processing Technology, 213,* 1166–1178.

Koyee, R. D., Heisel, U., Eisseler, R., & Schmauder, S. (2014, October). Modeling and optimization of turning duplex stainless steels. *Journal of Manufacturing Processes, 16,* 451–467.

Koyee, R. D., Schmauder, S., & Eisseler, R. (2013). Machining of stainless steels: A comparative study. In A. Maffei & A. Archenti (Eds.), *3rd International Conference on Advanced Manufacturing Engineering and Technologies* (pp. 53–62). Stockholm, Sweden.

Krolczyk, G., Legutko, S., & Gajek, M. (2013). Predicting the surface roughness in the dry machining of duplex stainless steel (DSS). *Metalurgija, 52,* 259–262.

Li, H. Y., Li, Y. H., Wang, X. F., Liu, J. J., & Wu, Y. (2013). A comparative study on modified Johnson Cook, modified Zerilli-Armstrong and Arrhenius-type constitutive models to predict the hot deformation behavior in 28CrMnMoV steel. *Materials and Design, 49,* 493–501.

Liu, R., Melkote, S., Pucha, R., Morehouse, J., Man, X., & Marusich, T. (2013). An enhanced constitutive material model for machining of Ti–6Al–4V alloy. *Journal of Materials Processing Technology, 213,* 2238–2246. doi:10.1016/j.jmatprotec.2013.06.015

Lorentzon, J., & Järvstråt, N. (2008). Modelling tool wear in cemented-carbide machining alloy 718. *International Journal of Machine Tools and Manufacture, 48,* 1072–1080.

Lorentzon, J., Järvstråt, N., & Josefson, B. L. (2009). Modelling chip formation of alloy 718. *Journal of Materials Processing Technology, 209,* 4645–4653.

Maranhão, C., & Paulo, D. J. (2010). Finite element modelling of machining of AISI 316 steel: Numerical simulation and experimental validation. *Simulation Modelling Practice and Theory, 18,* 139–156.

Nasr, M., Ng, E.-G., & Elbestawi, M. (2007). Effects of workpiece thermal properties on machining-induced residual stresses – Thermal softening and conductivity. *Proceedings of the Institution of Mechanical Engineers, Part B: Journal of Engineering Manufacture, 221,* 1387–1400. doi:10.1243/09544054JEM856

Niesłony, P., Grzesik, W., Chudy, R., & Habrat, W. (2014). Meshing strategies in FEM simulation of the machining process. *Archives of Civil and Mechanical Engineering.* Retrieved from http://dx.doi.org/10.1016/j.acme.2014.03.009

Nomani, J., Pramanik, A., Hilditch, T., & Littlefair, G. (2013). Machinability study of first generation duplex (2205), second generation duplex (2507) and austenite stainless steel during drilling process. *Wear, 304,* 20–28.

Odeshi, A. G., Al-ameeri, S., & Bassim, M. N. (2005). Effect of high strain rate on plastic deformation of a low alloy steel subjected to ballistic impact. *Journal of Materials Processing Technology, 162–163,* 385–391.

Olovsjö, S., Hammersberg, P., Avdovic, P., Ståhl, J.-E., & Nyborg, L. (2012). Methodology for evaluating effects of material characteristics on machinability – Theory and statistics-based modelling applied on Alloy 718. *The International Journal of Advanced Manufacturing Technology, 59,* 55–66.

Outeiro, J. C., Umbrello, D., & M'Saoubi, R. (2006). Experimental and numerical modelling of the residual stresses induced in orthogonal cutting of AISI 316L steel. *International Journal of Machine Tools and Manufacture, 46,* 1786–1794.

Paro, J., Hänninen, H., & Kauppinen, V. (2001). Tool wear and machinability of HIPed P/M and conventional cast duplex stainless steels. *Wear, 249,* 279–284.

Pu, Z., Umbrello, D., Dillon, O. W., Lu, T., Puleo, D. A., & Jawahir, I. S. (2014). Finite element modeling of microstructural changes in dry and cryogenic machining of AZ31B magnesium alloy. *Journal of Manufacturing Processes, 16,* 335–343.

Pujana, J., Arrazola, P. J., M'Saoubi, R., & Chandrasekaran, H. (2007). Analysis of the inverse identification of constitutive equations applied in orthogonal cutting process. *International Journal of Machine Tools and Manufacture, 47,* 2153–2161.

Puls, H., Klocke, F., & Lung, D. (2012). A new experimental methodology to analyse the friction behaviour at the tool-chip interface in metal cutting. *Production Engineering, 6*, 349–354.

Quiza, R., & Davim, J. P. (2011). Computational methods and optimization. In J. P. Davim (Ed.), *Mach hard mater* (pp. 191–208). London: Springer-Verlag.

Rao, R. V. (2013). *Decision making in manufacturing environment using graph theory and fuzzy multiple attribute decision making methods*. London: Springer.

Rech, J., Arrazola, P. J., Claudin, C., Courbon, C., Pusavec, F., & Kopac, J. (2013). Characterisation of friction and heat partition coefficients at the tool-work material interface in cutting. *CIRP Annals – Manufacturing Technology, 62*, 79–82.

Rhim, S. H., & Oh, S. I. I. (2006). Prediction of serrated chip formation in metal cutting process with new flow stress model for AISI 1045 steel. *Journal of Materials Processing Technology, 171*, 417–422.

Riedmiller, M., & Braun, H. (1993). A direct adaptive method for faster back propagation learning: The RPROP algorithm. In *Proceedings of IEEE International Conference on Neural Networks* (pp. 586–591), San Francisco, CA.

Rosochowska, M., Balendra, R., & Chodnikiewicz, K. (2003). Measurements of thermal contact conductance. *Journal of Materials Processing Technology, 135*, 204–210.

Samantaray, D., Mandal, S., & Bhaduri, A. K. (2009). A comparative study on Johnson Cook, modified Zerilli–Armstrong and Arrhenius-type constitutive models to predict elevated temperature flow behaviour in modified 9Cr–1Mo steel. *Computational Materials Science, 47*, 568–576.

Saunders, N., Guo, Z., Li, X., Miodownik, A. P., & Schille, J. P. (2004). Modelling the material properties and behaviour of Ni-based superalloys. *Superalloys, 2004*, 849–858.

SFTC. (2010). DEFORM v10.2.1 User's manual. Columbus, OH, USA.

Shrot, A., & Bäker, M. (2012). Determination of Johnson–Cook parameters from machining simulations. *Computational Materials Science, 52*, 298–304.

Sima, M., & Özel, T. (2010). Modified material constitutive models for serrated chip formation simulations and experimental validation in machining of titanium alloy Ti–6Al–4V. *International Journal of Machine Tools and Manufacture, 50*, 943–960. doi:10.1016/j.ijmachtools.2010.08.004

Smolenicki, D., Boos, J., Kuster, F., Roelofs, H., & Wyen, C. F. (2014). In-process measurement of friction coefficient in orthogonal cutting. *CIRP Annals – Manufacturing Technology, 63*, 97–100.

Song, W., Ning, J., Mao, X., & Tang, H. (2013). A modified Johnson–Cook model for titanium matrix composites reinforced with titanium carbide particles at elevated temperatures. *Materials Science and Engineering: A, 576*, 280–289.

Ulutan, D., & Özel, T. (2013). Determination of tool friction in presence of flank wear and stress distribution based validation using finite element simulations in machining of titanium and nickel based alloys. *Journal of Materials Processing Technology, 213*, 2217–2237.

Umbrello, D., M'Saoubi, R., & Outeiro, J. C. (2007). The influence of Johnson–Cook material constants on finite element simulation of machining of AISI 316L steel. *International Journal of Machine Tools and Manufacture, 47*, 462–470.

Vaziri, M. R., Salimi, M., & Mashayekhi, M. (2011). Evaluation of chip formation simulation models for material separation in the presence of damage models. *Simulation Modelling Practice and Theory, 19*, 718–733.

Wang, X., Huang, C., Zou, B., Liu, H., Zhu, H., & Wang, J. (2013). Dynamic behavior and a modified Johnson–Cook constitutive model of Inconel 718 at high strain rate and elevated temperature. *Materials Science and Engineering: A, 580*, 385–390. doi:10.1016/j.msea.2013.05.062

Xue, Q., Liao, X. Z., Zhu, Y. T., & Gray, G. T. (2005). Formation mechanisms of nanostructures in stainless steel during high-strain-rate severe plastic deformation. *Materials Science and Engineering: A, 410–411*, 252–256.

Yang, X., & Liu, C. R. (2002). A new stress-based model of friction behavior in machining and its significant impact on residual stresses computed by finite element method. *International Journal of Mechanical Sciences, 44*, 703–723.

Yang, X. S. (2010). *Nature-inspired metaheuristic algorithms*. London: Luniver Press.

Zorev, N. (1963). Inter-relationship between shear processes occurring along tool face and on shear plane in metal cutting. *ASME International Research in Production Engineering*, 42–49.

# 7

# Managing uncertainty through supply chain flexibility: reactive vs. proactive approaches

Reina Angkiriwang, I Nyoman Pujawan* and Budi Santosa

*Department of Industrial Engineering, Sepuluh Nopember Institute of Technology, Kampus ITS Sukolilo, Surabaya 60111, Indonesia*

The purpose of this paper is to obtain insights into the typology of uncertainty and relevant strategies adopted by manufacturing companies to achieve better supply chain flexibility. Strategies are classified into reactive (buffering) and proactive (redesigning). We develop a framework that links supply chain uncertainty, the two types of strategies for achieving supply chain flexibility and the relevant objectives to be achieved. Four case studies are compared in terms of uncertainty typology and strategies being adopted to improve supply chain flexibility. We present three propositions as a result of the study. First, to achieve better flexibility, companies have been focusing more on buffering than on proactive or redesign strategies. Second, companies tend to focus on internal operations rather than collaborating with external parties, and third, the power structure in a supply chain governs the type and configuration of supply chain flexibility.

**Keywords:** supply chain flexibility; uncertainty; strategy; proactive; reactive

## 1. Introduction

To remain competitive in business, manufacturers are required to produce products of a quality acceptable to customers and to deliver those products at competitive cost with highly reliable delivery times. Achieving high quality levels, timeliness of deliveries, and efficient processes along the supply chain cannot be reliant on a single organization, but should be ensured through collaboration and coordination with trading partners.

Increasing uncertainty has made the task of satisfying customers more challenging. Efficiency along the supply chain is important to maintain acceptable product prices, but flexibility to deal with time-varying or dynamic demand could be even more important nowadays. With the high probability that customers will suddenly increase, reduce, cancel, or move forward or backward their orders, supply chain players need to be more flexible in many respects. This may include the need to change capacity levels, to use different transportation modes, to switch supplier, to deal with small lot sizes and to have short or even negligible changeover times. Researchers have considered that flexibility, along with quality, cost, and speed of delivery, is critical for competitiveness (Avittathur & Swamidass, 2007; Bernardo & Mohamed, 1992; Gong, 2008; Shin, Collier, & Wilson, 2000) and should be viewed from the supply chain perspective in a collaborative way (Fantazy, Kumar, & Kumar, 2009; Kumar, Fantazy, Kumar, & Boyle, 2006; Pujawan, 2004; Zhang, Vonderembse, & Lim, 2006). As pointed out by Vickery,

---

*Corresponding author. Email: pujawan@ie.its.ac.id

Calantone, and Dröge (1999), supply chain flexibility has a positive impact on the bottom line of the companies. Their empirical study also reveals that volume flexibility is highly related to market share and market share growth.

Researchers have paid a great deal of interest to flexibility. Slack (1988) and Gupta (Gupta & Goyal, 1989; Gupta & Somers, 1992) are among the early authors that have brought up the issue of flexibility in the context of internal operations systems. Following their works, many authors have published papers on flexibility. Currently, flexibility is viewed not only from the perspective of internal operations systems; more and more authors are trying to link it to a supply chain context. Vickery et al. (1999), Duclos, Vokurka, and Lummus (2003), and Pujawan (2004) are among the authors that have extended flexibility from internal operations to the supply chain context. Seebacher and Winkler (2013) provide a comprehensive review and citation analysis of works on manufacturing as well as supply chain flexibility. Although there has been continuing interest in the flexibility issue within the context of the supply chain, further debate and discussion are required.

Although it is important to remain competitive in the market, supply chain flexibility is costly (Gupta & Goyal, 1989; Koste & Malhotra, 1999; Merschmann & Thonemann, 2011; Pujawan, 2004) and flexibility dimensions are not equally important for every supply chain (Sánchez & Pérez, 2005; Stevenson & Spring, 2007). Therefore, a supply chain should carefully assess how much flexibility is really needed (Pujawan, 2004) because increased flexibility does not always result in greater economic income (Bernardo & Mohamed, 1992; Gong, 2008). Hence, it is necessary to decide the right degree of supply chain flexibility and the appropriate strategy that should be adopted to respond to the need for flexibility. In this respect, companies may just reactively build up safety stock or add safety buffers to the lead time, or pursue more proactive paths such as redesigning products or supply chain networks. In spite of its importance, there have been very few works that attempt to classify flexibility strategies and to relate the typology of uncertainties to those strategies. The objective of this paper is to present a conceptual model and case studies that address uncertainty typology, strategies to improve supply chain flexibility, and the relevant objectives that the companies in a supply chain aim to achieve. In particular, this paper aims to address the following questions in an explorative manner:

(1) What strategies have companies pursued to improve flexibility in the supply chain?
(2) In dealing with uncertainty, do companies tend to reactively buffer themselves or use more proactive strategies?

## 2.  Literature review

The subject of flexibility has been present in the literature for more than two decades. This topic is highly related to uncertainty, both internal and external to a company or a supply chain. Supply chain flexibility is defined as the ability of a system or a chain to respond to unexpected and unpredictable changes due to uncertain environments to meet a variety of customer needs or requirements, while still maintaining customer satisfaction without adding significant cost. The type of strategies used to increase flexibility should be in line with the typology of uncertainty.

There have been changes in the direction of flexibility research. In the early stages, at the end of the eighties, flexibility was often discussed in the context of internal manufacturing systems. Cox (1989) stated that flexibility was a new concept with no acceptable measurement. Consequently, this triggered several researchers to develop an instrument to measure or assess flexibility (Bernardo & Mohamed, 1992; Gupta & Goyal, 1989; Gupta & Somers, 1992; Slack, 1988). Sethi and Sethi (1990) mentioned that flexibility is a complex, multi-dimensional, and hard-to-capture concept leading to several studies seeking to establish dimensions of flexibility during the 90s (Koste & Malhotra, 1999; Vokurka & O'Leary-Kelly, 2000). In the last few decades, intensifying market competition forced companies to change their orientation from internal operation systems to a supply chain context. Likewise, people extended the viewpoint of flexibility from the manufacturing to the supply chain context. Since then, a number of researchers have contributed to the literature on supply chain flexibility.

The major stream of work on flexibility has been related to the development of flexibility measures, as well as the conceptual framework that defines the relationships between various flexibility dimensions. Duclos et al. (2003) and Lummus, Duclos, and Vokurka (2003) developed a conceptual model and extended flexibility dimensions to the supply chain context. A framework for assessing supply chain flexibility has been developed by Pujawan (2004). The aim was to capture the relationship between supply chain flexibility drivers and flexibility dimensions, and ultimately to identify which dimensions have to be prioritized for improvement. Sánchez and Pérez (2005) built a hierarchy of supply chain flexibility dimensions based on bottom-up classification and tried to find a relationship between supply chain flexibility and firm performance. Kumar et al. (2006) developed a framework which recommended steps for implementing supply chain flexibility. It gives an overview of the relationship between each flexibility dimension in a supply chain.

As mentioned, a number of authors have tried to develop a framework to assess the importance of flexibility dimensions or relationships between flexibility dimensions and firm performance. However, little has been done to address and classify strategies to achieve the appropriate level of supply chain flexibility. In particular, no attention has been paid to relationships between the typology of uncertainty and the type of strategies chosen for better supply chain performance. Among the few authors to have addressed flexibility, Tachizawa (2007) and Tachizawa and Thomsen (2007) and Tachizawa and Gimenez (2010) developed models that relate between supply flexibility drivers and strategies to improve supply chain flexibility. However, the focus of their study is the purchasing or supply side, not the supply chain.

### 3. The conceptual model

In this paper, a conceptual model that links supply chain uncertainty, supply chain flexibility strategies, and the objectives pursued by companies along the supply chain is proposed. The elements of the conceptual model were synthesized from the literature. Figure 1 depicts the model, showing uncertainties as the factors that drive the need for supply chain flexibility (left), the two categories of strategy, i.e. four types of buffering (above), eight proactive strategies (below), and three objectives (right). More details will be given below.

### 3.1. Supply chain uncertainty

Uncertainty has been considered as a major factor behind the need for supply chain flexibility. We classify uncertainty in the supply chain context as upstream (supply) uncertainty, internal (process) uncertainty, and downstream (demand) uncertainty.

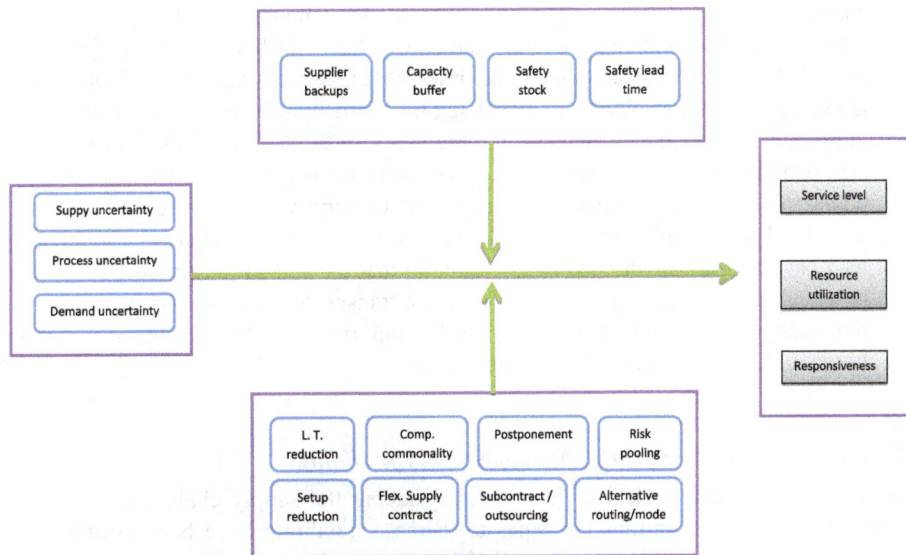

Figure 1.   The conceptual model.

- *Supply uncertainty*. This is related to the uncertainty of materials' supply. Supply uncertainty may be manifested in the form of uncertainty regarding material availability/supply capacity, material price (Tang, 2006; Zhang, Shang, & Li, 2011), alternative sourcing availability (Pujawan, 2004), and supply lead time (Mohebbi & Choobineh, 2005; Osman & Demirli, 2012; Ramasesh, 1991; Thomas & Tyworth, 2006). When supply is uncertain, a higher level of flexibility is required to achieve better customer service levels. For example, when the price of a certain material suddenly increases, the ability to change the production schedule in order to produce alternative products or to use a substitute material is helpful in maintaining the efficiency and effectiveness of the supply chain.
- *Process uncertainty*. Process uncertainty is related to the probabilistic nature of machine availability (Ho, 1989; Van Kampen, van Donk, & van der Zee, 2010), yield (Gurnani & Gerchak, 2007; Ho, 1989; Schmitt & Snyder, 2012; Tang, 2006), quality (Ho, 1989; Murthy & Ma, 1991; Wu, Blackhurst, & O'grady, 2007), and processing times (Cao, Patterson, & Bai, 2005; Schmitt, 1984). Uncertainty in internal operations may also be related to such factors as labor issue, instable availability of working capital, and problems with information technology. The more uncertain the internal processes, the higher the level of flexibility needed. For example, when the reliability of a manufacturing process is low, a capacity buffer is necessary to keep the system flexible.
- *Demand uncertainty*. This refers to the probabilistic nature of demand quantity, types, timing, and locations. Demand uncertainty could be in the form of errors in the demand forecast (Schmitt, 1984), changes in customer orders (Pujawan & Smart, 2012; Van Kampen et al., 2010; Wong, Boon-Itt, & Wong, 2011), uncertainty about the product specification/mix that the customers will order (Li, Sikora, Shaw, & Tan, 2006; Van Donk & van der Vaart, 2005), and competitor actions regarding

marketing promotion (Wong et al., 2011). An empirical study involved 106 manufacturing companies by Pujawan and Smart (2012) suggested that most manufacturing companies experience order volatility from customers. Manufacturing companies, especially those producing innovative products, are facing shrinkage in product life cycle and increasing competition in the market. This ultimately creates demand uncertainty and forcing supply chain players to markdown product prices. Supply chain flexibility is important to cope with the dynamic nature of demand. For example, when demand for a product is highly uncertain, companies may have to increase the inventory buffer so that the sudden increased demand can be satisfied. Alternatively, the use of multiple transportation modes to cope with differing degrees of demand urgency may be important to improve customer service levels under highly uncertain demand.

### 3.2.  Classification of strategies for supply chain flexibility

Flexibility is a strategy to deal with uncertainty facing the supply chain. As mentioned in the literature review, various strategies to increase flexibility have been evaluated by different authors. Koste and Malhotra (1999) mentioned that flexibility can be either reactive or proactive in nature. The reactive nature of flexibility addresses the environmental uncertainty, both internal and external, faced by an organization, while the proactive nature of flexibility allows an organization to redefine market uncertainties or influence what customers have come to expect from a particular industry. In the following subsections, we will list the strategies that belong to the above two classifications.

#### 3.2.1.  Reactive or buffering strategies

Basically, companies make no attempt to influence the level of uncertainty but, rather, react to it in an attempt to maintain their service level to customers or to maintain efficiency through, for example, better capacity utilization. These are the type of strategies which companies use to buffer themselves:

- *Safety stock.* Safety stock is one of the most common approaches to increase flexibility under the existence of demand and supply uncertainty. With safety stock, a company can reduce the probability of inventory shortage to an acceptable level. According to Van Kampen et al. (2010), safety stock also increases responsiveness. We argue that safety stock is a reactive strategy because this approach merely responds to the current level of uncertainty, without any attempt to proactively reduce it.
- *Capacity buffer.* One way to cope with uncertainty is to have flexible capacity. However, in many situations it is costly or even impossible to dynamically adjust the capacity level. Hence, to achieve flexibility, companies may set capacity higher than average demand so that they can avoid substantial shortages during peak periods. A capacity buffer could be used as a substitute of or a complement to safety stock. As mentioned by Manuj and Sahin (2011), some companies prefer to maintain surge capacity in the form of extra assembly lines that are staffed when necessary rather than maintain buffer inventory.
- *Supplier backups.* Working with a single supplier is risky. Companies often maintain multiple suppliers, which will guarantee availability but in most situations

will increase costs. The use of multiple suppliers to improve supply flexibility has been discussed in the literature, including for example Das (2011) and Manuj and Sahin (2011).

- *Safety lead times*. In dealing with uncertainty, companies often add a safety lead time to the actual cycle time (Raturi & Jack, 2004). While increasing inventory and adding costs, safety lead time enables companies to increase materials' availability and thus become more flexible in responding to demand. Van Kampen et al. (2010) suggested that adding safety lead time increases mix flexibility.

### 3.2.2. *Proactive or redesign strategies*

These are the type of strategies in which companies attempt to increase supply chain flexibility through proactively redesigning products, processes, and the supply chain network as well as proactively negotiating more effective relationships with trading partners. The following strategies belong to this classification:

- *Component commonality*: one of the obvious strategies related to product design that can improve flexibility is the use of component commonality. There has been wide discussion of the possibility that use of common component could increase efficiency as well as responsiveness when a manufacturing company offers a large product variety or when uncertainty in demand exists (Mohebbi & Choobineh, 2005). Large component commonality also increases supply chain flexibility. The use of component commonality to increase supply chain flexibility has been discussed by Pujawan (2004).
- *Postponement*: design or redesign of processes, both in terms of the manufacturing shop floor and in administrative processes, could significantly improve flexibility. Manufacturing or logistics postponement are examples of process design that improves supply chain flexibility. Increased component commonality and the use of postponement as strategies for better flexibility have been discussed in the literature (Barad & Sapir, 2003; Das, 2011; Kumar, Shankar, & Yadav, 2008; Pujawan, 2004; Tachizawa, 2007).
- *Risk pooling*: When demand is highly uncertain and involves multiple sales regions, a supply chain often designs a network for risk pooling. Reducing facilities or centralizing stocks to fewer facilities reduces risks as well as improves flexibility in allocating stock to multiple destinations or sales regions. In line with this idea, Stevenson and Spring (2009) argued that consortium purchasing among several independent firms or the use of centralized procurement among firms in the same group are among strategies to overcome the problem of buyers having minimum power and having to accept large minimum order quantities.
- *Subcontracting/outsourcing*: The use of external capacities through subcontracting and outsourcing is also a strategy used to be flexible in dealing with uncertainty. As suggested by Stevenson and Spring (2009), outsourcing is a tactic to obtain purchasing flexibility because it can reduce the risks of capacity utilization and amortization, especially when demand is uncertain, irregular, low, and/or temporary.
- *Flexible supply contract*: proactively negotiating supply contracts to alleviate minimum order quantities or to obtain a commitment from suppliers to supply materials or services in the case of a significant increase in demand are among the

strategies that could improve supply chain flexibility. Flexible procurement contracts can provide supply flexibility (Tang & Tomlin, 2008), ensure stability for the supplier, and help the buyer respond to demand fluctuations (Stevenson & Spring, 2007).

- *Lead time reduction*: reducing lead time is an important strategy to enable better flexibility. With a shorter lead time, companies will be able to better respond to demand uncertainty. Proactively reducing lead time may be done through redesigning procurement processes, changing supplier selection criteria from cost focus to speed focus, or developing suppliers for better lead time management.

- *Setup time reduction*: long production setup is a major cause of inability to create volume and mix flexibility. Significantly reducing setup time requires systematic effort within a manufacturing company. Vonderembse, Uppal, Huang, and Dismukes (2006) suggest that reduction of setup time will allow the economic production of small quantities, thereby achieving cost reductions, flexibility, and internal responsiveness. Having suppliers with short setup times would create supply flexibility and thus should be considered as an important criteria when selecting suppliers (Wu & Shen, 2006).

- *Alternative routing/mode*: when a customer places a sudden order, companies may have to use alternative transportation modes to speed up shipment. In another situation, when problems occur on a certain route, the company's flexibility to use alternative routes is important. The use of alternative transportation routing/modes to create flexibility has been mentioned in the literature (see e.g. Kumar et al., 2008; Morlok & Chang, 2004; Pujawan, 2004).

### 3.2. *Objectives of supply chain flexibility*

Although supply chain flexibility is often costly to achieve, it is rewarding to have appropriate levels of flexibility for companies working under uncertain environments. A study conducted by Merschmann and Thonemann (2011) involving German companies suggests that in uncertain environments companies with highly flexible supply chains perform better than companies with less flexible supply chains, while the opposite holds true under certain environments. As widely discussed in the literature, flexibility is viewed as a system's ability to cope with internal and external variation (Gong, 2008) or to cope with environmental uncertainty in an attempt to improve market responsiveness (Das, 2011), maintain customer service levels (Kumar et al., 2008), and achieve better resource utilization. Hence, we define here three objectives of supply chain flexibility that companies aim to achieve:

- *Higher service level*. Service level is a measure of a company's ability to satisfy customers' demand. It is normally measured in terms of percentage of customer demand satisfied without a backlog. When demand and supply are both uncertain, a flexible supply chain system is necessary to maintain a high level of service. As pointed out by Kumar et al. (2008), flexibility is the strategy by which the supply chain aims to maintain customer service levels by absorbing disturbances in supply and sudden changes in demand. There are many other works that address the linkage between flexibility and service level, including for example Alfredsson and Verrijdt (1999), Mohebbi and Choobineh (2005), and Van Hoek (2001).

- *Resource utilization*. Supply chain activities normally involve a variety of resources, such as those used for production, storage, material handling, transportation, and administrative activities. Resource utilization is generally a measure of how much of the available capacity of resources is being used for productive outputs. Under uncertain operating environments, it is often difficult to obtain high levels of resource utilization. However, when appropriate supply chain flexibility is in place, there is an opportunity to better utilize the resources along the supply chain. Some earlier publications mention how flexibility may improve resource utilization. Ndubisi, Jantan, Hing, and Ayub (2005) stated that there is a relationship between resource utilization and volume flexibility. Pramod and Garg (2006) suggest that as any flexibility level increases, the system utilization increases.

- *Responsiveness.* Responsiveness is a very important capability which companies need to have in doing business. In an uncertain environment, responsiveness can only be achieved if there is sufficient level of flexibility throughout the supply chain. Flexibility in responding to changes in demand reflects responsiveness to changing customer needs and wants (Duclos et al., 2003). Kumar et al. (2006) stated that there is a need to make operations more flexible in an environment with unpredictable changes, and thereby increase responsiveness. Further, Yi, Ngai, and Moon (2011) suggest that better supply chain responsiveness can be achieved in two ways: reducing uncertainties and improving supply chain flexibility.

## 4.  Case studies

We conducted a case study involving four manufacturing companies in Indonesia. According to Kähkönen (2011), case study is an appropriate research method for gathering rich empirical data and thereby to gain a deep understanding of the phenomenon in question. The objective of the case study was to gain more insights into supply chain flexibility in practice, and particularly the strategies which companies adopt to achieve better flexibility. We selected case companies with different supply chain characteristics, i.e. one is representing the engineering-to-order model, the other one is representing make-to-stock model, and the other two are somewhere in between of the above two models. A list of open-ended questions were prepared before interacting with each case company. In each case, we interviewed 3–5 people in 2–4 meetings involving those in charge of procurement, planning, manufacturing, product development, engineering, and marketing functions. In addition to the interview, we also collected and analysed pertinent data that could support our understanding of supply chain flexibility. We will provide details of each case in this section. Cross-case analysis will be conducted to compare uncertainty typology facing each case company and strategies they applied to create better supply chain flexibility.

### 4.1.  Case 1: company A

This company is in the business of producing new ships and repairing old ones. As a project-based company, design, procurement, and production respond to customer orders. The company serves many domestic customers, but only three of them are considered major customers. The production facility is located on the coast of East Java. The size of the production facility limits its capacity. They can produce up to two new ships and repair up to 100 ships each year. The company employs around 550 permanent and 600 subcontracted workers.

For every product, customers will be involved at the early stage of developing the product specification. The raw material will not be ordered until the design and specifications for the end product are agreed on by the customer. Each product is composed of more than 10,000 components. However, there are only four major types of component: plate, pipe, engine, and electricity. In each of these major types, there are about five different variants. The majority of materials, about 80%, are obtained from overseas suppliers with 2–6 months' lead time. To avoid lateness of material supply, the company sets a lead time buffer of about four weeks. The company has about three suppliers for each type of component. The relationship with customers and suppliers for each project is contractual in nature rather than a partnership type. If a supplier is unable to supply the quantity ordered, an alternative supplier should be used.

The production process consists of five stages. It begins with body construction, pipe fitting, installation of machinery or engine, electrification and, finally, running tests. The duration of a project building a new ship is somewhere between 4 and 20 months, depending on the product type, size, and complexity. Substantial changes in the design at the production stage rarely happen. There are usually only minor revisions which do not significantly affect the production schedule and costs.

Unavailability of material will lead to production rescheduling. Interestingly, delays in the supply of materials are not because of problems with the suppliers, but quite often because of poor cash availability at the company. For ship production, capacity constraints can be relaxed by hiring more subcontracted workers, although it is not always easy to adjust the level of the workforce. However, for ship repair, space availability is a major constraint which cannot be adjusted. The uncertain nature of internal operations and materials supply leads to poor levels of on-time delivery to customers.

### 4.2. Case 2: company B

Company B is a supplier of packing items such as plastic bottles and tubes for the consumer goods industry. The company operates three factories in Indonesia with fixed production capacities, employing around 1000 people in total. Currently, they supply packing items to about 10 major customers. They have two types of product, exclusive and V-item, with about 100 variants. Exclusive products are those specific to a certain customer, whose design or patent is owned by the customer. V-item is a nonpatented product and can be sold to any customer. However, most of the products produced by this company are exclusive to certain customers. For exclusive products, the customers are involved in product design and innovation. All changes are based on customer requests.

The production process is fairly simple, with only a few steps, including molding, printing, and stamping. Even though the process is very simple, the level of process uncertainty is high due to machine breakdowns, high setup time, and defective products. Materials can be categorized into resin, master bed, and art work. There are around 20 varieties for resin, 50 for master bed, and 100 for art work. Supply of materials is not considered a problem since they can be obtained easily from local or overseas suppliers. The company has about 10 active suppliers with three to eight weeks' lead time and 80–90% on-time delivery. The relationship with the customer is generally transactional in nature. However, the company collaborates with mold suppliers to ensure that new models are built according to the customer's requirements.

Customers have high bargaining power. Demand uncertainty is passed on to the company, where orders could fluctuate by up to 40% from the initial figures they submitted to the company. This is the major cause of uncertainty facing the company and triggers much of the production rescheduling. Some customers even change their orders one day before the delivery due date. The high flexibility needed with regard to customers' rescheduling or changing orders is possible because the company keeps high levels of inventory buffers and has relatively short supply lead times for most materials.

### 4.3.  Case 3: company C

This company produces circuit breakers for several industry sectors, such as telecommunication, transportation, watercraft and special vehicles, and industrial equipment. The company operates two factories in East Java with around 250 employees. The majority of products are sold to four major customers, all of which are located overseas. Customers' geographic area influences product variety. The majority of the materials are supplied by one main supplier. The company produces up to 4500 product variants; all of them belong to nine major product types. Variety is created through different product attributes such as ampere, mounting, number of poles, thread neck design, hardware, types of terminal, push button, color of push button, and auxiliary contact.

A representative of the company stated that this type of industry has very few competitors, because they deal in a very special product which needs extremely high performance and accuracy. Difficulty in obtaining replaceable materials and alternative suppliers, coupled with uncertain supply, results in a very high inventory buffer of materials.

Production processes for one type of product are different from those for other types and the processing times vary significantly. Even variants within the same product type could have different processes. Some types of product have less than 20 variants, but some could have thousands.

Each product consists of 11–45 components, with about 10–20% common components across different finished products of the same type. To speed up production time, subassemblies are prepared before the company receives customer orders. However, the final assembly is normally done after the receipt of orders from customers. The number of machines will determine production capacity and each machine is dedicated to a single product type only. This creates a low level of production capacity flexibility. On the other hand, the company absorbs demand fluctuation by working overtime hours.

Demand is lumpy for all product variants. The company sets no constraint on minimum order quantity. The order quantity from customers could be very large for some product variants; for other variants, customers sometimes order very small quantities, and even only one unit. This lumpiness creates much idle capacity and poor on-time delivery. To avoid having a large idle capacity, the company produces some regular products whose demand is quite certain in anticipation of customer orders. However, it results in high inventory of finished goods. The company did not set time-fencing mechanism for orders. As a consequence, there is frequent production rescheduling. On the positive side, the changeover time for producing different product variants within the same type is very short.

### 4.4.  Case 4: company D

Company D produces pharmaceutical products, operating three plants. All are located in the West Java Province of Indonesia, with about 400 workers. Every plant is dedicated to a certain type of product. Currently, they only serve the Indonesian market. The company supplies products to about 100 customers, but only three are considered major customers. There are four types of product – tablet, capsule, syrup, and dry syrup – with 64 product variants. The main product itself cannot be modified; however, the company usually changes the product package once a year.

Raw materials originate overseas but the import process is managed by a local distributor. Supply of materials from local distributors requires two to three days of lead time. It can take up to six months to receive special materials that do not have a local distributor. This company has around 30 suppliers for main materials and 28 for packaging items, many of which are considered as backup suppliers. The policy of the management is to have at least three suppliers for each type of material. Relationships with suppliers are more transactional than partnership in type. One supplier can supply several types of main materials for different finish goods.

Materials are categorized into four: main material, supporting material, primary packaging, and secondary packaging. In total, there are about 200 types of material where 50 are categorized as main materials, 100 are supporting materials, 9 are primary packaging items, and 64 are secondary packaging items. The product structure is generally very simple. A product consists of one to four main materials, one primary packaging item and one secondary packaging item. Product variations are created by different product weights, packaging, or product composition. Some materials can be used in more than one product variant within the same type.

There is high process similarity across different products. However, the batch quantity differs significantly. Production capacity for each product type is fixed, but there is a possibility to increase production volumes through the use of overtime hours. Even though the production lines are dedicated to certain product types, a changeover is needed when there is a change in product variants to be produced. The changeover time is about 2–3 h for cleaning and about half a day for changing a mold. The company responds to this long changeover time by freezing order and production schedule for one month.

In general, demand for most products is quite stable and little affected by segments and the geographical area of a market. On the other hand, there is uncertainty associated with supply of materials, mainly due to high competition to obtain them. In addition, the warehouse capacity is quite small, which limits the company's ability to do forward buying for materials. The company normally buys sufficient materials to satisfy the production schedule for a month, even though they can actually predict material requirements for three months. For materials that have long lead times, this creates a major material availability problem.

## 5.  Cross-case comparisons and analysis

In this section, we will compare the uncertainty typology of each case company and the strategies adopted to achieve better supply chain flexibility.

### 5.1.  Typology of supply chain uncertainty

Table 1 shows the comparisons between four companies in terms of uncertainty typologies. The four cases represent different types of operations: the first is a project-based or engineering-to-order company, and is followed by make-to-order, assembly-to-order, and make-to-stock companies. With the exception of Case 4, all companies are facing highly uncertain demand. Demand uncertainty may be associated with frequently revised orders

Table 1.  Typology of supply chain uncertainty.

| | Case 1 | Case 2 | Case 3 | Case 4 |
|---|---|---|---|---|
| Type of operations | Engineering to order | Make-to-order | Assembly-to-order | Make-to-stock |
| *Level of uncertainty* | | | | |
| • Demand | Highly uncertain | Highly uncertain, customers frequently change orders, high mix uncertainty | Highly uncertain product mix, high product variety, demand fluctuating over time | Low demand uncertainty (orders received well in advance) |
| • Supply | Highly uncertain supply lead time | Low supply uncertainty, relatively short lead time | Highly uncertain supply lead time and material price | Low supply uncertainty for most materials |
| • Process | Moderately uncertain, very high complexity | Highly uncertain in terms of cycle time, yield, availability | Moderately uncertain: low process similarity, high variability in processing time | Stable process, low uncertainty |

by customers (as in Case 2) or the inability of the company to forecast what products they are going to sell in the future (as in Cases 1 and 3).

From the supply side, Cases 1 and 3 are facing high supply uncertainty while the other two cases are fortunate to have low supply uncertainty. In Case 1, supply uncertainty is mostly due to the long lead times (4–6 months), but in Case 3, uncertainty is associated with lead time as well as the price. In both of these cases, most suppliers are located overseas, while for Cases 2 and 4 most materials are obtained from local suppliers.

In terms of operations processes, only Case 1 deals with highly complex processes; the other three involve simple production processes. On the other hand, process uncertainty in Case 2 is high where the company is unable to predict the cycle time accurately. In addition, yield and production system availability are uncertain. It is often the case that the production system fails and requires unplanned maintenance. In Case 3, process uncertainty is relatively lower, although processing time varies greatly depending on the types of product being produced.

From the typology of uncertainty of the four case companies, it is obvious that the company in Case 4 is operating under a stable and low level of uncertainty where not much flexibility is needed. The other three cases are facing different types of uncertainty and require flexibility in the supply chain.

### 5.2.  *Flexibility strategies adopted*

The way in which a supply chain creates flexibility to handle demand uncertainty differs from case to case. Table 2 shows which strategies are adopted by each case company. Low, moderate, and high refer to the intensity of the strategies being used. The ratings are given subjectively by the authors after obtaining information from each case company. An 'X' sign means there is no clear evidence of the use of that strategy by the respective case company. For example, the company in Case 3 uses safety stock with a high level of intensity but there is no evidence of it using backup suppliers. From the list of strategies presented in the framework of Figure 1, all four cases are more focused on buffering rather than the proactive or redesign strategies. For example, safety stock and safety lead time are used by all four cases, although with differing levels of intensity. On the other hand, the proactive strategies are less intensively adopted to improve supply chain flexibility. Based on this fact, we arrive at a proposition:

*Proposition 1*: *The flexibility of companies tends to be of a more reactive than proactive nature.*

The use of buffers to increase flexibility has been discussed extensively. For example, Caputo (1996) pointed out that buffers could be considered alternative or complementary factors to technological flexibility. Any strategy to increase flexibility has cost implications. The choice of positioning a buffer as an alternative or complementary way of achieving flexibility depends on the cost benefits of the two solutions. However, as suggested by Christopher and Lee (2004), buffers are built up as a result of no confidence or low visibility across the supply chain. Manuj and Sahin (2011) also warned that the use of buffers, especially capacity buffers, could be very costly if not planned carefully.

Although all four cases are using some type of proactive strategy, they are concentrating on internal operations rather than working with supply chain partners. For example, all four cases seem to use some level of component commonality. Increasing component commonality is possible with changes in product and process design. The objective is to reduce the number of components a company has to manage while maintaining the ability to produce a large product variety. All four case companies are adopting component commonality as a strategy to achieve better flexibility. The use of common components, along with the postponement strategy, has long been considered an effective way of increasing flexibility. Gunasekaran and Ngai (2005) pointed out the commonality of components is an important strategy to achieve flexibility, especially in the context of a built-to-order supply chain. Likewise, manufacturing postponement has long been used as a hedging strategy against the proliferation of product variety (Feitzinger & Lee, 1997; Yang, Burns, & Backhouse, 2004).

Table 2.   Flexibility strategies adopted.

|  | Case 1 | Case 2 | Case 3 | Case 4 |
|---|---|---|---|---|
| Type of operations | Engineering to order | Make-to-order | Assembly-to-order | Make-to-stock |
| *Strategies/enablers* |  |  |  |  |
| Safety stock | Low | High | High | Moderate |
| Safety lead time | High | Low | Low | Low |
| Capacity buffer | Low | High | Low | X |
| Backup supplier | High | Low | X | High |
| Lead time reduction | X | X | X | Low |
| Setup time reduction | X | X | High | X |
| Component commonality | Moderate | High | Moderate | Low |
| Postponement | High | High | Moderate | X |
| Risk pooling | X | X | X | X |
| Flexible supply contract | X | X | X | X |
| Subcontracting/outsourcing | High | X | X | X |
| Alternative routing/modes of transportation | X | X | Low | X |

Lead time reduction, flexible supply contracts, and subcontracting/outsourcing are not popular strategies among the four cases, although the company in Case 4 has been able to reduce supply lead time through the use of local suppliers. Obviously, when a company is undertaking any of these three strategies, the focus should be shifting from internal operations to supply chain collaboration. Supply lead time could not be effectively reduced without the involvement of suppliers. Likewise, flexible supply contracts would require the supplier to alleviate 'minimum order quantity' or to ensure security of supply in case of increasing demand and to enable speeding up delivery in case urgent deliveries are emerging. This then leads us to the following proposition:

*Proposition 2*: *The flexibility of companies tends to focus more on internal operations rather than on collaborating with external parties.*

As mentioned in the literature review, authors have recognized the importance of better coordination and collaboration to increase supply chain flexibility (Pujawan, 2004; Vokurka & O'Leary-Kelly, 2000). However, it should be noted here that Proposition 2 above suggests that efforts to increase supply chain flexibility have not been undertaken in accordance with supply chain principles. The move from manufacturing flexibility to supply chain flexibility should be interpreted as a need for collective efforts of supply chain members to create flexibility and, hence, responsiveness to end customers (Das, 2011; Kumar et al., 2008). Literature on supply chain flexibility suggests that reasons for the need for flexibility are both internal and external (Gong, 2008). Therefore, flexibility cannot be created without involving trading partners.

### 5.3. Flexibility attributes, problems, and objectivess affected

Table 3 presents the most obvious flexibility attributes, dominant problems that hinder flexibility and impacts of insufficient flexibility. In Case 1, as is usually the case for ETO companies, there is high flexibility in producing different products. However, the problem is mostly in the form of the supply of materials which have long lead times and the problem of cash availability that often results in delayed procurement of materials. In Case 2, the company attempts to provide very high flexibility to customers, so order revisions are allowed. With a long setup time, such volatile demand can only be satisfied with high inventory, both in the form of materials as well as finished products. In Case 3, the company also attempts to provide high flexibility to its customers, who may even order only a single unit. This is very well supported by the internal production system, where changeover time is very short on one side and there is moderate level of component commonality on the other side, but there is a big challenge from the supply side because of the long lead time. As a result, the company has to keep a high level of inventory buffers.

### 5.4. The impacts of power structure in supply chain relationships

Power structure in a supply chain seems to affect supply chain flexibility configuration. Although some studies have indicated that competition has shifted from company vs. company to supply chain vs. supply chain in the past two decades, each member in the chain is still exercising their power in the relationship (Crook & Combs, 2007). The power structure in a supply chain determines the parties that have to keep higher inventories, work with higher capacity flexibility and alleviate 'minimum order quantity'

Table 3. Dominant problem for each company that hinders supply chain flexibility.

| | Most obvious flexibility attributes | Problems | Impacts |
|---|---|---|---|
| Case 1 | • Ability to produce various types of products | • Problems with cash availability that prevents the company to obtain sufficient materials in due time | • Poor on-time delivery (only about 70%)<br>• Idle resources because of waiting for materials |
| Case 2 | • No frozen schedule, customers are allowed to change orders any time | • Long setup time<br>• Frequent breakdown | • Large inventory buffer (up to 6 weeks demand)<br>• Fluctuated capacity utilization |
| Case 3 | • No limit on minimum order quantity, some customer orders have a quantity of one unit<br>• Ability to produce a very large number of products variants | • There is only one very dominant supplier that limits supply flexibility<br>• Long supply lead time | • Poor on-time delivery reliability to customers<br>• High inventory of raw materials (about 3–4 months) |
| Case 4 | • None looks obvious in Case 4 | • Limited warehouse capacity that prevents forward buying for materials | On-time delivery of about 90% which is quite low for highly predictable demand Lost sales |

restrictions. The party that holds more power may require its trading partners to share information but reciprocate (Mishra, Raghunathan, & Yue, 2007), and hence tends to pose more uncertainties for others.

Differences in power structure and flexibility configuration are also exhibited by the cases observed in this study. In Case 2, customers are demanding high flexibility for changing orders, but the case company is unable to pass this risk on to the suppliers. The uncertainty on the part of the customers has to be absorbed by the case company. This is worsened by the fact that the case company is facing long setup times and uncertain processes. Although the supply lead time is relatively short, there is a need for the company to keep high material inventories and maintain a high level of capacity buffers. Figure 2 illustrates the power structure and flexibility configuration of Case 2. The solid arrows connecting the company with the suppliers and the customers indicate the power direction. The single direction arrow from customers to the company indicates the dominating power of the customers, while the two-way directions between suppliers and the company indicates balanced power between the two. The dashed

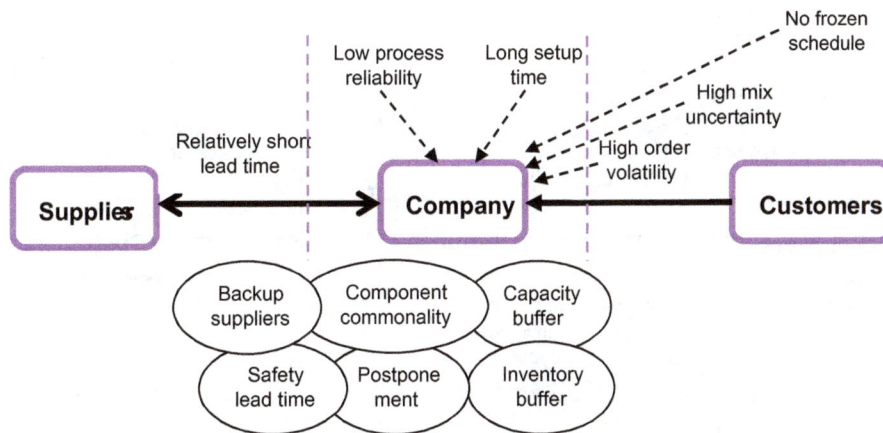

Figure 2.    Illustration of power structure and flexibility configuration for Case 2.

arrows represent challenges that foster the need for better flexibility and the circles are strategies for achieving better flexibility.

The company in Case 3 seems to be positioned between dominant, inflexible suppliers and demanding customers. As a consequence, the company has to absorb uncertainties regarding inventory of materials. As presented in the two tables above, the company in Case 3 is facing customers that require high product variety and a very wide range of order quantities, from as small as one unit to as large as thousands of units. Facing this uncertainty, the case company has to be flexible in producing different lot sizes and different product types. The company has been responding to this need by reducing setup costs, postponing the assembly process until orders are received, and increasing component commonality. On the other hand, the company is very much dependent on one very large, powerful supplier with long lead times, and hence there is little choice but to keep high inventory buffers, add buffers to the lead time and switch the transportation mode to air freight in case urgent delivery is needed. Figure 3 illustrates the power structure and flexibility configuration of Case 3. Hence, we suggest the following proposition:

*Proposition 3*: *The power structure and typology of uncertainty in the supply chain will govern the type and configuration of supply chain flexibility.*

## 6.   Concluding remarks

### 6.1.   Summary

In this paper, we have presented a conceptual model that links the uncertainties, the two classifications of strategies for improving supply chain flexibility and the pertinent objectives. We classified the strategies into reactive or buffering and proactive or redesign strategies. Four case studies have been analysed to obtain insights into uncertainty typology and the strategies each case company is adopting to have better supply chain flexibility. Our case studies suggest that each company may be facing different uncertainty typology and choosing different flexibility strategies.

We suggest three main propositions that need further testing. First, companies tend to use reactive rather than proactive strategies. This is somewhat understandable as

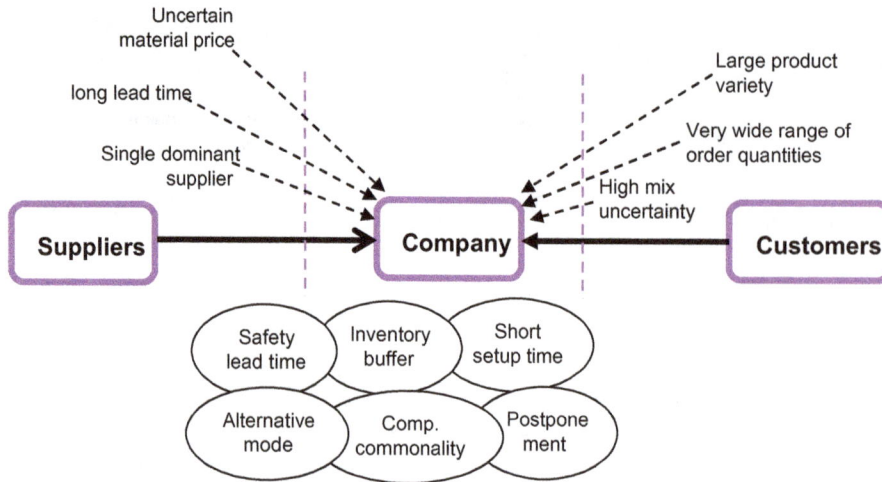

Figure 3.   Illustration of power structure and flexibility configuration for Case 3.

reactive strategies are easier to carry out and require less time and resource investment, although it may not be better in the long term. We suspect that such a pattern is quite common in the manufacturing industry in general, and hence a follow-up study involving a larger number of companies is necessary. Second, companies seem to be more inward than outward-looking when there is a need for higher flexibility. There is not much evidence of collaborative improvement of flexibility across the supply chain network through, for example, flexible supply contracts, outsourcing, or supply lead time reduction. Companies tend to work more on their internal systems. However, this might be largely dependent on the power structure in a supply chain. A manufacturing company positioned between powerful customers and suppliers is normally unable to influence trading partners, and thus uncertainty has to be absorbed internally through safety stock, safety lead times, extra capacity, and other internal reactions.

### 6.2.   Research implications

Obviously, there are a number of opportunities to extend this work. First, there is a need to do a larger scope study to confirm if the above three propositions hold true for a wider range of samples. In addition, uncertainty typology may affect the strategy applied. An interesting research question would be if any relationship exists between uncertainty typology and the flexibility strategies adopted by manufacturing companies. Finally, the strategies applied might be affected by common factors, leading to the following research question: would the choice between buffering and redesigning strategies be affected by any common factor?

Another avenue for future research is to find the balance between flexibility and schedule instability. Schedule instability is a subject that has been extensively discussed in the literature since as early as the 1970s (see e.g. an early paper by Steele, 1975) but remains a topic of ongoing research (see e.g. Pujawan & Smart, 2012 for a recent paper). A company offering high flexibility would not impose such policies as a long 'frozen schedule' and a large 'minimum order quantity'. However, being flexible with customers may also mean much disruption to the production schedule. Such disruptions are normally called schedule instability. Quantitative models or empirical research trying to find the balance between schedule instability and supply chain flexibility is an important topic for future research.

### 6.3. Managerial implications

Managers would benefit from our findings in a number of ways. First, the conceptual model can be used as a general guideline for achieving better supply chain flexibility when dealing with uncertainties. Managers could adopt various strategies to achieve supply chain flexibility, including four buffering strategies and eight redesign strategies. With this model, managers should be aware of various options, but the appropriateness of each strategy would depend on each specific situation.

Second, it should be quite obvious that too much focus on buffering strategies may be counterproductive in the long term. Buffering strategies are normally easier to execute and require less effort and risk. The move toward more proactive or redesigned strategies needs not only time and money, but also leadership that focuses on long-term benefits and willingness to take up short-term risks. Managers should be motivated to initiate a more long-term focus on flexibility rather than reactively responding to uncertainties with buffering strategies.

Finally, managers need to be motivated to explore possibilities to work with external parties to create better supply chain flexibility. The ideas behind supply chain management that fosters better collaboration with supply chain partners are much easier to say than to do. It is likely that most supply chain managers understand the importance of having close relationships with trading partners, but the reality, at least from our case study, suggests that such close collaboration is not easy to operate. Managers are more internal rather than external-looking. Potentials to improve flexibility together as a chain are little realized. This again implies the need to explore possibilities to work more closely with trading partners to improve supply chain flexibility.

### References

Alfredsson, P., & Verrijdt, J. (1999). Modeling emergency supply flexibility in a two-echelon inventory system. *Management Science, 45*, 1416–1431.

Avittathur, B., & Swamidass, P. (2007). Matching plant flexibility and supplier flexibility: Lessons from small suppliers of US manufacturing plants in India. *Journal of Operations Management, 25*, 717–735.

Barad, M., & Sapir, D. E. (2003). Flexibility in logistic systems – Modeling and performance evaluation. *International Journal of Production Economics, 85*, 155–170.

Bernardo, J. J., & Mohamed, Z. (1992). The measurement and use of operational flexibility in the loading of flexible manufacturing systems. *European Journal of Operational Research, 60*, 144–155.

Cao, Q., Patterson, J. W., & Bai, X. (2005). Reexamination of processing time uncertainty. *European Journal of Operational Research, 164*, 185–194.

Caputo, M. (1996). Uncertainty, flexibility and buffers in the management of the firm operating system. *Production Planning & Control, 7*, 518–528.

Christopher, M., & Lee, H. (2004). Mitigating supply chain risk through improved confidence. *International Journal of Physical Distribution & Logistics Management, 34*, 388–396.

Cox, T., Jr. (1989). Toward the measurement of manufacturing flexibility. *Production and Inventory Management Journal, 30*, 68–72.

Crook, T. R., & Combs, J. G. (2007). Sources and consequences of bargaining power in supply chains. *Journal of Operations Management, 25*, 546–555.

Das, K. (2011). Integrating effective flexibility measures into a strategic supply chain planning model. *European Journal of Operational Research, 211,* 170–183.

Duclos, L. K., Vokurka, R. J., & Lummus, R. R. (2003). A conceptual model of supply chain flexibility. *Industrial Management & Data Systems, 103,* 446–456.

Fantazy, K. A., Kumar, V., & Kumar, U. (2009). An empirical study of the relationships among strategy, flexibility, and performance in the supply chain context. *Supply Chain Management: An International Journal, 14,* 177–188.

Feitzinger, E., & Lee, H. L. (1997). Mass customization at Hewlett-Packard: The power of postponement. *Harvard Business Review, 75,* 116–123.

Gong, Z. (2008). An economic evaluation model of supply chain flexibility. *European Journal of Operational Research, 184,* 745–758.

Gunasekaran, A., & Ngai, E. W. (2005). Build-to-order supply chain management: A literature review and framework for development. *Journal of Operations Management, 23,* 423–451.

Gupta, Y. P., & Goyal, S. (1989). Flexibility of manufacturing systems: Concepts and measurements. *European Journal of Operational Research, 43,* 119–135.

Gupta, Y. P., & Somers, T. M. (1992). The measurement of manufacturing flexibility. *European Journal of Operational Research, 60,* 166–182.

Gurnani, H., & Gerchak, Y. (2007). Coordination in decentralized assembly systems with uncertain component yields. *European Journal of Operational Research, 176,* 1559–1576.

Ho, C. (1989). Evaluating the impact of operating environments on MRP system nervousness. *International Journal of Production Research, 27,* 1115–1135.

Kähkönen, A. K. (2011). Conducting a case study in supply management. *Operations and Supply Chain Management, 4,* 31–41.

Koste, L. L., & Malhotra, M. K. (1999). A theoretical framework for analyzing the dimensions of manufacturing flexibility. *Journal of Operations Management, 18,* 75–93.

Kumar, V., Fantazy, K. A., Kumar, U., & Boyle, T. A. (2006). Implementation and management framework for supply chain flexibility. *Journal of Enterprise Information Management, 19,* 303–319.

Kumar, P., Shankar, R., & Yadav, S. S. (2008). Flexibility in global supply chain: Modeling the enablers. *Journal of Modelling in Management, 3,* 277–297.

Li, J., Sikora, R., Shaw, M. J., & Tan, G. W. (2006). A strategic analysis of inter organizational information sharing. *Decision Support Systems, 42,* 251–266.

Lummus, R. R., Duclos, L. K., & Vokurka, R. J. (2003). Supply chain flexibility: Building a new model. *Global Journal of Flexible Systems Management, 4*(4), 1–13.

Manuj, I., & Sahin, F. (2011). A model of supply chain and supply chain decision-making complexity. *International Journal of Physical Distribution & Logistics Management, 41,* 511–549.

Merschmann, U., & Thonemann, U. W. (2011). Supply chain flexibility, uncertainty and firm performance: An empirical analysis of german manufacturing firms. *International Journal of Production Economics, 130,* 43–53.

Mishra, B. K., Raghunathan, S., & Yue, X. (2007). Information sharing in supply chains: Incentives for information distortion. *IIE Transactions, 39,* 863–877.

Mohebbi, E., & Choobineh, F. (2005). The impact of component commonality in an assemble-to-order environment under supply and demand uncertainty. *Omega, 33,* 472–482.

Morlok, E. K., & Chang, D. J. (2004). Measuring capacity flexibility of a transportation system. *Transportation Research Part A, 38,* 405–420.

Murthy, D. N. P., & Ma, L. (1991). MRP with uncertainty: A review and some extensions. *International Journal of Production Economics, 25,* 51–64.

Ndubisi, N. O., Jantan, M., Hing, L. C., & Ayub, M. S. (2005). Supplier selection and management strategies and manufacturing flexibility. *Journal of Enterprise Information Management, 18,* 330–349.

Osman, H., & Demirli, K. (2012). Integrated safety stock optimization for multiple sourced stock-points facing variable demand and lead time. *International Journal of Production Economics, 135,* 299–307.

Pramod, M., & Garg, S. (2006). Analysis of flexibility requirements under uncertain environments. *Journal of Modelling in Management, 1,* 196–214.

Pujawan, I. N. (2004). Assessing supply chain flexibility: A conceptual framework and case study. *International Journal of Integrated Supply Management, 1,* 79–97.

Pujawan, I. N., & Smart, A. U. (2012). Factors affecting schedule instability in manufacturing companies. *International Journal of Production Research, 50*, 2252–2266.

Ramasesh, R. V. (1991). Procurement under uncertain supply lead times – A dual-sourcing technique could save costs. *Engineering Costs and Production Economics, 21*, 59–68.

Raturi, A. S., & Jack, E. P. (2004). Creating a volume-flexible firm. *Business Horizons, 47*, 69–78.

Sánchez, A. M., & Pérez, M. P. (2005). Supply chain flexibility and firm performance: A conceptual model and empirical study in the automotive industry. *International Journal of Operations & Production Management, 25*, 681–700.

Schmitt, T. G. (1984). Resolving uncertainty in manufacturing systems. *Journal of Operations Management, 4*, 331–345.

Schmitt, A. J., & Snyder, L. V. (2012). Infinite-horizon models for inventory control under yield uncertainty and disruptions. *Computers & Operations Research, 39*, 850–862.

Seebacher, G., & Winkler, H. (2013). A citation analysis of the research on manufacturing and supply chain flexibility. *International Journal of Production Research, 51*, 3415–3427.

Sethi, A. K., & Sethi, S. P. (1990). Flexibility in manufacturing: A survey. *The International Journal of Flexible Manufacturing Systems, 2*, 289–328.

Shin, H., Collier, D. A., & Wilson, D. D. (2000). Supply management orientation and supplier/buyer performance. *Journal of Operations Management, 18*, 317–333.

Slack, N. (1988). Manufacturing systems flexibility – An assessment procedure. *Computer Integrated Manufacturing Systems, 1*, 25–31.

Steele, D. C. (1975). The nervous MRP system: How to do battle. *Production and Inventory Management, 16*, 83–89.

Stevenson, M., & Spring, M. (2007). Flexibility from a supply chain perspective: Definition and review. *International Journal of Operations & Production Management, 27*, 685–713.

Stevenson, M., & Spring, M. (2009). Supply chain flexibility: An inter-firm empirical study. *International Journal of Operations & Production Management, 29*, 946–971.

Tachizawa, E. M. (2007). *Uncertainty, integration and supply flexibility* (Unpublished dissertation). Universitat Pompeu Fabra, Barcelona, Spain.

Tachizawa, E. M., & Gimenez, C. (2010). Supply flexibility strategies in Spanish firms: Results from a survey. *International Journal of Production Economics, 124*, 214–224.

Tachizawa, E. M., & Thomsen, C. G. (2007). Drivers and sources of supply flexibility: An exploratory study. *International Journal of Operations & Production Management, 27*, 1115–1136.

Tang, C. S. (2006). Perspectives in supply chain risk management. *International Journal of Production Economics, 103*, 451–488.

Tang, C., & Tomlin, B. (2008). The power of flexibility for mitigating supply chain risks. *International Journal of Production Economics, 116*, 12–27.

Thomas, D. J., & Tyworth, J. E. (2006). Pooling lead-time risk by order splitting: A critical review. *Transportation Research Part E: Logistics and Transportation Review, 42*, 245–257.

Van Donk, D. P., & van der Vaart, T. (2005). A case of shared resources, uncertainty and supply chain integration in the process industry. *International Journal of Production Economics, 96*, 97–108.

Van Hoek, R. I. (2001). The rediscovery of postponement a literature review and directions for research. *Journal of Operations Management, 19*, 161–184.

Van Kampen, T. J., van Donk, D. P., & van der Zee, D.-J. (2010). Safety stock or safety lead time: Coping with unreliability in demand and supply. *International Journal of Production Research, 48*, 7463–7481.

Vickery, S. N., Calantone, R., & Dröge, C. (1999). Supply chain flexibility: An empirical study. *The Journal of Supply Chain Management, 35*, 16–24.

Vokurka, R. J., & O'Leary-Kelly, S. W. (2000). A review of empirical research on manufacturing flexibility. *Journal of Operations Management, 18*, 485–501.

Vonderembse, M., Uppal, M., Huang, S., & Dismukes, J. (2006). Designing supply chains: Towards theory development. *International Journal of Production Economics, 100*, 223–238.

Wong, C. Y., Boon-Itt, S., & Wong, C. W. Y. (2011). The contingency effects of environmental uncertainty on the relationship between supply chain integration and operational performance. *Journal of Operations Management, 29*, 604–615.

Wu, T., Blackhurst, J., & O'grady, P. (2007). Methodology for supply chain disruption analysis. *International Journal of Production Research, 45*, 1665–1682.

Wu, I. L., & Shen, Y. C. (2006). A model for exploring the impact of purchasing strategies on user requirements determination of e-SRM. *Information & Management, 43*, 411–422.

Yang, B., Burns, N. D., & Backhouse, C. J. (2004). Management of uncertainty through postponement. *International Journal of Production Research, 42*, 1049–1064.

Yi, C. Y., Ngai, E. W. T., & Moon, K.-L. (2011). Supply chain flexibility in an uncertain environment: Exploratory findings from five case studies. *Supply Chain Management: An International Journal, 16*, 271–283.

Zhang, G., Shang, J., & Li, W. (2011). Collaborative production planning of supply chain under price and demand uncertainty. *European Journal of Operational Research, 215*, 590–603.

Zhang, Q., Vonderembse, M. A., & Lim, J.-S. (2006). Spanning flexibility: Supply chain information dissemination drives strategy development and customer satisfaction. *Supply Chain Management: An International Journal, 11*, 390–399.

# 8

# Influence of subcontracting constraints on the performance of manufacturing industries in Nigeria

Victor Chukwunweike Nwokocha* and Iganatius Ani Madu

*Department of Geography, University of Nigeria Nsukka, Nigeria*

In this work, an attempt has been made to show the influence of subcontracting constraints on firm performance in Nigeria. The study in line with the literature identified a number of constraints hindering an effective subcontracting arrangement in the study area. While the constraints were found not to have affected the use of subcontracting in the country, low capital intensity, disclosure of commercial secrets, poor services and interest conflict were found to have restricted subcontracting arrangements in the study area to sharing of equipment and short-term contracts. These constraints however were found not have affected the performance of manufacturing industries in the study area. This paper keeping in mind the findings of this study suggested that manufacturing industries in Nigeria should invest more in machineries and tools so as to increase subcontracting co-operations among industries.

**Keywords:** subcontracting; firms; performance; manufacturing industries; contraints; Nigeria

## 1. Introduction

Industrialisation is one of the packages in the economic development process of any country that could be used to accelerate economic growth. It acts as a catalyst to transform the economic structure of countries, from simple, slow-growing, and low-value activities to more productive activities (manufacturing) that enjoy greater margins driven by technology, and have higher growth prospects (Logan, 2006).

The success of industrialisation or industrial production, however, largely depends on the establishment of useful linkages between industries (Nguyen and Khac, 2013). Industrial linkage is the interrelationship amongst various industrial activities through the input–output relationship or the economic value chain (Mayhew, 2009). Subcontracting which is a type of industrial linkage based on alliance is a work contract that externalise production processes from one firm to another through interfirm relationships (Ajayi, 2007). In developing countries, business linkages with local small and medium enterprises, including procurement, distribution and sales, offer large firms an avenue through which to address some of these concerns. These relationships can allow large firms to reduce input costs while increasing specialisations and flexibility.

In Nigeria, subcontracting; a strategic positioning of industrial activities started in the early 1960s, the post-independence period (Ajayi, 2007). The earliest stage in the adoption of production subcontracting as an industrial production technique in Nigeria

---

*Corresponding author. Email: chukwuweike_nwokocha@yahoo.com

was characterised by insignificant growth and rapid growth thereafter. Subcontracting became very important after the introduction of the Structural Adjustment Programme in 1986, and it is perceived by industrialist as very important in reducing the cost of production (Ajayi, 2007) and most recently following the works of Nwokocha, Madu, Ocheje, and Olerum (2015a), subcontracting is very important in enhancing the operational efficiency in industries, induce specialisation so as to gain professional resources and to strengthen cooperation amongst industries.

Firm performance on the other hand is a relevant construct in strategic management research and frequently used as a dependent variable. Despite its relevance, there is hardly a consensus about its definition, dimensionality, and measurement, what limits advances in research and understanding of the concept (Santos & Brito, 2012). The research into firm performance suffers from problems such as lack of consensus, selection of indicators based on convenience, and little consideration of its dimensionality (Combs, Crook, & Shook, 2005; Crook, Ketchen, Combs, & Todd, 2008; Richard, Devinney, Yip, and Johnson 2009). Many studies measure firm performance with a single indicator and represent this concept as one-dimensional, even while admitting its multidimensionality (Glick, Washburn, & Miller, 2005). If several dimensions exist, a researcher should choose the dimensions most relevant to his or her research and judge the outcome of this choice (Richard et al., 2009). Ray, Barney, and Muhanna (2004) stress this, warning against the difficulties of testing the resource-based theory using aggregated measures of performance and suggesting the use of indicators directly connected to the resources under analysis.

Consequently, this paper proposes to use a multidimensional measurement model of firm performance, structured in the Goal-Setting Theory (Chong, 2008; Locke & Latham, 2002) in order to assess the influence of subcontracting constraints on firm performance in Nigeria. Goal setting is the mechanism by which a firm delivers results against its strategy (e.g. subcontracting) on the extent to which there is clarity, challenge, commitment, feedback, and task complexity (Locke & Latham, 2002). It is explained using a combination of both financial and non-financial indicators. The financial indicators are: sales growth, growth in profits, changes in assets by gross value plant and machinery, return on assets to measure capital efficiency. The non-financial indicators were: growth in market share; product success; increase in number of employees and; labour productivity (Gakure, Kimemia, & Waititu, 2014; Hu, Zheng, & Wang, 2011; Kongmanilaa & Takahashib, 2009; Marimuthu, Arokiasamy, & Ismail, 2009; Ongonga & Abeka, 2011; Tuan and Yoshi, 2010).

Empirical researches on subcontracting have received growing attention in the academia during the past few years. This is evidenced by an increasing number of publications and studies on the topic (Ajayi, 2001, 2002, 2003, 2007; Berry, 1997; Diaz-Mora & Triguero-Cano (2007); Dahane, Clementz, & Rezg, 2008; Dahane, Dellagi, Clementz & Rezg, 2011; Deardorff & Djankov, 2000; Furlan, Grandinetti, & Camuffo, 2007; Grossman & Helpman, 2005; Gubik, 2005; Hajej, Rezg, & Gharbi, 2014; Kongmanila and Takahashi, 2009; Nwokocha et al. 2015a; Nwokocha, Madu, Ocheje, Olerum, & Nwosu, 2015b; Taymaz and Kiliçaslan, 2002; Tuan and Yoshi, 2010; Watanabe, 1971). According to these studies, subcontracting helps small and medium scale enterprises as well as large scale firms to stimulate, reduce, and control operation cost in manufacturing, improve company's focus, and to access world class capabilities, develop joint maintenance and production strategies amongst others.

Similarly, researches on subcontracting constraints have never fallen short in the academia. Some of these researches include the works of (Annim & Machethe, 1998;

Bbenkele, 1998; Dahane et al., 2008; Hajej et al., 2014; Kumar & Subrahmanya, 2007; Okatch, Mukulu, & Oyugi, 2011). While the works of Dahane et al. (2008), generalised subcontracting constraint in computational engineering, Hajej et al. (2014), developed optimal production plans of principal and subcontracting machines which minimises the total production and inventory cost for cases without and with returned products under service level and subcontracting transportation delay. These two studies however were structured to solve subcontracting constraint arising from transportation delay, joint maintenance, and production management. Similarly, while Annim and Machethe (1998), in a study on business linkages in Kwazulu-Natal, point out that the constraint on the expansion and improvement of subcontracting linkages is, according to suppliers: limited application of new technology, poor product quality, unreliable delivery of goods or services, and high products prices, the studies of Bbenkele (1998) and Okatch et al. (2011) posited that the constraints to subcontracting are inability of the SMEs to supply quality products to schedule, lack of local suppliers for certain parts, the proliferation of makes and models, and competition from imported second hand vehicles from Japan and Europe.

Conceding from the above, there is a widespread consensus on the constraints of subcontracting on industrial activities. These empirical findings, however, did not state the influence or contributions of these constraints to the performance of manufacturing industries. Following the findings of Berry (1997), Yasuda (2005), Marsall, Mcivor, and Lamming (2007), and Gakure et al. (2014)), which suggested that subcontracting has a positive impacts on firm's performance and the works of Hu et al. (2011), which observed that there is no significant relationship between subcontracting and firm performance, it will be difficult to make an informed assessment on the influence of subcontracting constraints on firm performance. This work is thus oriented to investigate the influence of subcontracting constraints on the performance of manufacturing industries in Nigeria.

## 1.1.   *Theoretical framework*

The theoretical framework for analysing subcontracting is hinged on three theories. These theories are the dualistic approach, development approach, and networking and clustering approach. While dualistic approach considers subcontracting as an unequal power relationship whereby large contractors realise benefit at the expense of small contractors, development approach formulated by development economists considers subcontracting as a relationship between large and small firms, but emphasises a positive role of it.

The networking and clustering approach, largely regarded as the modern theoretical foundation of subcontracting, supports networking initiatives and the development of industrial cluster (Pyke, 1992; UNCTAD, 1994). In the case of the cluster approach, subcontracting itself is one of the main types of networking on which clusters could be established. This approach looks at a group of firms cooperating (and competing) within a complex web of supportive institutions. Externalities, linkages, and economics of scale generated by this form of cooperation and competition are internalised by the network so that the collective efficiency and flexibility of the industry is enhanced (Ceglie & Dini, 1999).

From the fore going, the approaches to subcontracting have shown that subcontracting is a form of relationship established by firms not only to be more competitive in their production activities, but also to share mutual benefits. These approaches however failed to disclose the impact of subcontracting constraints on the performance of manufacturing industries. Based on this, three key questions are posed in this research in order to answer the above-stated question:

(1) What are the constraints hampering effective subcontracting arrangement?
(2) Have these constraints affected the use of subcontracting arrangement by firms?
(3) Have these constraints affected the performance of firms?

This work tends to explore the lacuna in the approaches to subcontracting in order to strengthen it.

## 1.2. Material and methodology

The purpose of this study was to establish the impact of subcontracting constraints on the performance of firms in Nigeria. The study adopted a number of methods comprising semi-structured interviews, field observations, reference to relevant literature, and questionnaire survey of 120 manufacturing industries as described elsewhere (Nwokocha et al., 2015a).

The appropriate sample size for the population-based survey was determined largely by the estimated prevalence of the variable of interest – subcontracting in this instance (estimated at 15%); the desired level of confidence, 95%; and the acceptable margin of error, 5% (standard value of .05). A pilot test on 20 firms helped to remove ambiguities and improve the instrument as well as test for its reliability and validity. Open- and close-ended questions asking respondents to rate various questionnaire items using a five-Likert scale of; 5 – very important, and 1 – not at all important. Likert-type ordinal scale representing a spectrum of subjective feelings and opinions were used to solicit specific responses from the industrialists.

### 1.2.1. Instrument/questionnaire validation

To establish the degree of reliability of the questions in the questionnaire, 20 manufacturing industries were made to rate each question in the questionnaire on a five-point scale. Principal Component Analysis method was used to extract the factors. The criteria states that: Cronbach's Alpha of a scale should be greater than .70 for items to be used together as a scale while factor loadings greater than .40 are considered statistically significant for studies with sample size less than 200. Therefore, in the present study, ±0.40 was used as the cut-off for loadings since the sample size of the study was less than 200. The higher the factor loadings were, the closer they were related to the variable.

### 1.2.2. Constraints to subcontracting

Constraints to subcontracting were the independent variable in this study. Following our field observations, fourteen (14) variables were identified as the constraints to subcontracting in the industrial sector of Nigeria. The constraints were analysed using a five-Likert scale of; 5 – very important, and 1 – not at all important. The pilot test result can be seen in Table 1.

Following the results in Table 1, the original 14 factors which constituted the constraints to subcontracting were reduced to 10 factors. This was because four factors with

Table 1.   Factor analysis of subcontracting constraints.

| Items | Factor loading |
| --- | --- |
| Proliferation of makes | 0.67 |
| Increased demand for specialised technology | 0.29* |
| Low visibility | 0.56 |
| Poor services | 0.70 |
| Limited application of new technology | 0.83 |
| Competition from imported parts | 0.86 |
| Low capital intensity | 0.81 |
| Lack of local suppliers for certain parts | 0.58 |
| Unfulfilled order from subcontractors | 0.13* |
| Interest conflicts with subcontracting partners | 0.66 |
| Disclosure of commercial Secrets | 0.55 |
| Legal disputes | 0.11* |
| Weaken Culture | −0.02* |
| Decrease compatibility of innovation | 0.75 |
| Number of Items | 14 |
| Cronbach's Alpha | 0.91 |

Notes: Authors Computation, 2015.
NB: The factors marked with (*) were eliminated from further analysis.

loadings less than .40 were discarded as shown in Table 1. Thus, ten (10) factors with factor loadings between .55 and .93 were subsequently considered valid as the constructs to represent the constraints to subcontracting. The Cronbach's Alpha for these 10 factors was .91, which exceeded the reliability cut-off value of .70.

### 1.2.3.   Firm performance

Firm performance was the dependent variable in the present study and in accordance with the literature, a combination of both financial and non-financial indicators led to a balanced performance measurement. The pilot test was conducted using a five-Likert scale of; 5 – significantly increased, and 1 – significantly decreased.

Firm performance measures had a total of 8 items generated from literature comprising both the financial and non-financial indicators. The results indicate that with factor loadings of between .70 and .90, the construct of the 8 measurement items was valid for firm performance (see Table 2).

The Cronbach's Alpha coefficient for the 8 indices was =0.92, which means the instrument had an excellent level of consistency, and fit for use in data collection.

Table 2.   Factor analysis of firm performance.

| | |
| --- | --- |
| Growth in sales (volume) | 0.90 |
| Growth in profit | 0.88 |
| Gross value of capital (machinery) | 0.81 |
| Return on assets | 0.78 |
| Growth in market share | 0.87 |
| Product success | 0.76 |
| Labour productivity | 0.71 |
| Increase in workers | 0.70 |
| No. of indices | 8 |
| Cronbach's Alpha | 0.92 |

Note: Author's Computation, 2015.

### *1.3.  Data analysis*

Ordinal multinomial logit model was used to test for the effect of explanatory variables on those responses that had more than two categories. Similarly, ordinary logit model was used for binary responses. Relevant statistical techniques were used to analyse the data. All analyses were carried out with the aid of Statistical Packages for Social Sciences version 17 (SPSS 17).

## 2.  Results

The result of the study is discussed below. The focus is on the impact of subcontracting constraints on the performance of manufacturing industries in Nigeria.

### *2.1.  Analysis of respondents*

Out of the 120 questionnaires administered, 80 (66.7%) were considered valid, with no missing data. The response rate of 66.7% was, therefore, considered adequate for the study. The targeted enterprises were aged 10 years for the firms and 45 years and above for the managers on the average. About 64.5% of the firms were managed by men. The preliminary findings showed that 80.4% of the respondents had attained secondary level of education while 50.6% had attained product-related skills training through apprentice-ship and learning on the job. From our observations, industrial products manufactured were Plastics and rubber items (ranging from gallons, buckets, paper and plastic bags, etc. 25%), metallic windows and doors including aluminium roofing sheets (28%) food and Beverage products (28%), and foams and chemical products (ranging from detergents, engine oil, bed forms, etc. 19%). About 85.6% of the investigated firms were engaged in subcontracting arrangements.

### *2.2.  Constraints to subcontracting*

Despite the benefits of production subcontracting, there are constraints limiting its effectiveness in the industrial sector. These constraints based on the field observation have been synthesised into ten (10) factors, and how they appeal to the industries were analysed using a five-point Likert-type scale ranging from 1 = 'not at all important' to 5 = 'very important' in relation to how they affect the industrialist (see Table 3).

The analysis shown in table 3 reveals that the majority of the industries believed that engaging in production subcontracting has a number of constraints associated with it. From the analysis, most of the industries identified: disclosure of commercial secrets, low capital intensity, interest conflicts, decrease compatibility of innovation, importation of parts and poor services with mean and standard deviation values of 4.23 (STD = .77), 4.23 (STD = .77), 4.18 (STD = .72), 3.52 (STD = .50), 3.52 (STD = .50), 3.52 (STD = .50), and 4.13 (STD = .74), respectively, as the major constraints affecting the effectiveness of subcontracting in Nigeria. Most of the industrialists interviewed have experienced one of these problems at one time or the other making them either to change their subcontracting partner or acquire the capacity to handle the activity (ies) in house. While 'Low capital intensity' was found to have made most of the firms in the study area to invest very little in machineries, reducing subcontracting strategies predominantly to sharing of equipment and short-term contracts, 'Poor services, Disclosure of commercial secrets, importation of parts and interest conflict' were found

Table 3.  Mean distribution of responses to subcontracting constraints.

| Variables | N statistic | Range statistic | Minimum statistic | Maximum statistic | Mean statistic | Std deviation |
|---|---|---|---|---|---|---|
| Proliferation of makes | | 0.00 | 2.00 | 2.00 | 2.00 | 0.00 |
| Limited application of advance technology | | 0.00 | 2.00 | 2.00 | 2.00 | 0.00 |
| Decrease compatibility of innovation | | 1.00 | 3.00 | 4.00 | 3.52 | 0.50 |
| Disclosure of commercial secrets | | 3.00 | 2.00 | 5.00 | 4.23 | 0.77 |
| Interest conflicts | | 3.00 | 2.00 | 5.00 | 4.18 | 0.72 |
| Low visibility | | 0.00 | 2.00 | 2.00 | 2.00 | 0.00 |
| Poor services | | 2.00 | 1.00 | 3.00 | 4.13` | 0.74 |
| Imported goods | | 1.00 | 3.00 | 4.00 | 3.52 | 0.50 |
| Low capital intensity | | 3.00 | 2.00 | 5.00 | 4.23 | 0.77 |
| Lack of local suppliers for certain parts | | 0.00 | 2.00 | 2.00 | 2.00 | 0.00 |
| Valid number | 80 | | | | | |

Source: Field Work and Author's Computation, 2015.

to have forced firms to frequently change their subcontracting partners or acquire the capacity to handle the activity (ies) in house.

## 2.3.  Firm performance

Firm performance, the dependent variable in this study was measured in both financial and non-financial indicators. The respondents were asked to evaluate their firm's performance by rating various indicators of their business performance in the last 10 years on a scale of 1–5, where 1 represented – significantly decreased and 5 represented – significantly increased. Following the result in Table 4, 7.90% of the respondents had experienced significant increase while 35.31% had seen relative increase in firm performance in the last 10 years. Furthermore, while 18.12% of the manufacturing industries had experienced relative decrease in their business performance, 33.13% had experienced significant decrease. The result also revealed that 33.96% of the SMEs were static and recorded no changes (see Table 4).

The large number of firms that experienced significant decrease with reference to the little number that experience significant increase in performance is an indication that the Nigerian Industrial sector have underperformed. Similarly, the overall mean and standard deviation score of 2.38 and .71 shows that majority of the firms experienced some increase in performance. The respondents indicated that they have achieved highest performance in Profit Growth with an average mean value of 2.84 and standard deviation of .80 and least performance in return to assets with an average mean of 2.04, and standard deviation value of .54 (see Table 4).

Furthermore, in order to show the impact of these constraints on subcontracting activities in the study area, the share of subcontracting for all the industries was computed using a ten years (10 years) time frame. The share of subcontracting according to Taymaz and Kiliçaslan (2002) and Holl (2007) is subcontracted input – the share of a plant's inputs subcontracted to supplier plants, in total inputs and subcontracted output – the share of output subcontracted by plants, in total output. Subcontracted input share

Table 4.  Percentage and mean distribution of responses to firm performance.

| Item | Significantly decreased (1) % | Relatively decreased (2) % | Static (3) % | Relatively increased (4) % | Significantly increased (5) % | Mean | Standard deviation |
|---|---|---|---|---|---|---|---|
| Growth in sales | 2.30 | 28.30 | 10.00 | 40.40 | 19.00 | 2.15 | 0.46 |
| Profit Growth | 3.50 | 29.30 | 16.60 | 50.60 | 0.00 | 2.82 | 0.80 |
| Return on assets | 5.60 | 16.70 | 43.10 | 24.60 | 10.00 | 2.04 | 0.54 |
| Gross value of capital | 5.50 | 19.00 | 34.30 | 21.00 | 20.20 | 2.64 | 0.89 |
| Profit success | 0.60 | 12.60 | 39.40 | 45.10 | 2.30 | 2.15 | 0.36 |
| Growth in market share | 10.00 | 13.40 | 25.20 | 50.30 | 1.10 | 2.44 | 0.90 |
| Increase in workers | 5.00 | 10.90 | 48.10 | 35.40 | 0.60 | 2.16 | 0.96 |
| Labour productivity | 5.00 | 14.90 | 55.00 | 15.10 | 10.00 | 2.64 | |
| Total average | 33.13 | 18.12 | 33.96 | 35.31 | 7.9 | 2.38 | 0.71 |

Note: Field work and Author's Computation, 2015.

Table 5.	Subcontracted input and output share over a period of ten (10) years.

| Years | Subcontracted input share (%) | Subcontracted output share (%) |
|---|---|---|
| 2000–2001 | 6.31 | 6.56 |
| 2001–2002 | 5.22 | 5.30 |
| 2002–2003 | 4.54 | 4.80 |
| 2003–2004 | 4.41 | 4.52 |
| 2004–2005 | 5.38 | 5.40 |
| 2005–2006 | 4.41 | 4.52 |
| 2006–2007 | 4.41 | 4.00 |
| 2007–2008 | 4.50 | 4.87 |
| 2008–2009 | 5.36 | 5.50 |
| 2009–2010 | 5.77 | 5.99 |

Note: Fieldwork Author's Computation, 2015.

can also be referred to as the proportion of subcontracted input in total inputs while sub-contracted output share refers to the proportion of subcontracted output in total output (Morrison & Yasar, 2008). These definitions were based on the survey definition of income from subcontract as income generated from the processing of materials provided by the firm offering the subcontract, see Table 5.

The results in Table 5, Figures 1 and 2 shows that the subcontracted input share and output share of the firms have been relatively stable. This relative stability shows that most firms still make use of this process despite the constraints associated with subcon-tracting as was noted earlier. This lays credence to our earlier finding which noted that about 85.6% of the surveyed firms were engaged in subcontracting arrangements in the study area. This result also reveals that the constraints to subcontracting have no impact or influence on the performance or underperformance of firms. This is because the subcontracted input and output share shown in Table 5 have remained relatively stable and therefore cannot account for the performance or underperformance of firms in the study area.

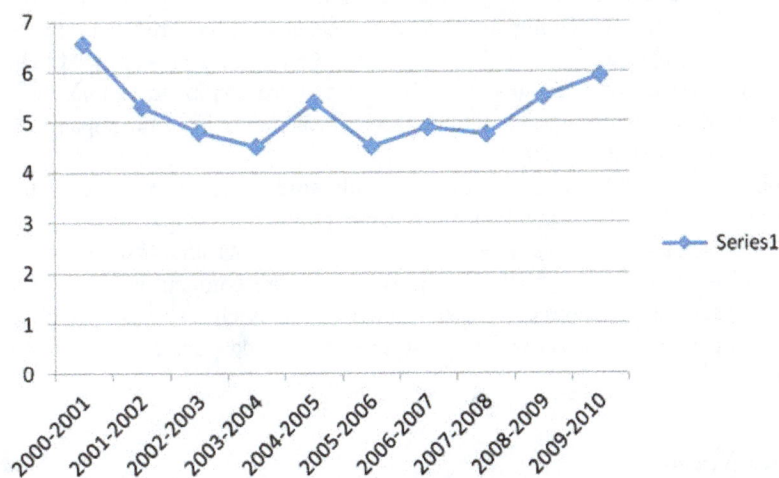

Figure 1.	Subcontrated input share (time series analysis).

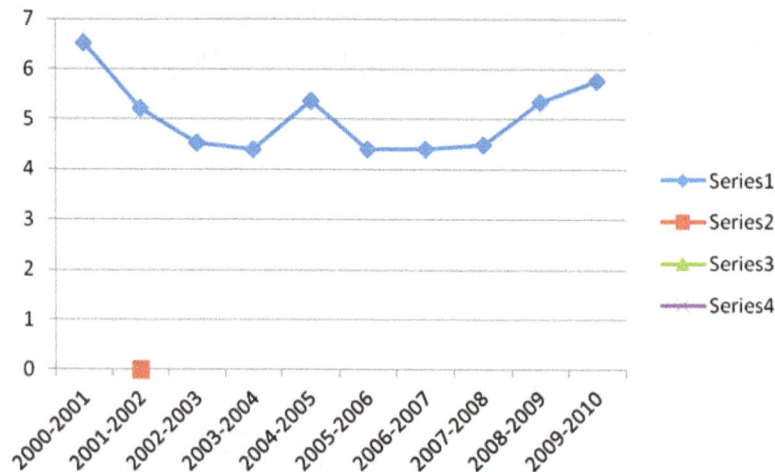

Figure 2.   Subcontrated output share (time series analysis).

## 2.4.  Summary of finding and conclusion

This paper in line with the literature identified a number of constraints to subcontracting arrangement in the study area. These constraints ranges are disclosure of commercial secrets, poor quality of service, interest conflicts, decrease compatibility of innovation, importation of parts, and unfulfilled orders from subcontractors. While 'Low capital intensity' was found to have made most of the firms in the study area to invest very little in machineries and tools thereby making their subcontracting strategies in the study area to sharing of equipment and short-term contracts, 'Poor services, Disclosure of commercial secrets, importation of parts and interest conflict' was found to have forced firms to frequently change their subcontracting partners or acquire the capacity to handle the activity (ies) in house. Similarly, the analysis of firm performance also showed that a majority of the firms have experienced some increase in performance. The firms indicated that they have achieved highest performance in Profit Growth and least performance in return to assets. Going forward, the paper also discovered that subcontracted input and output share of the firms have been relatively stable. This showed that despite the constraints associated with subcontracting, firms have continued to make use of this production process, signifying that subcontracting constraints have no impact on the use of subcontracting strategy by the firms.

Keeping in view the findings of the study and conclusion drawn, this paper recommends that manufacturing industries in Nigeria should invest more in machineries and tools so as to increase subcontracting co-operations amongst industries in the study area. This will not only help the firms to be more competitive, it will also help them to engage in subcontracting more intensely as well as building up superior capabilities in order to bypass some of the aforementioned constraints associated with subcontracting.

### Disclosure statement

No potential conflict of interest was reported by the authors.

# References

Ajayi, D. D. (2001). Industrial subcontracting linkages in the Lagos Region, Nigeria. *The Nigerian Journal of Economic and Social Studies (NJESS), 43*, 265–277.

Ajayi, D. D. (2002). Temporal pattern of production subcontracting in Nigeria. *Annals of the Social Science Academy of Nigeria, 14 & 15*, 67–81.

Ajayi, D. D. (2003). Nature and scope of production subcontracting in Nigeria. *Africa Development, XXVIII*, 89–111.

Ajayi, D. D. (2007). Recent trend and patterns in Nigeria's industrial development. *Africa Development, 32*, 139–155.

Annim, F. D. L., & Machethe, C. (1998). *Promoting the growth of MSEs through business linkages in the Northern Province of Kwazulu-Natal.* South African Province: Research Report: United States Agency for International Development Bureau for Africa.

Bbenkele, E. C. (1998). Enhancing economic development by fostering business linkages between the pharmaceutical companies and the traditional medicines sector. Research report. Center for Partnerships in Enterprise Research and Technology Transfer. University of Natal, Pietermaritzburg South Africa.

Berry, A. (1997). *SME competitiveness: The power of networking and subcontracting* in R. W. Gakure, P. N. Kimemia, & G. A. Waititu. Influence of Subcontract Offering on the Performance of Manufacturing Micro and Small Enterprises in Kenya. *Journal of Humanities and Social Science (IOSR-JHSS), 19*, Ver. II, 37–46.

Ceglie, G., & Dini, M. (1999). *SME cluster and networking development in developing countries: The experience of UNIDO* (Working Paper No. 2). Vienna: Private Sector Development Branch, UNIDO.

Chong, H. (2008). Measuring performance of small-and medium sized enterprises: The grounded theory approach. *Journal of Business and Public Affairs, 2*(1), 1–10.

Combs, J. G., Crook, T. R., & Shook, C. L. (2005). The dimension of organizational performance and its implications for strategic management research. In D. J. Ketchen & D. D. Bergh (Eds.), *Research methodology in strategy and management* (pp. 259–286). San Diego, CA: Elsevier.

Crook, T. R. Ketchen, D. J., Combs, J. G., & Todd, S. Y. (2008) Strategic resources and performance: A meta-analysis. *Strategic Management Journal, 29*, 1141–1154.

Dahane, M., Clementz, C., & Rezg, N. (2008). Analysis of joint maintenance and production policies under a subcontracting constraint. *International Journal of Production Research, 46*, 5393–5416.

Dahane, M., Dellagi, S., Clementz, C., & Rezg, N. (2011). Development of joint maintenance and production strategies in a subcontracting environment. *International Journal of Production Research, 49*, 6937–6961.

Deardorff, A., & Djankov, S. (2000). Knowledge transfer under subcontracting: Evidence from Czech firms. *World Development, 28*, 1837–1847.

Diaz-Mora, C., & Triguero-Cano, A. (2007). *Why do some firm contract out Production? Evidence from the firm-level panel data.* Albacete: Department of International Economics, Faculty of Juridical and Social Science, University of Castilla-La Mancha Albacete Spain.

Furlan, A., Grandinetti, R., & Camuffo, A. (2007). How do subcontractors evolve? *International Journal of Operations & Production Management, 27*, 69–89.

Gakure, R. W., Kimemia, P. N., & Waititu, G. A. (2014). Influence of Subcontract Offering on the Performance of Manufacturing Micro and Small Enterprises in Kenya. *IOSR Journal of Humanities and Social Science (IOSR-JHSS), 19*, Ver. II, 37–46.

Glick, W. H., Washburn, N. T., & Miller, C. C. (2005). *The myth of firm performance.* Proceedings of the Annual Meeting of American Academy of Management, Honolulu, Hawaii, USA, in J. B. Santos & L. A. L. Brito (Eds.), *Towards a subjective measurement model for firm performance* (pp. 95–177). Brazilian Administrative Review.

Grossman, G. M., & Helpman E. (2005). Outsourcing in a global economy. *Review of Economic Studies, 72*, 135–159.

Gubik, A. (2005). New opportunities for SMEs founded by cooperation. *European Integration Studies, Miskolc, 4*, 25–36.

Hajej, Z., Rezg, N., & Gharbi, A. (2014). Forecasting and maintenance problem under subcontracting constraint with transportation delay. *International Journal of Production Research, 52*, 6695–6716

Holl, A. (2007). *Production subcontracting and location.* Madrid: FEDEA Foundation for Applied Economics Studies.

Hu, Z., Zheng, J., & Wang, J. (2011). Impact of industrial linkages on firm performance in Chinese development zones, Yangtze River Delta, Jiangsu Province, China. *The Chinese Economy, 44,* 78–105.

Khac, N. N. (2013). *Demand creation and competition effect of export-platform FDI on backward linkages; evidence from panel data analysis of Vietnames supporting industries.* Paper provided by Centre d'Études des Politiques Économiques (EPEE), Université d'Evry Val d'Essonne in its series Documents de recherche with number 13-02, 1–42.

Kongmanilaa, X., & Takahashib, Y. (2009). Determinants of subcontracting and firm performance in Lao PDR: Evidence from a Garment industry cluster. *Asia Pacific Management Review, 15,* 97–112

Kumar, R. S., & Subrahmanya, B. (2007). Subcontracting relationships of Indian SMEs with global TNCs: Do SMEs gain, How. *Journal of Asian Economics, 5,* 2–35

Locke, E. A., & Latham, G. P. (2002). Building a practically useful theory of goal setting and task motivation: A 35-year odyssey, American Psychologist. *American Psychological Association, 57*(9), 705–717.

Logan, M. I. (2006). The shift from informal to formal: Some cost and benefits of large scale industrialization in Asian Countries. *Singapore Journal of Tropical Geography, 3*(2), 170–176

Marimuthu, M., Arokiasamy, L., & Ismail, M. (2009). Human capital development and its impact on firm performance: Evidence from developmental economics. *The Journal of International Social Research, 2/8*(Summer 2009), 265–272.

Marsall, D., Mcivor, R., & Lamming, R. (2007). Influences and outcomes of outsourcing: Insights from the telecommunications industry. *Journal of Purchasing & Supply Management, 13,* 245–260.

Mayhew, S. (2009). *A dictionary of Geography.*(4th ed.), Oxford: Oxford University Press.

Morrison, C. J., & Yasar, M. (2008). *Outsourcing, productivity, and input composition at the plant level. Canadian Journal of Economics, 42*(2), 422–439.

Nwokocha, V. C., Madu, I. A., Ocheje, J. F., & Olerum, V. N. (2015a). Production subcontracting: A strategy for the survival of small and medium scale industries in Nigeria. *Mediterranean Journal of Social Sciences, 6,* 641–651.

Nwokocha, V. C., Madu, I. A., Ocheje, J. F., Olerum, V. N., & Nwosu, I. G. (2015b). Production subcontracting: A policy issue for small and medium scale manufacturing industries in Nigeria. *Academic Journal of Interdisciplinary Studies, 4,* 375–385.

Okatch, B. A., Mukulu, E., & Oyugi, L. (2011). Constraints to subcontracting arrangements between SMEs and large firms in the motor vehicle industry in Kenya. *International Journal of Business and Social Science, 2*(15), 208–223.

Ongonga J. O., & Abeka E. O. (2011): Networking in the Kenyan informal sector: An attempt to manage the market failures. *African Journal of Business Management, 5,* 11323–11334.

Pyke, F. (1992). *Industrial development through small-firm co-operation.* Geneva: ILO.

Ray, G., Barney, J. B., & Muhanna, W. A. (2004). Capabilities, business processes, and competitive advantage: Choosing the dependent variable in empirical tests of the resource-based view. *Strategic Management Journal, 25,* 23–37.

Richard, P. J., Devinney, T. M., Yip, G. S., & Johnson, G. (2009). Measuring organizational performance: towards methodological best practice. *Journal of Management, 35*(3), 718–804.

Santos, J. B., & Brito, L. A. L. (2012). Towards a subjective measurement model for firm performance [Special issue]. *Brazilian Administrative Review, 9,* 95–117.

Taymaz, E., & Kiliçaslan, Y. (2004). *Determinants of subcontracting and regional development: An empirical study on Turkish textile and engineering industries.* Ankara: Department of Economics Middle East Technical University

Tuan, N. P. & Yoshi, T. (2010). Vertical Linkage and Firm's Performance in Supporting industries in Vietnam. *Asian Journal of Management Research, 1*(1), 1–14.

UNCTAD. (1994). *Technological dynamism in industrial districts: An alternative approach to industrialization in developing countries.* New York, NY: United Nations.

Watanabe, S. (1971). Subcontracting, industrialisation and employment creation. *International Labour Review, 104,* 51–76.

Yasuda, T. (2005). Firm growth, size, age and behaviour in Japanese manufacturing. *Small Business Economics, 24,* 1–15.

# 9

# Production scheduling for continuous manufacturing systems with quality constraints

Jalel Ben Hmida[a]*, Jim Lee[a], Xinchun Wang[a] and Fathi Boukadi[b]

[a]*Systems Engineering, University of Louisiana at Lafayette, P.O. Box 44170, Rougeou, Hall Room # 244, Lafayette, LA 70504, USA;* [b]*Department of Petroleum Engineering, University of Louisiana at Lafayette, P.O. Box 44408, Madison Hall, Room # 134, Lafayette, LA 70504, USA*

This research is motivated by a real world production scheduling problem in a continuous manufacturing system involving multiple objectives, multiple products and multiple processing lines with various inventory, production and quality constraints. Because of the conflicting objectives, a global optimization approach is considered as not feasible by the plant management. Given a customer demand forecast, two practical heuristic or sequential optimization algorithms are developed to generate daily production schedules for two primary objectives: minimize shipment delays (pull-backward procedure) and minimize average inventory levels (push-forward procedure). A third heuristic algorithm (reduce switch-over procedure) which is based on the current management practice is also developed to serve as a benchmark. A factorial experiment was performed to evaluate the performance of the heuristic procedures and to identify factors that might affect the performance differences among heuristics. Since each heuristic is designed to give priority to one of the three conflicting objectives, none of them is absolutely superior to the other algorithms in all aspects. However, the first two heuristic procedures performed better than the current management practice in shipment delays and average daily inventory. The production schedules generated by the two procedures also satisfy the quality constraints. The experimental results also showed that the performance of the algorithms is significantly affected by product mix, inventory levels, and demand pattern.

**Keywords:** production scheduling; continuous manufacturing system; product quality; multiple objectives; heuristic algorithms

## 1. Introduction

Scheduling is concerned with efforts to see that the resources of a manufacturing system are well utilized so that the products are produced within reasonable accord with the customer demand. Numerous planning and scheduling methods have been developed for discrete part manufacturing systems, and published research has shown that the approach for scheduling production depends on the type of manufacturing system involved. Reviews of the characteristics of continuous manufacturing systems in Vollmann, Berry, and Whybark (1992), Askin and Standridge (1993) suggest that the continuous manufacturing systems require different modeling approaches.

---

*Corresponding author. Email: jalel.benhmida@louisiana.edu

This research is motivated by an actual production scheduling problem in a continuous manufacturing system of a large producer of chemical products. On average, the chemical plant receives more than 200 customer orders in a month, totaling about four million pounds of chemical products. The production system is capable of producing a large number of different types of products through multiple processing lines. A large number of process constraints exist due to varied capabilities of the production lines and the processing requirements. Most of the products can be produced on more than one line and some of the processes require the sharing of special tools. Some products have precedence constraints, and some products have similar production conditions that should be scheduled for production consecutively.

Currently, scheduling daily production in the system is subjectively based on the management's experience. With an increasing emphasis on the multiple objectives of on-time shipment, low inventory, and production quality; the management of the plant needs a scheduling tool to improve the production scheduling for better system performance. The three objectives identified by the plant management are described as below:

(1) *On-time shipments*. The current overdue shipment rate (OSR) is unacceptable by the top management. The lack of on-time delivery may prevent the plant from being considered as a major supply candidate by the downstream industries. A measure of on-time shipment is commonly considered as the most important performance indicator of a continuous manufacturing system.
(2) *Low inventory*. To reduce overdue shipments and if processing lines are available, finished products are normally produced before the due dates. This results in high level of inventory. While "zero inventory" may not be practically feasible, efforts must be made to reduce the overdue shipments and the inventory level at the same time.
(3) *Product quality*. Production run length is defined as the number of days a processing line is scheduled to produce the same product type. Longer runs can produce finished products with consistent quality. Frequent product switchovers in the processing line can result in quality problems. However, a run length larger than necessary can increase the inventory level.

The management of continuous manufacturing systems of this type faces the dilemma of meeting customer delivery dates while operating the system efficiently. This involves conflicting objectives. The conflict arises because improvement in one objective can be made to the detriment of one or more of the other objectives. Moreover, various production and quality constraints need to be satisfied. All of these concerns are addressed in this research. Currently the approach used by the plant management gives priority to product quality where the product switchover is minimized. A scheduling tool, which can generate several daily production schedules based on algorithms with priorities given to different objectives, would be helpful to the plant management.

## 2. Literature review

Production scheduling, perhaps because of its considerable quantifiable structure and richness of virtually endless variations, continues to attract many researchers for scholarly work. Patterson, Slowinski, Talbit, and Weglarz (1989) stated that most of the realistic scheduling problems could not be solved in a computationally efficient manner. Linear programming (LP) is commonly used to solve production planning and scheduling problems optimally, but the problem formulation is normally over-simplified and the solution is used only at the strategic level. Heuristic rules have been developed to deal with practical scheduling problems on the shop floor.

Deepark and Grossmann (1990) proposed a multi-period linear programming model for the simultaneous production planning and scheduling of multi-product batch manufacturing systems that may consist of several non-identical parallel lines. The model involves a simplistic representation of the scheduling problem where many possible constraints are not considered. Gupta, Bector, and Gupta (1990) provided an approach to determine the optimal sequence of products with due date in a single machine system. Linear programming models were developed for three different due date assignment methods. Minton, Johnston, Phillips, and Laird (1992) described a simple heuristic approach to solve large-scale constraint satisfaction and scheduling problems. Anwar and Nagi (1997) formulated an integrated scheduling and lot-sizing problem. An efficient heuristic is developed that schedules operations by exploiting the critical path of a network and iteratively groups in order to determine the lot-size that minimizes the make span as well as set-up and holding costs. Certain aspects of scheduling problems such as operational pre-conditions (e.g. product sequencing and cleaning requirements) are important to avoid compromising final product quality, safety and feasibility of operations. (Amaro & Barbosa-Póvoa, 1999) developed a Mixed Integer Linear Programming (MILP) formulation so as to account for operational conditions. The formulation is solved using a Branch and Bound (B&B) procedure. In some production systems, early shipments are forbidden. Scudder, Hoffmann, and Rohleder (1993) explored various policies on shipments using a simulation model and also came up with best policies.

Many scheduling problems are multi-objective that can be efficiently solved by a heuristic approach. Daniels (1994) described a multi-objective scheduling situation where information concerning the relative importance of the criteria is available through interaction with a given decision maker. Yim and Lee (1996) used Petri nets and heuristic search for multiple objectives scheduling in a flexible manufacturing systems (FMSs). Tung, Lin, and Nagi (1999) presented a hierarchical approach to scheduling in a FMS that pursues multiple performance objectives. Viana and de Sousa (2000) used heuristics to solve multi-objective resource constrained project scheduling problem. Lou, Liu, Zhou, Wang, and Sun (2012) present a method for developing a pro-active scheduling model that anticipates known uncertainties and a reactive model that accounts for and handles unknown uncertainties.

One approach to avoid frequent product switchover in production is to assign a long setup time between product switchovers. He and Kusiak (1992) considered a mixed-integer formulation of the single-machine scheduling program with sequence-dependent setup times and precedence constraints. Seki and Kogure (1996) developed a stochastic-demand model to deal with lot scheduling problem for multi-item continuous demand utilizing a single machine with setup times. A recent study by Wagner and Smits (2004) used local search heuristic for stochastic economic lot scheduling problem that considered multiple product in a single facility with limited capacity and significant setup times and costs.

Recent research in scheduling has also incorporated various management concepts. Modarress, Ansari, and Willis (2000) presented a technique using Just-In-Time (JIT) and Statistical Process Control (SPC) methods to minimize the short run production in an environment with demand uncertainties. Bose and Pekny (2000) presented Model Predictive Control (MPC) concept for planning and scheduling problem. A forecasting model was developed to calculate target inventory. The importance of Supply Chain (SC) production planning was studied in Caramanis, Anli, and Paschalidis (2003). Their model produces weekly production schedule that minimizes inventory and backlog costs subjected to non-linear constraints on production imposed by weekly varying dynamic lead times and inventory hedging policies.

Giannelos and Georgiadis (2003) investigated scheduling in a continuous manufacturing process. They developed a model for scheduling fast consumer good manufacturing process using MILP techniques combined with continuous representation of time so as to alleviate some of the computational problems incurred by discrete time-index formulations. Kogan, Leu, and Perkins (2002) suggested polynomial-time algorithm for the continuous-time scheduling problem in a manufacturing system where multiple product types are produced on a set of parallel machines. The main objective of the study was to schedule the production so that inventory, production, and planning horizon costs are minimized.

Review of scheduling literature indicates that most of the realistic scheduling problems are solved by efficient heuristic methods. These methods are developed primarily for discrete manufacturing systems and do not consider the effects of scheduling on product quality. In this paper, mathematical models which formulate the problem in continuous manufacturing systems with multiple objectives are presented first. Several heuristic algorithms are then proposed to solve the scheduling problem with priority given to different objectives. The management can determine an acceptable production schedule based on the proposed decision support procedures. The performance of the heuristic procedures is evaluated based on a simulation experiment.

## 3.  Problem description

The scheduling problem addressed in this research can be stated as follows: $A$ continuous manufacturing system can produce $M$ types of products. There are $N$ processing lines, each being capable of producing a subset of $M$ products. If finished products are produced earlier than the due date, they are stocked in the warehouse where each product has a limited storage area. The production rate varies by products on a line, and only one product can be produced by one line at any time.

Table 1 shows a sample of the data of maximum inventory level allowed and production rate of each processing line by product. For the purpose of modeling, the

Table 1.  Sample data for inventory levels and processing line capabilities.

| Product | Maximum inventory ($\times 1000$ lb) | Production rate by line ($\times 1000$ lb/day) | | | | |
|---|---|---|---|---|---|---|
| | | $A$ | $B$ | $C$ | $D$ | $E$ |
| 1SB | 150 | 25 | 25 | 0 | 0 | 0 |
| 1SC | 262 | 25 | 25 | 0 | 0 | 0 |
| 2SB | 80 | 24 | 0 | 24 | 0 | 0 |
| 2SC | 126 | 24 | 0 | 24 | 0 | 0 |
| 3SB | 150 | 23 | 0 | 0 | 23 | 0 |
| 3SC | 189 | 23 | 0 | 0 | 23 | 0 |
| 4B | 150 | 19 | 0 | 0 | 0 | 19 |
| 4C | 112 | 19 | 0 | 0 | 0 | 19 |
| 5 | 180 | 34 | 0 | 0 | 0 | 0 |
| 6B | 120 | 0 | 29 | 0 | 0 | 0 |
| 6C | 180 | 0 | 29 | 0 | 0 | 0 |

setup time is included in our estimation of production rate by processing line. For example, if line $A$ is used to produce product 1SB and the minimum run length is 4 days. The total production in this run of 4 days is estimated to be 100 (thousand pounds), and the daily production is considered to be $100/4 = 25$ (thousand pounds). The real production time of the last day in a run is normally shorter than the other days, as the line must be cleaned and adjusted (i.e. setup time) for another run with a different product. As a result, the total estimation is normally slight lower than the actual production to cover the loss of setup time. When the run length is long, it is possible that the production run can be completed 20–30 min early. The effect of the difference does not significantly affect the estimation on the production rate.

For example, if line $A$ is used to process product 1SB for one day, then 25 thousand pounds of 1SB can be produced. While most of the products can be produced on more than one line, some of the products can be produced only on one line (e.g. product 5 can be produced on Line $A$ only). A sample of customer orders by days is shown in Table 2. Based on the customer orders, the scheduling problem is to determine the daily production schedule for each processing line with production, inventory and quality constraints.

To formulate the problem we require the following notation:

$T$          planning horizon

$t$          day in planning horizon $(t = 1, \dots, T)$

$M$       number of product types

N        number of processing lines

$i,j$       product index $(i, j = 1, \dots, M)$

$k$         processing line index $(k = 1, \dots, N)$

$L_i$       waiting period for product $i$ in days. Waiting period is the number of days that a product needs to stay in warehouse before shipment for quality assurance

$U_i$      actual due date of product $i$

$E_i$      adjusted due date of product $i$ based on waiting period

$HL_i$    minimum run length (in days) of product $i$

$IH_i$     maximum inventory allowed for product $i$

$I_{it}$       inventory of product $i$ in day $t$

$D_{it}$      demand of product $i$ in day $t$

$Q_{ikt}$     production quantity of product $i$ on line $k$ in day $t$

$QO_i$    quantity of overdue product $i$

$W_{kt}$     times of product switchovers on line $k$ in day $t$

$C_{ik}$      daily production rate for product $i$ on line $k$

$P$         set of jobs $[i, j]$, where product $i$ precedes product $j$

$G_g$      set of product group which should be produced sequentially

$g$         represents the type of groups

$R_r$      set of products that require the same special tool (resource) $r$ for processing

$r$         represents the type of resource

$RA_r$    number of units of resources available for products in set $R_r$

$H_{ik}$     production run length (in days) of product $i$ on line $k$

$B_{ik}$     production starting day of product $i$ on line $k$

$O_{it}$      demand of product $i$ that cannot be produced on time in day $t$

$X_{ikt}$     a zero-one variable. $X_{ikt}$ is 1 if product $i$ is scheduled on line $k$ in day $t$

Table 2.   Sample monthly customer order (in thousand units).

| Due day | 1SB | 1SC | 2SB | 2SC | 3SB | 3SC | 4B | 4C | 5 | 6B | 6C | 7 | 8B |
|---------|-----|-----|-----|-----|-----|-----|----|----|---|----|----|---|----|
| 1  | 0  | 15 | 20 | 0  | 0 | 0  | 0 | 0  | 35 | 0  | 0  | 0 | 0  |
| 2  | 0  | 0  | 0  | 0  | 0 | 0  | 0 | 10 | 0  | 0  | 35 | 0 | 0  |
| 3  | 20 | 10 | 0  | 20 | 0 | 30 | 0 | 0  | 0  | 20 | 0  | 0 | 0  |
| 4  | 0  | 30 | 0  | 0  | 0 | 0  | 0 | 0  | 0  | 0  | 0  | 0 | 0  |
| 5  | 5  | 0  | 0  | 0  | 0 | 0  | 0 | 0  | 0  | 0  | 0  | 0 | 0  |
| 6  | 10 | 0  | 15 | 0  | 0 | 0  | 0 | 0  | 0  | 0  | 0  | 0 | 0  |
| 7  | 0  | 0  | 0  | 0  | 0 | 0  | 0 | 0  | 5  | 0  | 0  | 0 | 0  |
| 8  | 0  | 30 | 0  | 0  | 0 | 0  | 0 | 0  | 0  | 0  | 0  | 0 | 0  |
| 9  | 20 | 0  | 0  | 0  | 0 | 0  | 0 | 0  | 0  | 10 | 35 | 0 | 40 |
| 10 | 0  | 0  | 30 | 0  | 0 | 0  | 0 | 0  | 0  | 0  | 0  | 0 | 0  |

The three objectives to be considered in the production scheduling are: minimizing overdue shipment, minimizing inventory, and minimizing product switchovers. They are discussed below:

(1) OSR measures the amount of customer demands that cannot be met by a production schedule. In this research, it is expressed as ratio of overdue shipment amount to the total demands in the planning horizon. The overdue shipment of production day $t$ ($O_{it}$) is demand ($D_{it}$) minus the inventory $I_{i,t1}$ and minus the production of product $i$ on all lines ($Q_{ikt}$).

$$OSR = \sum_i QO_i / \sum_t \sum_i D_{it} \qquad (1)$$

where $QO_i = \sum_t O_{it}$ and $O_{it} = D_{it} - I_{i,t-1} - \sum_k Q_{ikt}$

If $O_{it} < 0$, let $O_{it} = 0$

(2) Average daily inventory (ADI) reflects the average daily amount of finished products produced before the due date. The inventory of product $i$ on day $t$ ($I_{it}$) is the initial inventory plus production on day $t$ and minus the demand on day $t$. The ADI is determined by the total inventory divided by the planning horizon $T$. It is shown as follow:

$$ADI = \sum_t \sum_i I_{it} / T \qquad (2)$$

where $I_{it'} = I_{io} + \sum_{t=1}^{t} \sum_k Q_{ikt} - \sum_{t=1}^{t} D_{it}$

If $I_{it'} < 0$, let $I_{it'} = 0$

(3) Total Switch Times (TST) of the product is used to check the quality of a product as frequent product switchovers could lead to poor product quality. For everyday $t$ and every line $k$, product switchover count $W_{kt}$ is 1 if the product scheduled on day $t$ is different from $t-1$. The TST can be defined as:

$$\text{TST} = \sum_k \sum_t W_{kt} \tag{3}$$

where $W_{kt} = 1$ if the product scheduled on day $t$ is different from $t - 1$.

Several types of production, inventory and quality constraints are considered in scheduling the daily produbelow.

(1)    *Production capacity and capability constraints.* The quantity of product $i$ that can be produced on line $k$ in day $t$, $Q_{ikt}$ should be less than the capacity $C_{ik}$ of line $k$. The capacity of line $k$ is a function of product type $i$. For every $i$ and every $k$,

$$Q_{ikt} \leq C_{ik} \tag{4}$$

where $C_{ik}$ is 0 if line $k$ is not capable of producing product $i$.

(2)    *Inventory constraints.* Each product is allocated limited storage space in the warehouse. A maximum inventory level $\text{IH}_i$ is imposed on each product $i$.

$$I_{io} + \sum_t \sum_k Q_{ikt} - \sum_t D_{it} \leq \text{IH}_i \tag{5}$$

(3)    *Minimum run length constraints.* To achieve the desired quality level, some products have minimum run length requirements as defined by $HL_i$. For every $i$, $H_{ik}$ represents the actual run length in days of product $i$ on line $k$:

$$H_{ik} \geq HL_i \tag{6}$$

(4)    *Precedence constraints.* If product $i$ must be produced before product $j$, then the production starting date of product $i$ on line $k$ ($B_{ik}$) should be completed before the line starts to process product $j$.

$$B_{ik} + H_{ik} \leq B_{jk} \tag{7}$$

In our example, product 2SB must be produced prior to product 4B and 4C.

(5)    *Group constraints.* Products with similar production conditions should be produced sequentially to ensure product quality. If product $i$ and $j$ are in the same group, it is desired to produce these product back to back on line $k$.

$$B_{ik} + H_{ik} = B_{jk}, \text{ or } B_{jk} + H_{jk} = B_{ik} \tag{8}$$

For example, Product 17B, 18B, 20 require similar production conditions. Then they should be produced sequentially in the production schedule.

(6)    *Resource sharing constraints.* The processing operations of some products require the use of special equipment with capacity constraints. Let $RA_r$ represent the number of tools available and $R_r$ is a set of products required in the production. At any time, no more than $RA_r$ products in $R_r$ can be produced.

$$\sum_{i \in R_r} \sum_k X_{ikt} \leq RA_r \tag{9}$$

where $X_{ikt} = 1$ if $i$ is scheduled on $k$.

For example, no more than two S type products can be produced at one time on all lines as the processing of a S type product requires the use of a special tool 1, and $RA_1 = 2$.

(7)    *Preferable product-line assignment.* Since most of the products can be produced on more than one processing line, priorities on the product-line assignment can be established by the management. The assignment may affect the production efficiency and

the quality of the product in some systems. For example, products 2SB and 3SB are preferably produced on line B though they can also be produced on line A.

The number of constraints for this multi-line, multi-product facility is more than one hundred, and the very challenging task is how to satisfy these constraints in the scheduling process while considering the conflicting objectives. This research proposes the development of practical heuristic scheduling algorithms to generate various production schedules based on the demand data and the constraints. With the scheduling tool, the management will be able to generate several feasible production schedules with estimated overdue shipment, daily inventory levels and the number of product switchovers. This information will allow the management to determine the actual daily production schedule that balances the three objectives.

## 4. Heuristic scheduling algorithms

The research problem is to construct production schedules that minimize (1) OSR, (2) ADI and (3) TST while satisfying the various production, inventory and quality constraints. With three conflicting objective functions, it is infeasible to develop a global optimization model which optimizes all three functions. Therefore, three heuristic or sequential optimization algorithms are developed in this research to optimize each of the three objectives. The Pull-Backward Heuristic Procedure (PB) first produces an initial production schedule to minimize the OSR. The initial schedule is then adjusted to reduce ADI without affecting the OSR. This procedure is basically a sequential optimization method which optimizes the OSR and ADI in two steps. The Push-Forward Heuristic Procedure (PF) first produces an initial schedule which minimizes the ADI. The initial schedule is then adjusted to reduce OSR without affecting the ADI. Similar to the PB schedule, the PF is a sequential optimization method which optimizes the ADI first then OSR in two steps. Finally, the Reduce Switchover Heuristic Procedure (RS) is used to produce an optimal solution on TST first. Then the OSR is considered next, followed by ADI.

The methodology underlying the three procedures is similar and follows the three basic steps: problem decomposition, product sequencing, and demand shifting. First, the planning horizon is broken into several smaller periods. If the planning horizon is one month, the length of a period can be about one week. All orders of the same product in a period are lumped together to form one product demand. The production for each of the periods is scheduled according to the lumped demand. After all the period schedules are obtained, they are merged together to form the schedule for the whole planning horizon.

In creating a period schedule, two sequencing rules are used to create priorities. The Earliest Due Date (EDD) rule is chosen as the primary rule, where the choice is made based on the product that has the earliest due date. The Shortest Processing Time (SPT) rule, where the choice is made based on the processing operation with the shortest duration, is used as a secondary rule. For example, if 1SB, 1SC and 2SB are produced in line A (see Table 2) in the first 7-day period, 1SB will be produced last as it has an order of 20 due on day 3 compared to the due date of 1 for 1SC (with order quantity 15) and 2SB (with order quantity 20). Since 1SC and 2SB have the same due date, they are sequenced based on SPT rule. The processing priority is given to 2SB because the demand of 35 thousand pounds of 2SB can be produced with a shorter processing time than the time required to produce 55 thousand pounds of 1SC.

When the production cannot be scheduled for on-time delivery due to capacity constraints, the demand can be shifted to another processing line according to line capability. If other lines are also occupied, the demand can be shifted to the next period. To satisfy precedence constraints, group constraints, and resource constraints; the product demand can be shifted to other dates based on the three objectives of the scheduling. Different approaches of shifting are employed for different heuristic procedures. However, the demand of a product is always shifted forward to satisfy the maximum inventory constraints. The procedures are summarized below.

### 4.1. Pull-Backward Heuristic Procedure (PB)

The PB procedure is basically a sequential optimization method which optimizes the OSR and ADI in two steps. If the demand is greater than the production capacity in a given period, then not all production runs can be scheduled before the due dates. In this case, an attempt is made to schedule the production in the previous period to avoid shipment delay. However, the ADI may go up.

Step 0. Initialization

Adjust the due date of customer order $U$ according to waiting period $L$ to form an adjusted demand date $E$ in planning horizon. Given a product $i$, let $E_i = U_i - L_i$ for all products with a non-zero waiting period. Break planning horizon $T$ into PN scheduling periods; let $v$ be the length of the period and $n$ be the $n$th period ($n = 1, \ldots, $ PN) in the planning horizon, Then PN $= \frac{T}{v}$. If PN is non-integer, it is rounded down to the nearest integer.

Step 1. Product-processing line assignment

Assign products to processing lines based on preferable product-line assignment, subject to capability constraints. One product is assigned to one line.

Step 2. Period schedule generation

Construct a trial production schedule based on demand for period $n$ subject to production capacity constraints and minimum production run length.

(2.1.) Compute the total demand due in period $n$ by product and by line. Let $DD_{ikn}$ represents this demand for product $i$ assigned to processing line $k$. For each $k$, sort due day $E_i$ for all products in ascending order, resulting in $k$ sorted lists. If a product has several orders in period $n$, $E_i$ is computed based on the order with the earliest due date. Products with smaller $E_i$ values have higher priorities to be scheduled for production. If two or more products have the same due day, the product with the smallest processing time (SPT) has a higher priority to be scheduled than other products.

(2.2.) For each processing line k, sequence products for the line based on the priority established in the sorted list from step 2.1.

(2.3.) Determine production run length $H$ for each of product scheduled in the period. Run length of a product in a period is determined according to the total quantity of demand, capability and minimum run length constraints:

$$H_{ik} = \begin{cases} DD_{ikn}/C_{ik} & \text{for} \quad H_{ik} > HL_i \\ HL_i & \text{for} \quad H_{ik} \leq HL_i \end{cases} \qquad (10)$$

where $DD_{ikn}$ is the demand of product $i$ on line $k$ in period $n$. $H_{ik}$ is rounded up to the next integer if $DD_{ink}/C_{ik}$ is not an integer. Repeat this step for all $k$. (For example, product 1SC in the first 5-day period in Table 2 has $DD_{ik1} = 55,000$ and $C_{ik} = 25,000$. $DD_{ink}/C_{ik}$ is 55,000/25,000, or 2.2. Therefore, $H_{ik}$ would be rounded up to 3.)

(2.4.) Shift the unfilled demand of a product to another processing line in accordance with capability constraints. If $H_{ik}$ cannot be totally assigned to line $k$ in the period, shift the unscheduled demand to another line which is not fully utilized. If product $i$ has an unfilled demand in period n and cannot be assigned to any other processing line, shift the unfilled demand to period $n+1$.

Step 3. Period schedule revision

(3.1.) Modify the schedule to meet the precedence constraints. If product j has been scheduled at time $t_1$ before product $i$ at time $t_2$ where $[i, j]$ has precedence constraint $P$, then re-schedule $i$ at $t$, where $t < t_1$. Adjust production schedule on the processing line accordingly. Repeat this step for all processing lines.
(3.2.) Modify the schedule to meet the group constraints. If production of a set of products $[i, j]$ in $G_g$ is separated in the schedule and product $i$ is in the earliest position, shift all other products in the same group next to and immediately following the product $i$. Shift the production of products without group constraints to a later day in the schedule if needed.

Step 4. Period iteration
Repeat step 1 to step 3 until PN schedules $S_1, S_2, \ldots, S_{PN}$ are obtained for each of the PN periods.
Step 5. Schedule Combination
Merge period schedules and check constraints.

(5.1.) Merge subsets of schedules $S_1, S_2, \ldots, S_{PN}$ to form a complete schedule $S$.
(5.2.) Exchange the position of products scheduled at the same line if precedence constraints are violated.
(5.3.) Check resource constraints for the production schedule across all processing lines. If a constraint is violated in day $t$, shift the demand of the product with lower $DD_{ikn}$ to an earlier day if possible. If the product cannot be produced on an earlier day, shift the demand to day $t+1$. Adjust the production schedule on the same processing line correspondingly.
(5.4.) Satisfy the maximum inventory constraints. Assume the production on product $i$ in line k can be shipped in the same day to meet the demand. Determine the inventory level for all products. If the inventory of product is greater than the maximum level allowed, shift that demand to day $t+1$ and then check the resource constraints. The production schedule will be shifted to the future if the new schedule violates the resource sharing constraints.

### 4.2. Push-forward heuristic procedure (PF)

The PF is a sequential optimization method which optimizes the ADI first then OSR in two steps. If the demand is greater than the production capacity in a given period, however, the demand is shifted to the next period to avoid inventory. Based on the previous description, only two steps need to be replaced. They are described below.

(3.1.) Modify the schedule to meet the precedence constraints. If product $j$ has been scheduled at time $t_1$ before product $i$ at time $t_2$ where $[i, j]$ has precedence constraint

$P$, then re-schedule $j$ at $t$, where $t > t_2$. Adjust production schedule on the processing line accordingly. Repeat this step for all processing lines.

(3.2.) Modify the schedule to meet the group constraints. If production of a set of products $[i, j]$ in $G_g$ is separated in the schedule and product $i$ is in the earliest position, shift all other products in the same group to be produced immediately before the product $i$. Shift the production of products without group constraints to a later day in the schedule if needed.

(5.3.) Check resource constraints for the production schedule across all processing lines. If a constraint is violated in day $t$, shift the demand of the product with lower $DD_{ikn}$ to day $t + 1$. Adjust the production schedule on the same processing line correspondingly.

### 4.3. Reduce Switchover Heuristic Procedure (RS)

The RS procedure focuses on minimizing product switchovers on processing lines which is similar to the current practice of the plant management. It is used to produce an optimal solution on TST first. Then the OSR is considered next, followed by ADI. Based on the current practice, once a processing line is assigned, the production run is continued until the total demand for the planning horizon (normally one month) is satisfied. The advantage of the current practice is that the method is easy to implement. The disadvantage is that both the OSR and ADI are not acceptable.

To implement the RS algorithm, the number of days in a period used in the previous procedures must be increased to allow a longer run. A test on various period lengths shows that the length of the period should be doubled or the number of periods (PN) should be reduced by about 50% in this procedure to observe a statistically significant different number of product switchovers. With RS, all the steps remain unchanged as procedure PB. Only the PN is changed. Other steps remain unchanged as procedure PB.

From the above discussions, it is noted that different production schedules can be obtained by using different heuristic procedures. A computer program written in C language is then implemented to include the three heuristic procedures. Many variables, such as the initial assignment of products to the processing lines and the number of periods in the planning horizon can be changed in the program. Consequently, alternative schedules can be produced for the management to make scheduling decisions.

## 5. Simulation experiments

Simulation experiments were carried out to evaluate the performance of the three heuristic procedures. The purpose of the experiment was to compare the performance of the procedures by answering the following questions:

(1) What is the magnitude of the relative performance of one procedure over another?
(2) If the performances of the two procedures are statistically significant different, then what are the factors related to the problem attributes affecting their differences?

To answer these questions, four factors that may have influence on the performance measures are identified. They are as follows:

(1) *Group ratio (G)*. The group ratio is expressed as a ratio of the number of

products with group constraints to the total number of products produced in the system. For example, if the total number of products is 50, and five of them are involved in group constraints, then the group ratio is 5/50 or 10%. Three levels are tested: 10, 25, and 50%.

(2) *Precedence strictness ratio (P).* The precedence constraints specify the products that must be produced before others. The precedence strictness ratio is defined as the number of precedence requirements among the products over the number of feasible pairs of products. For example, in a system producing five products, the maximum number of precedence constraints is 5!/2!3! or 10. If only one constraint exists for a pair of products, then the precedence strictness ratio is 1/ 10 or 0.1. Two levels are tested in the experiment: 0.1 and 0.2. These two ratios are relatively low because the type of products in the continuous manufacturing system we studied does not have too many precedence constraints.

(3) *Maximum inventory (I).* The maximum finished-goods inventory allowed in the system is basically limited by the space in the warehouse. To reduce the inventory cost, it is desired to reduce the maximum inventory level. Two levels of maximum inventory are tested. A level of 100% represents the actual maximum inventory allowed in a chemical plant, which is about 4500 thousand pounds of products in total. The level of 67% means that the maximum inventory allowed is reduced to $4,500 \times 67\%$ or 3000 thousand pounds in total.

(4) *Demand (D).* Three sets of real demand data were collected from a chemical plant. They are referred to as D1, D2, and D3. Each data set contains the daily demand for over 60 product types over a month period. The average total daily demands for D1, D2 and D3 are 96.6, 108.1 and 109.4 thousand pounds respectively. The standard deviations of the total daily demand of the three demand patterns are 93.90 for D1, 74.63 for D2, and 75.24 for D3.

Since the production scheduling issues are unique, a well-known data is not available in the literature. Using the three data sets from industry, a factorial experiment was designed and performed to evaluate the performance of the three heuristic procedures and to identify the factors affecting the three objectives scheduling.

Table 3 summarizes the factors and associated levels tested in the experiment. A full factorial experimental design is used, so the number of basic experiments designed for each heuristic procedure is $3 \times 2 \times 2 \times 3 = 36$. In order to account for variations in the experiments, five replications are created for each of the basic experiments. Considering three heuristic procedures and five replications per experiment, the total number of experiments would be $36 \times 5 \times 3 = 540$.

The method to replicate the experiment is based on a 10% random perturbation on demands. For every customer order in the planning horizon, a uniformly distributed random number between (0, 1] is generated to determine the direction of the perturbation. If the random number is between (0, 0.25], then the value is decreased by 10%. If the random number is between (0.25, 0.5], the order is decreased by 5%. If the random number is between (0.5, 0.75], the corresponding value will be increased by 5%. Finally, the value is increased by 10% if the random number is between (0.75, 1). This allows the demand data of replications to be varied within a 20% range. For example, if the original demand for a product is 25, it may be changed to 22.5 (i.e. $25 \times .90$), 23.75 (i.e. $25 \times 0.95$), 26.25 (i.e. $25 \times 1.05$) or 27.5 (i.e. $25 \times 1.10$) in the replications.

Three performance measures corresponding to each of the three scheduling objectives are used in the experiments: (1) OSR, (2) ADI and (3) Total switch times (TST).

Table 3.  Experiment factors and levels.

| Factor | Level |
|---|---|
| Group (G) | 10%; 25%; 50% |
| Precedence (P) | 10%; 20% |
| Maximum Inventory (I) | 100%; 67% |
| Demand (D) | D1; D2; D3 |

The three performance measures have been defined in Equations (1)–(3). Based on OSR, ADI and TST, a total cost index (TCI) can be estimated to represent an overall economical performance for a schedule where:

$$TCI = (400\ QO + ADI + 2500\ TST)/100 \qquad (11)$$

The quantity QO is the total amount (in thousand pounds) of overdue shipment products, ADI is the ADI in thousand pounds, and TST is the number of times the switchover occurs in the production schedule. Note that the coefficients of the three parameters are estimated to present a certain relation between parameters, and the values vary in accordance with different production systems.

## 6.  Analysis of experimental results

Based on the experimental design described in the previous section, a total of 540 experiments were performed. The performance measures of the 180 experiments for each of the heuristic procedures were averaged and presented in Table 4.

Among the three objectives of the scheduling, OSR and ADI can be considered as the primary or key performance measures for comparing different production schedules. As the product quality is critical to the success of the business, every production schedule developed by management must satisfy the run length constraints to insure that the quality of products is acceptable. Therefore, TST is simply used as a benchmark for schedule comparison purpose. In the real world a smaller TST is easier to implement and control, so it can be also used by management to evaluate the feasibility of schedule implementation. With the consideration of the minimum run length constraints in our scheduling problem, all three algorithms proposed can schedule production with acceptable product quality.

Table 4 shows that each heuristic procedure performs best for one of the three objectives as expected. Procedure RS results in an average of 12% OSR for the test problems. If the reduction of OSR is critical, then procedure PB can reduce the rate from 12% to 9%, and procedure PF is also slightly better than RS by 1%. For the ADI, PB is better than RS by 1%, and PF can reduce the inventory by more than 5%.

Table 4.  Summary performance by Heuristic.

| Performance measure | Heuristic procedure | | | % Difference | |
|---|---|---|---|---|---|
| | PB | PF | RS | PB-RS | PF-RS |
| OSR | 0.09[a] | 0.11 | 0.12 | 3 | 1 |
| ADI | 73,164 | 69,883[a] | 73,852 | 1 | 5 |
| TST | 42.40 | 40.33 | 36.51[a] | −14 | −9 |
| TCI | 3032[a] | 3233 | 3211 | 6 | 1 |

[a]Indicates the best performer by performance measure.

However, the product switchover (TST) for PB and PF will go up by 14 and 9%. Based on the TCI with the coefficients defined in equation (11), procedure PB is the most economical procedure among the three. However, this superiority can be diminished if the coefficients in equation (11) are changed.

The summary results shown in Table 4 reveal that the performance measures of the three heuristic procedures are different. To determine whether the differences are significantly away from 0, $t$ tests were performed. Under the null hypothesis, there should be no difference between the performance measures obtained from each procedure. If the null hypothesis is rejected at a selected level of significance, it is concluded that the difference exists. The assumption of normality on the data was first tested and verified. The test was set at 0.01 level of significance.

Four types of variables were used to measure the relative difference of the heuristic procedures. Variables DOSR$(i, j)$ measures the difference in OSR between two heuristic procedures $i$ and $j$:

$$\mathrm{DOSR}(i,j) = \mathrm{OSR}(i) - \mathrm{OSR}(j) \tag{12}$$

Variables for measures of ADI, TST and TCI are expressed as percentage difference between two procedures $i$ and $j$:

$$\mathrm{PADI}(i,j) = [\mathrm{ADI}(i) - \mathrm{ADI}(j)]/\mathrm{ADI}(j) \tag{13}$$

$$\mathrm{PTST}(i,j) = [\mathrm{TST}(i) - \mathrm{TST}(j)]/\mathrm{TST}(j) \tag{14}$$

$$\mathrm{PTCI}(i,j) = [\mathrm{TCI}(i) - \mathrm{TCI}(j)]/\mathrm{TCI}(j) \tag{15}$$

The reason of using only difference in OSR is that the OSR is already a percentage value. Moreover, the value for measure of percentage difference between two procedures sometimes might be infinite since OSR$(j)$ could be near zero.

The difference of the four statistics (OSR, ADI, TST and TCI) and the results of the $t$ tests appear in Table 5. A '*' is marked in the "Significance" column if the hypothesis that the difference is 0 cannot be rejected at the 0.01 level of significance.

Table 5.   $t$-test results by performance measure.

| Variable | No. Obs | $M$ (%) | SD | Significance |
|---|---|---|---|---|
| DOSR(PB,RS) | 180 | −2.24 | 0.036 | 0.0001 |
| DOSR(PB,PF) | 180 | −1.96 | 0.052 | 0.0001 |
| DOSR(PF,RS) | 180 | −0.28 | 0.064 | 0.554[*] |
| PADI(PF,RS) | 180 | −5.43 | 0.017 | 0.0001 |
| PADI(PF,PB) | 180 | −4.77 | 0.019 | 0.0001 |
| PADI(PB,RS) | 180 | −0.94 | 0.015 | 0.0001 |
| PTST(RS,PB) | 180 | −14.80 | 0.081 | 0.0001 |
| PTST(RS,PF) | 180 | −9.71 | 0.032 | 0.0001 |
| PTST(PF,PB) | 180 | −7.69 | 0.192 | 0.0001 |
| PTCI(PB,RS) | 180 | −5.17 | 0.214 | 0.0002 |
| PTCI(PB,PF) | 180 | −4.70 | 0.214 | 0.0037 |
| PTCI(RS,PF) | 180 | −4.43 | 0.428 | 0.071* |

[*]Indicates that the difference is insignificant at level 0.01.

Table 6.   ANOVA results by performance measure.

| Variable | Significant effects |
|---|---|
| DOSR(PB,PF) | G, I, D, G*D, I*D, G*I*D |
| DOSR(PB,RS) | G, I, D, G*D, I*D |
| PADI(PB, PF) | G, I, D, G*D, G*I*D |
| PADI(PB, RS) | G, I, D, G*D |
| PADI(PF, RS) | G, I, D, G*D, G*I*D |
| PTST(PB, PF) | G, D, G*D |
| PTST(PB, RS) | G, D, G*D, I*D |
| PTST(PF, RS) | G, D, G*D |
| PTCI(PB, PF) | G, I, D, G*D, I*D, G*I*D |
| PTCI(PB, RS) | G, G*D, I*D |

The $t$ tests show that procedure PB significantly outperforms PF and RS, but the difference between PF and RS is not significant. The OSR generated by procedure PB is about 2% less than that of procedure PF and is 2.24% less than that of procedure RS. Procedure PF can generate the lowest level of ADI, and procedure PB ranks second in ADI. The $t$ tests indicate that all the differences are significant, although the percent deviations are not very large. Procedure RS is the best on TST as the approach gives priority to reduce the number of product switchovers. The differences among the three procedures are statistically significant different. The statistics of the TCI show that procedure PB is significantly superior to the other two procedures for a system with cost coefficients similar to Equation (11). The difference between PF and RS is not significant. The results show that no procedure is absolutely superior to other procedures in every aspect. On the other hand, the experiment shows that the three procedures perform as desired since each of them is given a different priority.

The preceding observation suggests that further statistic analyses can be done to evaluate the effects of various system factors and their interaction effects on the difference between the procedures. To identify important factors affecting the performance of the heuristic procedures, an analysis of variance (ANOVA) was done using the SAS software system to evaluate the effects of the four factors and all their possible interactions. The ANOVA results on the four performance measures are summarized in Tables 6.

The variable column identifies the data set tested in the ANOVA model. For example, DOSR(PB, PF) represents the set of experimental data showing DOSR between procedures PB and PF. The level of significance is set at 0.05. All significant effects are listed in the tables.

The observations from ANOVA results are summarized as below:

(1) Group ratios (G) and demand patterns (D) significantly affect the performance differences in most of the cases.
(2) The interaction effects of G*D, I*D and G*I*D can also significantly affect the difference in performance measures.
(3) Three of the four factors (G, I, D) affect the performance. It seems that the performance of the heuristic strongly depends on the operating conditions. Only the precedence constraints (P) do not significantly affect the difference in performance measures.

## 7. Conclusion

This research presents a multiple objective formulation of production scheduling problem in continuous manufacturing systems and the development of three heuristic procedures to schedule daily production. Each heuristic procedure gives priority to one of three objectives. A computer program was developed to implement the heuristics. Many variables such as assignment of product to processing lines, coefficients of the cost function and the number of periods in the planning horizon can be changed in the program. Consequently, these heuristic procedures can be considered as "decision support" rules because the management will still need to make decisions themselves based on the various schedules produced.

A factorial experiment was designed and performed to evaluate the performance of the three heuristic procedures and to identify the factors that affect the performance differences. The analyses of the results show that no procedure is absolutely superior to the other procedures in all aspects. This is not a surprise because each of the three algorithms produces a local optimal solution. However, both the PB and PF algorithms can generate schedules with better OSR and ADI than the current management approach modeled by the RS algorithm. To provide a composite performance measure, a TCI was suggested. It is noted that the rank of the heuristics can be changed as a function of the cost coefficients in TCI.

Based on the ANOVA results, the differences between procedures are significantly affected by group ratio, maximum inventory and demand pattern. This information is of important value because it suggests that if one of the above factor changes in the manufacturing system, the method to schedule production needs to be re-evaluated. Further research is currently underway to study the sensitivity of the resource availability to the performance measures, and the development of a decision support system based on the heuristics proposed in this research.

## References

Amaro, A. C. S., & Barbosa-Póvoa, A. P. F. D. (1999). Scheduling of industrial distribution manifolds with pre-conditions. *European Journal of Operational Research, 119*, 461–478.

Anwar, M. F., & Nagi, R. (1997). Integrated lot-sizing and scheduling for just-in-time production of complex assemblies with finite set-ups. *International Journal of Production Research, 35*, 1447–1470.

Askin, R. G., & Standridge, C. R. (1993). *Modeling and analysis of manufacturing systems.* New York, NY: John Wiley.

Bose, S., & Pekny, J. F. (2000). A model predictive framework for planning and scheduling problems: A case study of consumer goods supply chain. *Computers & Chemical Engineering, 24*, 329–335.

Caramanis, M. C., Anli, O. M., & Paschalidis, I. C. (2003). Supply chain (SC) production planning with dynamic lead time and quality of service constraints. *Proceedings of the IEEE Conference on Decision and Control, 5*, 5478–5485.

Daniels, R. (1994). Incorporating preference information into multi-objective scheduling. *European Journal of Operational Research, 77*, 272–286.

Deepark, B. B., & Grossmann, I. E. (1990). Simultaneous production planning and scheduling in multiproduct batch plants. *Industrial & Engineering Chemistry Research, 29*, 570–580.

Giannelos, N., & Georgiadis, M. (2003). Efficient scheduling of consumer goods manufacturing processes in the continuous time domain. *Computers & Operations Research, 30*, 1367–1381.

Gupta, Y. P., Bector, C. R., & Gupta, M. C. (1990). Optimal schedule on a single machine using various due date determination methods. *Computers in Industry, 15*, 245–254.

He, W., & Kusiak, A. (1992). Scheduling manufacturing systems. *Computers in Industry, 20*, 163–175.

Kogan, K., Leu, Y., & Perkins, J. (2002). Parallel-machine, multiple-product-type, continuous-time scheduling: Decomposable cases. *IIE Transactions, 34*, 11–22.

Lou, P., Liu, Q., Zhou, Z., Wang, H., & Sun, S. X. (2012). Multi-agent-based proactive–reactive scheduling for a job shop. *The International Journal of Advanced Manufacturing Technology, 59*, 311–324.

Minton, S., Johnston, M. D., Phillips, A. B., & Laird, P. (1992). Minimizing constraint satisfaction and scheduling problems. *Artificial Intelligence, 58*, 61–205.

Modarress, B., Ansari, A., & Willis, G. (2000). Controlled production planning for just-in-time short-run suppliers. *International Journal of Production Research, 38*, 1163–1182.

Patterson, J. H., Slowinski, R., Talbit, F. B., & Weglarz, J. (1989). *An algorithm for a general class of research and resource constrained scheduling problems. Advances in Project Scheduling*. Amsterdam: Elsevler Science.

Scudder, G. D., Hoffmann, T. R., & Rohleder, T. R. (1993). Scheduling with forbidden early shipments: Alternative performance criteria and conditions. *International Journal of Production Research, 31*, 2287–2305.

Seki, Y., & Kogure, K. (1996). Lot scheduling problem for continuous demand. *International Journal of Production Economics, 44*, 7–15.

Tung, L., Lin, L., & Nagi, R. (1999). Multiple-objective scheduling for the hierarchical control of flexible manufacturing systems. *International Journal of Flexible Manufacturing Systems, 11*, 379–409.

Viana, A., & de Sousa, J. P. (2000). Using metaheuristics in multiobjective resource constrained project scheduling. *European Journal of Operational Research, 120*, 359–374.

Vollmann, T. E., Berry, W. L., & Whybark, D. C. (1992). *Manufacturing planning and control systems*. Boston, MA: Irwin.

Wagner, M., & Smits, S. (2004). A local search algorithm for the optimization of the stochastic economic lot scheduling problem. *International Journal of Production Economics, 90*, 391–402.

Yim, S., & Lee, D. Y. (1996). Multiple objective scheduling for flexible manufacturing systems using Petri nets and heuristic search. *Proceedings of the IEEE International Conference on Systems, Man and Cybernetics, 4*, 2984–2989.

# 10

# On decoupling points and decoupling zones

Joakim Wikner[a,b]*

[a]Department of Management and Engineering, Linköping University, Linköping, Sweden; [b]School of Engineering, Jönköping University, Jönköping, Sweden

In a market with demand surplus, it is possible to compete with standard products available in finished goods inventory. Sooner or later, the products will be sold and mass production can prevail. Competition is however increasing and to strike a competitive balance between cost efficiency and market responsiveness, it is becoming ever more important to establish a competitive level of customer-order-based management (COBM). This paper outlines a framework for this management approach based on content, represented by four key decision categories, and an overview of a process for applying the content. The content is based on a generic decision-based decoupling theory that is used for deriving the decision categories; flow driving, flow differentiation and flow delimitation. The derivation of these decision categories is based on analysis of strategic lead-times. Thereafter, the decision category flow transparency is included as the fourth content cornerstone of the framework. A process is then outlined for application of the framework. A basic bill-of-material is used as an illustration of applying the framework for COBM.

**Keywords:** decoupling points; postponement; decision-making; decision categories; operations strategy

## 1. Introduction

Decoupling points (see e.g. Blackstone, 2008) have played a crucial role in production and logistics management since the infancy of materials management. The objective of decoupling points, usually associated with stock points, has traditionally been to disconnect the material flow into sub-flows and thus enabling more focused and local flow-management. It is well known that if the flow is not decoupled, it becomes more sensitive to disturbances since the disturbances then easily propagate through large parts of the flow due to the dependencies (Goldratt & Cox, 1984). Managing this combination of uncertainties (referred to as fluctuations by Goldratt and Cox) and dependencies is the key challenge to flow management. In isolation, local uncertainties can be handled by probability theory and dependencies without uncertainties are suitable for tools from optimization theory. However, when uncertainties are combined with dependencies, these tools are challenging to apply.

Henry Ford showed the potential of tightly coupled flows under certainty when he developed the Ford production system (Ford & Crowther, 1988). A major drawback of this type of system is its inability to handle variations in volume or mix, which can be seen as disturbances that affect the continuous one-piece flow. GM and other companies

*Email: joakim.wikner@decouplingpoints.org

introduced frequent changes to the product lines which forced the system developed by Ford beyond its limits (Jones, 2001). Ford realized that one way to handle this challenge was to introduce a functional organization and stock points to decouple and reduce the dependency between different steps of the flow. GM went down a similar path but to a larger extent using arguments from cost accounting for decoupling the flow and this approach was adopted by many western companies for decades (Waddell & Bodek, 2005). This approach was also intrinsically connected to materials management systems such as material requirements planning (MRP). MRP is basically a system that is designed to maintain a target stock level, i.e. safety stock level, at the decoupling points of the materials flow. In general, this approach to decoupling points was indiscriminate to prioritizing different decoupling points even if the introduction of master scheduling and VAX-profiles provided some support (see e.g. Plossl, 1985, p. 177). The letters V, A, and X represent different types of material profiles, where the 'waist' of the profile is associated with a suitable level for performing master scheduling. The waist does not only provide an indication of suitable level for master scheduling but is also associated with the decoupling of customer-order-driven flow from forecast-driven flow (see e.g. Smith, 1989, p. 193), and hence a strategic location for inventory.

The importance of positioning decoupling points was further highlighted by the theory of constraints (TOC) (Goldratt, 1990) that emphasizes the importance of constraints for positioning of decoupling points. This approach proposes a design that can handle disturbances and dependencies by carefully positioning and dimensioning a combination of capacity buffers and material buffers. TOC uses time buffers, which are not in themselves decoupling points but rather an abstraction of flow segments related to non-constraints where the detailed flow analysis is replaced by earliest start time and latest finishing time for the time buffer (Stein, 1996). As a consequence, the time buffer will result in a physical buffer and hence a decoupling point. The most critical decoupling point is related to the bottleneck of the system. This is a different perspective compared to the VAX approach, which is material based. Olhager and Wikner (1998) showed how these two perspectives can be combined in support of master scheduling but provided limited support in terms of how to operationally combine them.

At the same time as MRP was established, Ford's original approach was further developed by Toyota that created Toyota production system (TPS) (see e.g. Ōno, 1988). TPS is the backbone of lean, which is a system that from a flow management perspective uses small decoupling points, referred to as supermarkets and FIFO-lanes, to handle product-mix demand uncertainties, and continuous improvement to reduce supply uncertainties. In all these approaches described above, the focus is on managing operational and tactical decoupling points (Wikner, Johansson, & Persson, 2009) when planning and controlling the materials flow. The lean approach also carries many similarities to the system simplification approach developed by the Cardiff group (see e.g. Wikner, Naim, & Towill, 1992). Both the lean and the Cardiff system approach emphasize simplification through elimination of unnecessary activities, i.e. waste. Lean is mainly concerned about waste at an operational level whereas the system's approach is concerned with waste at a structural level by creating effective control structures. The concept control structures as used here focuses on decision-making for creating effective flows whereas the lean approach has mainly been successful as an approach for creating efficient flows. The control structures of the system's approach can be described in terms of three components related to feedback and feed forward of information, and key decoupling points. These decoupling points are here referred to as strategic decoupling points (Wikner et al., 2009) due to their critical impact on competitiveness. In addition, there are tactical decoupling points related to items and item stock points, and operational decoupling points related to queues and buffers in the flow. Below the focus is exclusively on strategic decoupling points even if they are referred to simply as decoupling points.

The interest in decoupling points has developed in parallel streams of research related to, e.g. lead-time relation (see e.g. Mather, 1984; Shingō, 1989), decoupling point (see e.g. Hoekstra & Romme, 1992), customer order decoupling point (CODP) (see e.g. Bertrand, Wortmann, & Wijngaard, 1990), postponement (see e.g. Schmenner, 2001; van Hoek, 2001), order penetration point (Sharman, 1984), supply chain segmentation (see e.g. van der Veeken & Rutten, 1998), customization (see e.g. Graca, Hendry, & Kingsman, 1999), services (Fließ & Kleinaltenkamp, 2004; Wikner, 2012b) and leagility (Naim & Gosling, 2011; Naylor, Naim, & Berry, 1999). These different streams emphasize slightly different aspects of decoupling points but a fundamental property is the explicit focus on customers from a lead-time perspective. The main focus has been on the customer as the driver of the process but the literature on postponement (see e.g. García-Dastugue & Lambert, 2007), as well as the literature on CODP (see e.g. Hoekstra & Romme, 1992; Olhager & Östlund, 1990) have emphasized product differentiation as a separate aspect. Even if these streams have many similar properties, they have also to a large extent developed in parallel. They do however provide the decision-maker with a similar kind of decision support. In addition, a discussion on 'multiple decoupling points' has emerged in this context (see e.g. Banerjee, Sarkar, & Mukhopadhyay, 2011) where these issues are targeted to some extent but with a disperse foundation. An overall common structure is therefore needed that highlights key decisions related to decoupling points and thus enabling the development of a more cohesive theory related to decoupling points. The intention here is to identify a set of decision categories that outline the key decisions to be made when developing a competitive strategy for customer-order-based management (COBM), which here is defined as:

*Customer-order based management* (COBM) is a management approach that emphasizes the individual customer's demand as a key input to flow-based decision making in the supply network.

COBM is a management approach and based on decision-making from a generic process perspective in contrast to, e.g. customer-order-based production (Borgström & Hertz, 2011), customer-driven strategy (Wallace, 1992), and customer-driven manufacturing (Berry, Hill, & Klompmaker, 1995; Wortmann, Muntslag, & Timmermans, 1997) that are explicitly targeting manufacturing. In complex decision-making, the sheer number of options makes the situation difficult to embrace for the decision-maker and the decision problem must be organized in a structured manner. One approach is to identify so-called decision categories by disintegrating the decision problem in relatively independent categories. One of the earliest, and probably the most referenced set of decision categories, was defined by Hayes and Wheelwright (1984). They divided decisions into structural and infrastructural decision categories based on a rather resource-oriented approach in that it focuses on the preconditions in terms of invested capital for the value-adding flow rather than the core properties of the value-adding flow per se. With a more process-oriented and lead-time-based approach, the focus shifts from different functional areas of the business to how and where customer value is created by processes in the flow. From a supply perspective, this approach puts the customer in focus and the extent of customer influence on supply management is here referred to as decisions concerning the level of COBM that should be implemented. High level of COBM means that the customer can be offered a product that to a large extent is unique and may consist of services, but at the same time it also means that the customer must wait for delivery, i.e. for demand to be fulfilled. In contrast to this, a low level of COBM implies little customer influence, and that products are standardized goods with the possibility to be delivered with short lead-time.

In some cases, the decision on level of COBM is simplified to a decision about positioning of the CODP. This may be a sufficient description of the decision problem in some cases, but in many others it provides a too simplistic picture of the decision for segmentation of the supply chain from a customer perspective. Competition from low-cost countries has also put increasing pressure on companies in other countries to be more customers oriented. COBM is appropriate in this context as it involves more explicit customer focus both in terms of understanding specific customer requirements and cooperation with customers in different contexts. In this way, COBM rewards closeness to customers from a geographical perspective (due to costs and lead-times) as well as cultural/social perspective (due to the higher level of interaction between the parties) wherefore local suppliers, with high level of COBM, have advantages in this context compared to traditional goods focused providers.

Presently, there is very limited support available in terms of comprehensive frameworks concerning decision support for COBM. One reason is probably that COBM exists in a borderland between different challenges related to traditional manufacturing as well as service, distribution, and engineering activities. Due to these circumstances, it is interesting and important to investigate the challenges facing decision-makers in COBM to identify similarities and differences between these different types of businesses, and how synergies can be exploited to improve our ability to understand and manage these kinds of enterprises. It is also important to avoid adopting assumptions from specific industries. By keeping the approach focused on fundamental and generic concepts, the result can be employed in different kinds of industries. A consequence is however that industry-specific properties of importance may have to be added to the resulting framework.

The purpose of this paper is therefore to outline a framework that captures key aspects of COBM and in particular the role of lead-times and decoupling points from a decision-maker's perspective. As outlined above, the real-world relevance of this purpose is based on that the different widely applied management approaches all would benefit from a more explicit recognition of COBM. The research objectives targeted in this paper are therefore focusing on the foundation of flow-based COBM:

- Define a generic framework for lead-time-based analysis of strategic decoupling points.
- Identify the lead-times of strategic importance for decoupling.
- Outline key decision categories of a flow-based framework for COBM.

To achieve these objectives, the paper initially identifies the so-called logical entities which are defined from a management perspective. Decision domains are then introduced and used to define decoupling points and resource-based and process-based decoupling zones which are positioned using strategic lead-times. Thereafter, more general compounded decision domains are established. Based on this flow-based theory, the three decision categories flow driving, flow differentiation, and flow delimitation are defined based on strategic lead-times. These decision categories are complemented with flow transparency and altogether they constitute the four decision categories of the framework for COBM. Finally, the process for applying the content is outlined.

## 2. Research approach

The framework for COBM outlined below is the result of many lines of research related to both a deductive theoretical approach to theory development and an inductive empirical approach to both theory development and theory testing.

The theoretical baseline is a set of concepts developed in the literature related to decoupling points such as the CODP, postponement, customization, and leagility. The literature has mainly focused on the application of decoupling points in different areas and not so much on conceptual development. In particular, discussions on positioning of decoupling points (see e.g. Hoekstra & Romme, 1992) have gained much interest but less effort has been spent on investigating the more fundamental aspects of decoupling points and common properties of different kinds of decoupling. This is also the gap targeted here and the main contribution lies in the holistic perspective on decoupling points and the explicit flow-based approach.

The literature provided a set of theoretical concepts that were used in two research projects involving 5-6 companies. As the number of concepts used in the projects increased, a more general theoretical foundation of the decoupling-based concepts was established resulting in the generic framework presented below. The generic framework was then used as a platform for defining decoupling-oriented concepts providing a comprehensive and conceptual identification of COBM. The resulting framework for COBM therefore rests on a combined theoretical and empirical foundation. The actual cases from the research projects are rather complex and still not each covering all aspects of the framework. Hence, a fictitious example is used below that provides an opportunity to illustrate the application of the framework without introducing the complexity of the real cases.

## 3. Management perspectives

Cost efficiency has for many decades been the most important driver for enterprise development. This approach has turned out to result in focus on local efficiency where customers' needs are of lesser importance. The development of lean thinking (see e.g. Womack & Jones, 1996) as a framework for flow-based enterprise management has however put the customer in focus and as a consequence the enterprise's ability to create customer value becomes decisive. In this context, time is maybe the most important resource and the time that is available should be used to create customer value, otherwise the time is considered as wasted. This puts strain on the management system since time should be used for value creation at the same time as uncertainties must be considered. Before the importance of time in terms of lead-times is discussed, the context of lead-times is defined from a system's perspective.

### 3.1. Transformation-based system perspective

A lead-time analysis is performed in some kind of context in terms of a system's perspective. The context can, for example, be different aggregation levels or hierarchical control structures. The analysis performed here is however more management oriented and based on the transformation process.

Analysis of supply chains is often based on the interaction between different actors (see e.g. Harland, 1996). The actors may however be viewed from different perspectives (Wikner, 2012a). From an overall perspective, the actors are usually companies (legal entities) and the challenges are handled within *business management.* These *legal entities* interact through e.g. customer orders and purchase orders. In this way, the legal perspective handles issues associated with who is ultimately funding different types of transactions, i.e. who is acting as the *sponsor of transformation*, see Figure 1.

The financial and contractual dimension, represented by the legal perspective, is however only a reflection of processes being performed by resources at the physical entities in terms of geographical places or organizational units such as departments, production sites or distribution centres. To define actors from this perspective as *physical entities* is important when analysing localization issues or how different units collaborate from a production perspective. This also involves how different strategies for managing the supply chain are applied related to e.g. leagility and postponement. The management approach is therefore focused on logistics activities in a broad sense, including production, and can be referred to as *supply chain management.* A physical entity is hence a value-adding node in a physical network that performs transformation in terms of form, place or time and is associated with *type of transformation.*

The division into physical entities is not that obvious from an enterprise management perspective since the business information systems, such as ERP systems, now offer the possibility to manage multiple physical units as one integrated network, which may also be referred to as a virtual supply chain (Chandrashekar & Schary, 1999). The idea of virtual supply chains dates back to virtual inventory management and e.g. different types of base stock systems (see e.g. Clark & Scarf, 1960) which is interesting since inventory management in essence is concerned with the management of tactical decoupling points. The concept of virtual supply chains is emphasizing a transient property with short-term orientation as supply chains may change frequently based on market requirements. In contrast, the framework introduced here makes no assumption on short term vs. long term but rather emphasizes a structural management perspective. The key property here is that this type of network, from a planning and control perspective, is a system of multiple geographically or organizationally dispersed units that are managed as one entity and it is here referred to as a *logical entity.* Logical entities are not concerned with the type of transformation that is performed but defines the flow in terms of processes and objects and is thus associated with a generic *transformation* of input to output from a *process management* perspective. The concept logical entity is a fundamental construct of COBM and is hence explicitly defined here:

> A *logical entity* is a network of one or more physical entities that from a management perspective can be considered as one integrated network, offering the same level of controllability in all its parts.

The three perspectives legal, physical, and logical are summarized in Figure 1.

### *3.2. Legal and logical entities*

The three system's perspectives introduced above can be combined in different ways. A company (legal perspective) can be divided into manufacturing and distribution/transportation with different functional belongings (physical perspective) but due to integrated business information systems, they can be managed as one virtual integrated unit

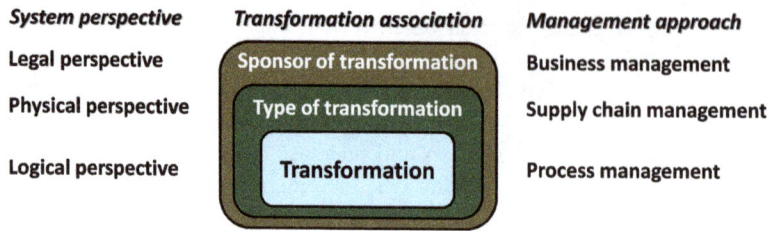

Figure 1.   System perspective, transformation association and management approach.

(logical perspective). In particular, it is important to separate between the legal perspective and the logical perspective when performing flow analysis. Even if the economic consequences of the transformation in the flow is related to the legal entities, the mission for effective flow is created by considering the parts of the flow that can be managed as an integrated unit, i.e. as a logical entity. Figure 2 illustrates this with four fundamental configurations that can be identified by combining legal entities with logical entities. More general networks may of course contain multiple legal and logical entities but this would basically be a generalization of the simple configurations of Figure 2.

The configuration with one legal entity and one logical entity corresponds to two legal entities and two logical entities since in both cases there is one logical entity per legal entity, i.e. each company can be managed as one integrated system. This is probably the most common scenario since legal entities with one production unit usually can be managed as one integrated system and hence constitutes one logical entity. When there is one logical entity for two legal entities, it means that the two legal entities are integrated and managed as one unit. If, on the other hand, one legal entity has two logical entities, it corresponds to a situation where one legal entity has not been able to establish one integrated control approach but instead is divided in, e.g. different departments that are managed as separate, non-integrated units.

Below, the point of departure is different logical entities. How these logical entities relate to legal entities is of less importance for this framework since the focus here is on efficient and effective flows rather than who is responsible for different parts of the flow and economic evaluation of the flow. An implicit assumption below is therefore that one logical entity may correspond to a part of a legal entity, a complete legal entity or a multiple legal entities that are managed as one unit. Since the logical entity is defined from a flow and control perspective, it is also important for flow-based decision-making. By defining the specific preconditions that are valid for decision-making

Figure 2.   Four fundamental configurations of legal entities (actors) and logical entities.

in different parts of the logical entity, the key decision categories for COBM can be identified.

## 4. Framework for flow-based decision-making

Flow is a general concept that basically implies a 'change', or in other words a transformation, and a rate of transformation. The transformation can take place in the form, place, or time dimension (see e.g. Bucklin, 1965). Being a rate-based concept means that flow is closely associated to time and understanding implications of time in this context is critical. A detailed flow analysis does however require a more elaborate view in terms of the types of flow involved. Depending on the context of the analysis, different types of flow are significant but in most cases the five types identified by Forrester (1958) (information, materials, money, manpower, and capital equipment) suffice. At a higher level of abstraction, the concept of flow can be defined from a customer perspective as:

> A *flow* represents the value-adding for customers through a set of transformation processes.

In this case, the concept of flow is used to represent that value adding is taking place, where value is added in a number of transformation processes (of course also non-value adding activities may be included but the emphasis is on value adding) which is in the spirit of, e.g. swift and even flow (Schmenner & Swink, 1998) and simplified material flow (Childerhouse & Towill, 2003). In every such process, a number of flow-based decisions are made that are decisive for how value adding is performed. The detailed design of the processes and the value-added flow may of course vary but here a common denominator is that for illustrative purposes, the flow direction is assumed to be from left to right and that the flow rate is directly or indirectly based on customer demand. The flow may hence be associated with materials flow but it is important to also recognize that the suggested framework does not exclude services since service can be defined as a process-based concept (Vargo & Lusch, 2004).

### 4.1. Decision categories and decision domains

Decision-making in COBM is comprehensive and covers a wide range of aspects to be effective. Providing a comprehensive framework for all types of decisions is both challenging to assemble due to the sheer number of different aspects to cover and would also be of limited value to the decision-maker due to the complexity provided by the comprehensiveness. To reduce the complexity due to a too high level of details, the decisions that should be covered can be classified according to their impact. Hayes and Wheelwright faced a similar challenge when outlining their framework for manufacturing strategy: 'Because of the diversity of manufacturing decisions made in different businesses, an organizing framework that groups them into major categories is a useful tool in both identifying and planning the functional strategy for manufacturing'. (Hayes & Wheelwright, 1984, p. 84). The major categories for COBM are defined later but the decision categories, as a concept that were used by Hayes and Wheelwright, are also suitable for this purpose and is here defined as:

> A *decision category* groups decisions into major categories that are fundamental to decision making for a particular purpose.

The scope of a decision category (DC) is based on some fundamental aspect of decision-making, which is here referred to as a decision criterion. Thus, decision categories are a way of classifying decisions and the decision criterion is what the classification is based on:

> A *decision criterion* is a standard on which a decision may be based.

To manage an enterprise involves making numerous decisions of different character but all related to transformation processes. Some decisions are of more simple character and do not affect the flow to a large extent while others are of decisive importance to the performance of the process, such as initiating a process or not. Effective process management therefore requires that the critical decisions can be isolated and classified to enable the design and application of a management framework with decision categories. These categories should then be applicable on a wide spectrum of processes with similar preconditions. In support of this classification of processes, the concept of decision domain is introduced:

> A *decision domain* is characterized by consistent preconditions for decision making related to a specific decision criterion.

A decision domain thus identifies what is common to a number of processes from a decision perspective and can hence be perceived as sub-categories within a DC (Wheelwright, 1984). A decision domain might, for example, define that the processes related to a decision domain are performed based on that a customer order has been received and hence that the process can be classified as customer-order-driven. This decision domain would then belong to the DC 'flow driving' and be based on the decision criterion 'flow driver' with the property 'driven by customer order' (see e.g. Figure 20 for an overview of decision categories). From this perspective, it would be reasonable to illustrate a decision domain as if the processes are within a decision domain. But for practical reasons, this relation is illustrated with a decision domain whose extension along the flow determines the processes associated with a particular decision domain, as shown in Figure 3. This means that the processes U1 and U2 in Figure 3 are classified as belonging to Upstream decision domain (from a flow perspective) and processes D1 and D2 as belonging to Downstream decision domain.

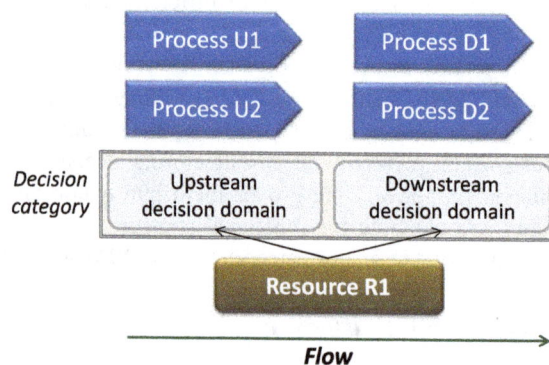

Figure 3.    Relation between processes, resources and decision domains.

Resources that perform processes have a relation to decision domains that follow a different pattern compared to processes. Since a resource can perform more than one process, it can also be related to more than one decision domain. Resource R1 in Figure 3 is, for example, related to both process U1 and process D2. Process U1 may be forecast-driven and process D2 customer-order-driven wherefore they are classified differently in terms of the DC flow driving. But, since both these processes are performed by the same resource, the capacity of the resource is required by both forecast-driven and customer-order-driven processes. As Figure 3 shows, a decision domain is here illustrated by a rectangle with rounded corners and with black text on grey background. Processes are depicted with traditional arrows, resources with rectangles with rounded corners and white text on dark background.

## 4.2. Decision domains and decoupling points

The analysis above defines decision domains as a collecting concept for processes with a specific property in common. The decision domains can be relatively independent of each other in the sense that they can be based on different decision criteria that highlights different process perspectives, such as focusing on credit risk or representing disassembly decisions. From a flow perspective, it is of particular interest to focus on decision criteria that separates the flow into two separate parts. A specific example of this is processes that are customer-order-driven and processes that are forecast-driven. For the DC flow driving, these processes are completely different but still they are connected to the same decision criterion, i.e. what decides that a process is initiated. From a flow perspective these two decision domains can be seen as being in a sequence in the flow and characterized as one upstream decision domain and one downstream decision domain that from a decision perspective are disconnected from each other.

The decision criterion is decisive for the type of property that can be classified as either upstream or downstream. To emphasize this difference, a decoupling point is introduced that indicates that the decoupled decision domains are related to the same decision criterion:

> A *decoupling point* separates decisions that are made under different consistent properties related to a specific decision criterion.

The decision domains are disjunctive in the sense that they represent completely different properties, such as forecast-driven or customer-order-driven, considering the same decision criterion. This relation is depicted in Figure 4 where a general decoupling point is illustrated with an ellipse which symbolically connects the two decision domains of the DC.

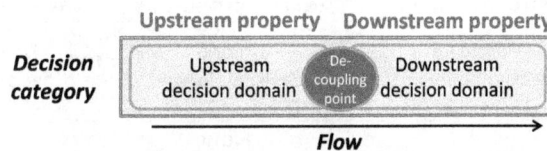

Figure 4.   Decision domains with disjunctive properties.

However, in many cases, the decision support is not homogenous and based on only one property at each side of the decoupling point. In some cases, the transition is not instantaneous along the flow, i.e. an upstream property is not changed to a downstream property at one point, but rather can be seen as a gradual transition consisting of a mix of properties from each side of the decoupling point of Figure 4. A process driver may, for example, be a combination of forecast and customer order. This type of mixed properties is here associated with a hybrid decision domain, see Figure 5. It is then logical to model this as three decision domains and consequently two decoupling points for each DC.

Hybrid decision domains represent more complex decision-making where the criterion has mixed properties. In some cases, it can be an advantage to use both decoupling points but this approach also increases the complexity. An alternative is to emphasize only one decoupling point and this is a useful approach when the decision criterion is intrinsically linked to one of the decision domains. A typical example is the flow driver criterion of COBM, which is customer focused and hence the logical place to decouple is upstream of the customer-order-driven decision domain. This domain is customer focused, whereas the other decision domains also have other properties. Since the focus below is on COBM, only decoupling point 1 will be considered and below simply be denoted 'decoupling point'. The mixed properties that the hybrid decision domain relate to represent, from a flow perspective, a transition from the properties valid upstream to the properties valid downstream of the hybrid decision domain. This transition with a hybrid decision domain can be seen as a zone that decouples the two single-property decision domains.

Decoupling points by definition mean, as described above, that different preconditions exist upstream and downstream from the decoupling point. From a material flow perspective, this type of discontinuity corresponds to that the preconditions for the flow changes significantly which leads to disturbances of the flow. Different kinds of buffers are used to reduce the impact of these disturbances. Capacity buffers, related to resources, offer interesting possibilities since they increase the capability to be agile in processes performed to, e.g. customer order or backorders at stock points. A drawback with capacity buffers is however that they do not provide instant reaction but requires a lead-time before a result can be delivered (since the capacity is used to perform some kind of transformation which takes time). The material buffers do not have this drawback since they provide immediate availability of what is demanded even if this type of buffer provides less flexibility since decisions, about what to include in the material buffer, are made before the requirement is identified. In this way, the material buffer is a more lead-time responsive type of buffer that is suitable in relation to decoupling points. The capacity buffer does, however, provide important support downstream from the

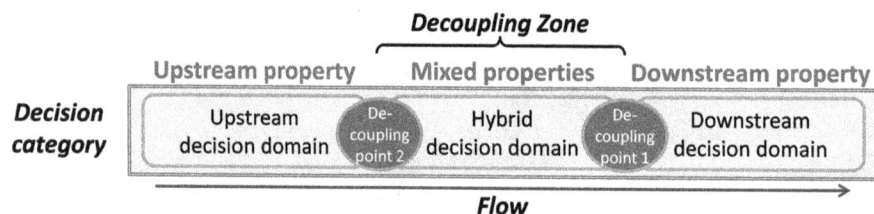

Figure 5.   Decision domains and strategic decoupling points.

decoupling point as it enables the resources to be more flexible. The material buffer is here referred as a stock point and later illustrated in the figures by a black triangle pointing downwards representing physical inventory in the flow.

The significance of so-called discontinuities can be reduced, or even eliminated, by introducing decoupling zones that enable a softer transition from the conditions at the upstream decision domain to the conditions at the downstream decision domain. If discontinuities are completely eliminated, there is also no need for a corresponding buffer at a stock point.

### 4.3. Decision domains and decoupling zones

Decoupling points are positioned at an instantaneous transition along the flow from upstream property to downstream property. This represents a 'black or white' scenario which in some cases is too simplistic. The transition may be gradual and hence represents a zone with different 'shades of grey' related to the hybrid decision domain as described above.

> A *decoupling zone* covers decisions that are made under mixed properties related to one or more decision criteria.

There are mainly two different perspectives that require these 'shades of grey'. The first perspective is the gradual transition from upstream property to downstream property and it is here called *process-based decoupling zone* since it exists along one individual flow and corresponds to one single decision domain. The other perspective exists when multiple flows are considered in parallel, from an aggregate perspective, and is in particular related to when one resource is involved across multiple flows at the same time. This scenario has its roots in how resources are positioned in relation to decision domains in Figure 3 and is therefore here referred to as *resource-based decoupling zone* that corresponds to an aggregation of two or more different decision domains.

### 4.3.1. Process-based decoupling zones

The concept decoupling zone refers to a mix of different properties related to decision domains. In case there is a gradual transition along a flow, a number of attributes related to the property can be affected. The properties can, for example, be associated with uncertainty or certainty concerning customer orders. The certainty can however be based on different types of attributes related to the customer order such as requested delivery date or requested quantity. When deciding on the number of attributes to use, a balance between the complexity of using many attributes and not challenging the relevance of the attributes, by using too few, must be established. In this framework for COBM, two attributes are used for each DC related to a process-based decoupling zone, but from a more general perspective, one or more than two attributes can also be used.

> A *process-based decoupling zone* is based on the gradual transition between two properties of one decision criterion related to one flow.

The process-based decoupling zone (PB-DZ) is in Figure 6 illustrated by a rectangle placed above the hybrid decision domain. The zone is expanded to a two-dimensional surface that represents two attributes that each can take values on a scale from upstream

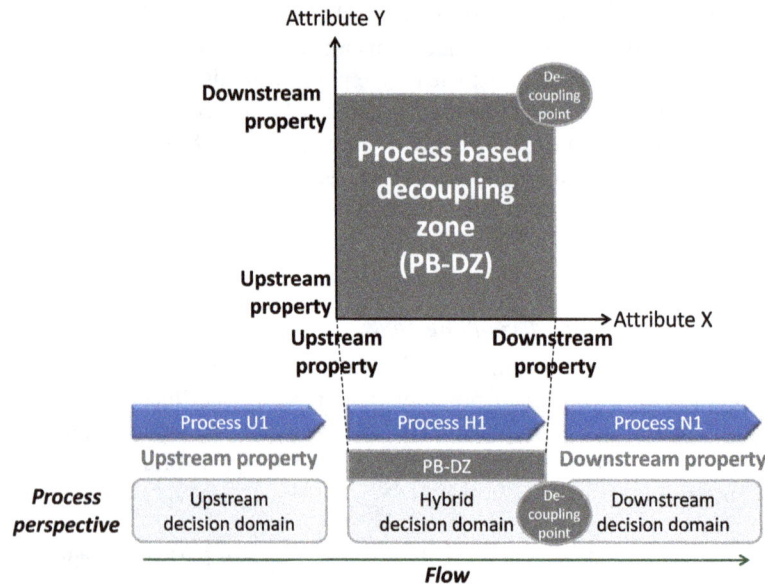

Figure 6.    Process-based decoupling zone along one flow.

to downstream property. The end-points (the corners) where the attributes have the same 'value' represent the limit of upstream and downstream decision domain, respectively. Based on the definition of decision domain and how it is related to processes means that there are now, at least, three types of processes along the flow; in Figure 6, this is illustrated with the processes U1, H1 and N1. At the beginning of the flow (on the left in the figure), the upstream property is of relevance to decisions. When the flow reaches the PB-DZ, the decision preconditions correspond to the lower left corner of the two-dimensional surface. The flow through the PB-DZ corresponds to that the preconditions change as the value of the two different attributes changes along a trajectory from the lower left corner to the upper right corner of the surface, i.e. when the decoupling point is reached. Thereafter, the downstream property is relevant for decision-making until the end of the flow is reached.

### 4.3.2.   Resource-based decoupling zones

From a resource perspective, it looks different compared to the process perspective since resources can be connected to multiple decision domains with different properties as shown in Figure 3. This multi-property load on a resource is related to resource-based decoupling zones.

> A *resource-based decoupling zone* is based on the intersection between different properties of one decision criteria across one or more flows.

In Figure 7, this is illustrated with the two flows, Flow A and Flow B, where each consists of two processes related to one upstream and one downstream decision domain, respectively, and thus has one decoupling point each. One individual flow is therefore relatively simple from a process perspective with two basic decision domains both based

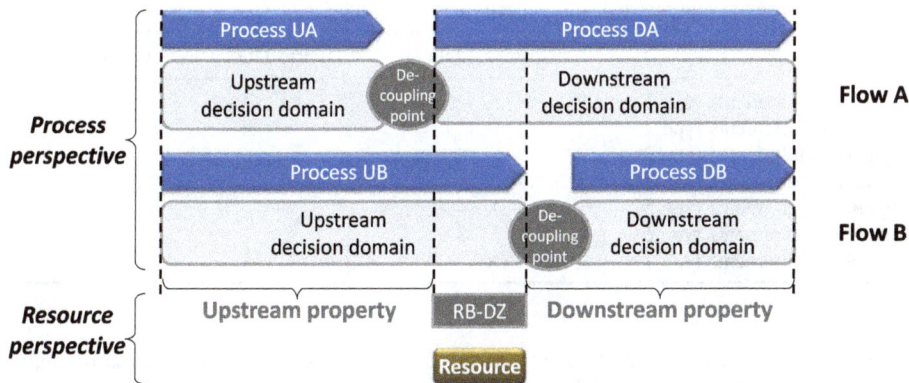

Figure 7.   Resource-based decoupling zone across multiple flows.

on the decision criterion in question. In Figure 7, one resource is identified and from the resource's perspective, the situation is more complex since it is involved in process DA and process UB, i.e. two processes that are related to different decision domains with different properties. This type of scenario is here referred to as a resource-based decoupling zone (RB-DZ) and could, for example, mean that some of the load on the resource is forecast-driven (based on UB) and some is customer-order-driven (based on DA). This means that the load analysis becomes considerably more complex compared to when only UA and UB or only DA and DB are involved, where the preconditions would be more homogenous what concerns the actual DC. Note that the assumption is that the same decision criterion is in focus for all processes, i.e. they belong to the same DC. In addition, the individual flows may not contain a process-based decoupling zone but in aggregation a resource-based decoupling zone can be identified. In case the decoupling points of the individual flows are positioned at the same place also, the aggregate would not contain a decoupling zone and the decoupling point could be referred to as a 'resource-based decoupling point'.

The aggregation discussed above is across multiple flows. The corresponding situation can occur in the case of aggregation in time where, for example, weeks are used as time periods rather than days. With a resolution of days, a more detailed analysis can be performed resulting in that resources are either loaded with forecast-driven or customer-order-driven activities. Using the aggregation level of weeks on the other hand might result in a mix of different types of load for a particular resource since the timing of load cannot be determined as exactly as in the case of time periods based on days. Also, resource aggregation may create a similar problem as individual resources are not loaded, but instead groups of resources are loaded. Hence aggregation in terms of flows (i.e. products), time, or resources can generate similar challenges from a capacity requirement perspective.

In summary, the capacity requirement for a resource may be created by seven different combinations of processes related to three different decision domains of one single decision criterion. Note that Figure 7 is based on that each flow consists of two decision domains whereas Table 1 is based on the existence of also a hybrid decision domain, i.e. PB-DZ. In Table 1, these combinations are illustrated and the first three cases (type 1–3) in the table are based on one decision domain each with capacity requirement from

Table 1.  The seven different types of capacity requirement.

|  | Upstream DD | Hybrid DD | Downstream DD |
|---|:---:|:---:|:---:|
| Capacity requirements type 1 | X |  |  |
| Capacity requirements type 2 |  | X |  |
| Capacity requirements type 3 |  |  | X |
| Capacity requirements type 4 | X | X |  |
| Capacity requirements type 5 | X |  | X |
| Capacity requirements type 6 |  | X | X |
| Capacity requirements type 7 | X | X | X |

the respective decision domain. Three of the combinations originate in the combination of two types of decision domains. Capacity requirement type 7, finally, is based on three different types of decision domains, i.e. one resource would be used in processes of three different decision domains. The example of Figure 7 can, based on this, be classified as a resource with capacity requirement type 5 since the capacity requirements originate from a upstream DD and a downstream DD.

For both the PB-DZ and the RB-DZ, the 'length' of each decision domain is important since it determines the position of both decoupling point and decoupling zone. The decision domain is based on one decision criterion; for example, flow driver, which means that the length of the decision domain and the priority of the chosen decision criterion are different sides of the same coin. The definition of important decision categories for COBM can therefore be based on the identification of important lead-times.

### 4.4.  Positioning of decoupling point and decoupling zone

The extension of decision domains have so far been based on a relative perspective. A decoupling point has been identified that separates the flow into two distinct parts related to one decision domain each. In addition, a hybrid decision domain may be positioned between these two decision domains resulting in a sequence of three decision domains along the flow. Identifying the decision domains is a first step but it is also important to determine the actual extension of each decision domain and as a consequence also the position of the decoupling point and the decoupling zone.

A useful measure for the extension would operationalize the concepts introduced so far. A number of candidates exist for acting as a point of reference such as the physical extension of the flow, organizational properties such as functions involved, or cost aspects related to e.g. resource ownership. It is however important to have a baseline that is absolute in the sense that it is not something that can easily be manipulated to suit different purposes. It should rather be something that can act as a point of reference that is important and transparent to all concerned interest parties. Considering the history of operations and supply chain management, a number of cases have shown that the key resource to manage is time as time lost can never be recovered. In addition, time has also been shown to be a key driver of competitiveness. Henry Ford was one of the first to explicitly highlight time as one of the pillars of his management philosophy (Ford & Crowther, 1988) and time has continued to play a key role in, e.g. Time-based management (Stalk & Hout, 1990), BPR (Hammer & Champy, 1993) and lean thinking (Womack & Jones, 1996). Presently, the main proponent of time as a key resource is probably lean in all its flavours such as lean thinking, lean production, lean services, lean administration, etc. At the core of all these, lean 'flavours' is the continuous improvement to reduce waste, and waste is basically different incarnations of lost time. For example value stream mapping (Rother & Shook, 1998), the most highlighted part of the analysis is the timeline capturing the total lead-time as well as the value adding and non-value adding time. From a time perspective, the timeline positions the different parts of the value stream in relation to the customer, i.e. the end-point of the value stream.

The decision criterion and the related decision domains provide a conceptual division of the flow from a decision perspective. The decision in itself and what it involves are important to define the decision situation, but the timing of the decision is absolutely critical in defining the preconditions for the decision-maker. The timing is defined in relation to when the result of the sub-flow, corresponding to the decision domain, is required. With this as a point of departure, three lead-times can be defined as of Figure 8, related to one decision domain each: upstream (US), hybrid (H), and downstream (DS). Accumulating these lead-times the total lead-time for the flow $i$ ($F_i$) is obtained, where: $FLT_i = USLT_i + HLT_i + DSLT_i$. The decoupling zone highlighted here is related to a decision domain and hence of type PB-DZ. The RB-DZ can be positioned in a similar way based on that the decoupling zone is related to an intersection of a set of flows. DSLT would in this case correspond to $DSLT = \min_{i \in Flows} \{DSLT_i\}$ where Flows is the set of all flows where the resource is involved. In a corresponding way the lead-time of the RB-DZ can be defined as: $HLT = \max_{i \in Flows} \{HLT_i + DSLT_i\} - DSLT$. USLT can then finally be defined as: $USLT = \max_{i \in Flows} \{FLT_i\} - HLT - DSLT$.

Based on the argument earlier, where the second decoupling point was considered as of less interest, the decoupling point upstream of the hybrid decision domain is not analysed further. The positioning of the decoupling point is hence based on the lead-time of the downstream decision domain, DSLT, in relation to a reference lead-time such as the total flow lead-time, FLT. The position of the decoupling point can hence be expressed in two different ways:

- The lead-time DSLT, which is useful in e.g. analysis based on the time-phased bill-of-material to calculate the position of the decoupling point. This is the 'absolute' lead-time expressed in suitable time units.
- The 'relative' lead-time results in a percentage and expresses how long a lead-time is in relation to a 'benchmark' lead-time and in particular we are interested in the fraction of the benchmark lead-time that is related to downstream property. For positioning of the decoupling point the key lead-time is the relation between the lead-times DSLT and FLT, which is called the DSLT:FLT relation. If DSLT:FLT = 1 then the decoupling point is at the beginning of the flow. If DSLT:FLT $\ll$ 1

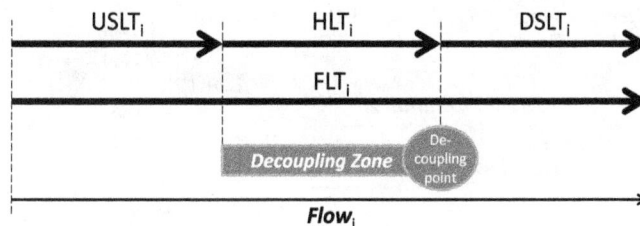

Figure 8.   Decision domains and lead-times.

(or DSLT:FLT = 0) then the decoupling point is at the end of the flow and finally if DSLT:FLT < 1 the decoupling point is positioned within the flow.

The 'relative' lead-time is measured in relation to a reference value and is thus of more practical use in COBM. From a general perspective it can be defined as:

A *lead-time relation* is a ratio of two strategic lead-times.

Positioning of the decoupling point also provides information about the extension of the upstream and the downstream decision domains in case there is no hybrid decision domain. If also a hybrid decision domain is present additional lead-time analysis is required to position also this decision domain. This additional analysis can be based on either USLT or HLT. Once the decoupling point and decoupling zones are positioned, the different decision categories can be combined resulting in compounded decision domains.

### 4.5.  Compounded decision domains

In an actual decision situation, multiple decision criteria may be important to consider at the same time. Still each decision criterion can be defined separately as in Figure 9 where this is illustrated on the left side with decision criteria 1 and 2. In most cases, the different decision criteria are independent of each other and capturing different aspects of the challenges facing the decision-maker. Therefore the two decision criteria may be modelled as two separate dimensions as shown on the right of Figure 9. The resulting four decision domains are referred to as compounded decision domains (CDD):

A *compounded decision domain* covers the intersection of two decision domains related to two separate decision criteria.

Each CDD is a combination of upstream (U) and/or downstream (D) decision domains based on decision criteria 1 or 2. For example, D1, D2 is a CDD where the decision criterion is a combination of decision criteria D1 and D2. Depending on the actual decision criteria used some of the combinations might not be possible, or at least not competitive, and hence a strategy for that CDD might not be necessary. However, from a generic point of view all CDDs are assumed to be valid and hence all possible CDDs are included.

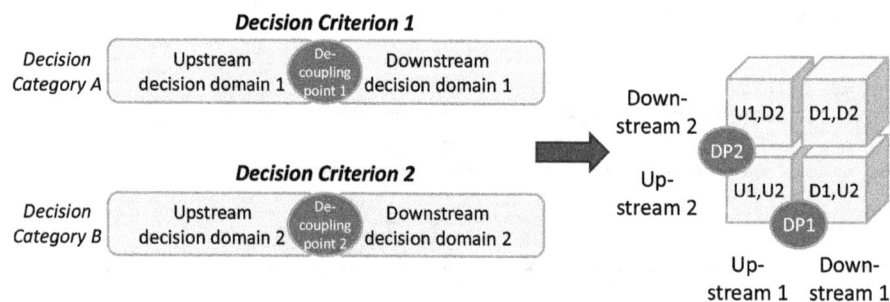

Figure 9.   Example of compounded decision domains.

The CDDs add complexity since they introduce a multi-dimensional perspective on decision domains. The example in Figure 9 is based on two flows with decoupling points only. By also including decoupling zones there are some more challenging issues to consider.

### 4.5.1. Compounded decision domains and process-based decoupling zones

The concept of CDD can easily be expanded to also include hybrid decision domains and PB-DZs. The example of Figure 9 represents the possible combinations when no hybrid decision domains are included. By allowing hybrid decision domains each dimension would be extended to three blocks and in total nine CDDs would be included on the right in Figure 9. This scenario, with nine CDDs, is somewhat similar to the analysis of the resource-based decoupling zone of Table 1 (with seven possible combinations). It is however important to note that in this case the two dimensions are based on different criteria whereas in Table 1 the two dimensions were based on the same criterion. Since the CDD here is based on two criteria the type 7 case cannot occur and the criteria are different for types 4–6 which means that it would result in another three types (for example, U1, D2 does not correspond to D1,U2 when different criteria are used). In summary there would be four CDDs only involving upstream and downstream decision domains and an additional five that include at least one hybrid decision domain. This would result in three decision domains in each dimension of Figure 9 and, consequently, in total nine different CDDs (cf. Figure 16 for an example).

### 4.5.2. Compounded decision domains and resource-based decoupling zones

The CDDs creates additional complexity in terms of RB-DZs since the number of possible combinations increases significantly. The example of Figure 7 resulted in one additional scenario (totally three) compared to when one flow was analysed. The number of possible combinations of decision domains, $C$(Number of flows, Number of decision domains per flow) affecting a resource when each flow can be divided into $n$ different decision domains is a combinatorial problem. For the case in Figure 7, which is based on that the decision domains belong to the same DC, the number of combinations would be: $C_{\text{One DC}}(2,2) = 2 + 1 = 3$ different combinations of decision domains. In case each flow would contain a hybrid decision domain the number of combinations would be: $C_{\text{One DC}}(2,3) = 3 + 3 + 1 = 7$ different combinations (as shown in Table 1). When the CDDs are introduced the possible combinations for a resource to be involved in more than one decision domain increases dramatically for RB-DZs. Consider, for example, the case introduced above in Figure 9 with two dimensions and two decision domains in each dimension resulting in a total of four CDDs. One case could be that a resource is acting in the CDD U1, U2 for one flow and the CDD D1, D2 in the other flow. By definition the RB-DZ is related to when a resource is involved in two CDDs of different properties which means that if a resource is involved in CDD U1, U2 of two different flows it is not defined as a RB-DZ. For this case with two flows with four CDDs each there are a total of $C_{\text{Two DCs}}(2,4) = 12$ possible combinations. Introducing PB-DZs in each flow of course increases the complexity further and the number of combinations fast becomes hard to manage. It is therefore important to identify the critical intersections and to exclude the less important from the analysis.

### 4.6.  Summary of framework for flow based decision-making

The generic framework for decoupling points and decoupling zones provides the means for identifying some key decision categories related to supply chain and operations management. In summary, the generic framework for flow based decision-making contains ten key concepts (which have been explicitly defined above):

(1) *Logical entities* are the platform for decisions on flow based management.
(2) *Flow* provides the point of reference for identifying lead-time based decision categories.
(3) *Decision categories* are defined based on decision criteria.
(4) *Decision criteria* are used to define decision domains within a decision category.
(5) *Decision domains* for the same decision criteria are separated by decoupling points.
(6) *Decoupling points* are positioned at the interface between two decision domains.
(7) *Decoupling zones* correspond to mixed properties in a decoupling point scenario.
(8) *Process-based or resource-based* perspective can be applied on decoupling zones.
(9) *Lead-time relations* position decoupling points/zones and decision domains.
(10) *Compounded decision domains* reflect the complexity of an actual decision problem.

The generic framework with these ten key concepts is next used to define three lead-time relations and thereafter four decision categories for COBM.

## 5.  Lead-time based flow analysis

Lead-time analysis is used in different contexts to e.g. reduce lead-times, identify wastes or simply to identify the magnitude of the lead-times. This type of lead-time analysis can be categorized as 'absolute' lead-time analysis since focus is on the length of the lead-times. It is however 'relative' lead-time analysis that is of main interest as an indicator of positioning of decoupling points and decoupling zones since it involves the relation between different lead-times. Before these lead-time relations can be defined it is necessary to identify the key 'strategic' lead-times.

### 5.1.  Lead-time definitions

Balancing the requirements and availability of material and capacity of operations are important challenges in logistics flow. Common to these challenges is that timing is of critical importance. The relationship between available capacity and capacity requirements should be investigated with respect to a timeline and also the balancing of demand and supply of materials requires a time-phased approach (however less significant in a rate-based context). The comprehension of lead-times and how they should be managed is therefore of utmost importance to enterprise management.

The standard item lead-times ($L$) are fundamental, but even more important from a competitiveness perspective are the five strategic lead-times (Wikner, 2011). From a supply perspective the cumulative lead-time (Blackstone, 2008) is critical since it represents the lead-time of the product including all items. The cumulative lead-time is then split in two parts representing items provided by the logical entity in focus (internal lead-time) and items provided by the upstream logical entities (external lead-time). Another key aspect is the level of possible customization involved and this is here referred to as adapt from a supply perspective. Customization is based on customer requirements and adapt may hence also be seen as based on customer requirements from a demand perspective. Finally, a key aspect from a customer perspective is also the requested delivery lead-time as it also frequently defines a key characteristic of the order winners.

Five strategic lead-times (referred to as SEIAD based on their variable names) are investigated here and further elaborated on below:

- Supply lead-time ($S$) is the cumulative lead-time for the product through the whole (extended) logical entity.
- External lead-time ($E$) can in many cases be seen as related to the (purchased) component and is therefore usually associated with purchase orders.
- Internal lead-time ($I$) corresponds to the controllable part of the bill-of-material and is usually associated with the own provisioning lead-time.
- Adapt lead-time ($A$) is related to the customer order but based on when the supply performed is actually customer order unique. The possible level of customization is related to a supply perspective ($A_S$) and the requested level of customization is related to a demand perspective ($A_D$). This lead-time could also be referred to as the customization lead-time but to avoid confusion with different types of customization strategies the more neutral term adapt is used. The adapt lead-time is important in e.g. the context of mass customization but it is also a key component in other cases such as e.g. postponement strategies.
- Delivery lead-time ($D$) is based on market requirements for delivery and directly associated with demand/customer order.

### 5.2.  Example: lead-times and time-phasing

The different types of lead-times defined above are related to each other and a simple example is introduced as an illustration of the theory. Some of the concepts introduced here have also been applied on actual cases (Bäckstrand et al., 2013) but in general, actual bill-of-material that can be used as an illustration of several concepts are rather complex. The sheer complexity of the bill-of-material in these cases and their area of application diverge the focus from the illustration of the concepts. The example introduced below is therefore fictitious and designed to be simple but still sufficient to illustrate the key concepts used here. The example is shown in Figure 10 in terms of a traditional material based bill-of-material (product structure) to the left and a time-phased bill-of-material to the right (see e.g. Bäckstrand & Wikner, 2013; Clark, 1979; Wikner & Rudberg, 2005b). The material based bill-of-material includes where-used relations and lead-times. For example, item $Y$ consists of item $X$ and item $U$, and item $Y$ has a lead-time ($L_Y$) of 3 periods. On the right in Figure 10 this bill-of-material is instead represented by a time-phased bill-of-material where the horizontal distance between two filled circles corresponds to each item's lead-time ($L$). In this case the supply lead-time $S = 12$ periods (the cumulative lead-time for the bill-of-material) and

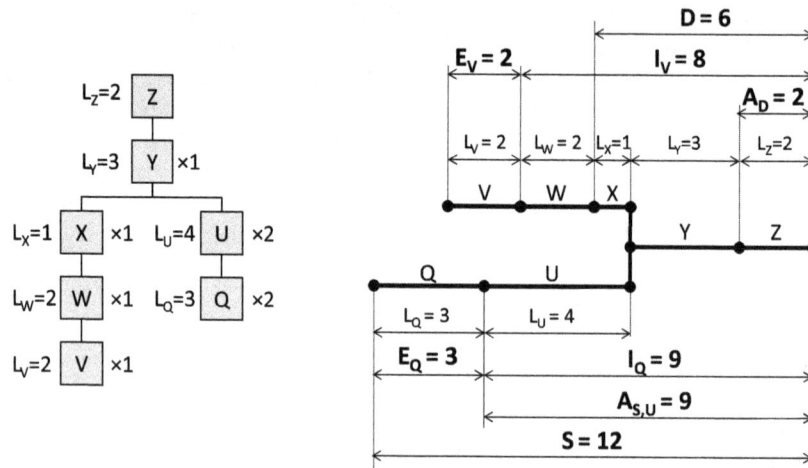

Figure 10.   Example: Material-based and lead-time-based bill-of-materials.

the delivery lead-time to the customer is given as $D = 6$ periods. (The estimation of the delivery lead-time $D$ will be more elaborated on later, see e.g. Figure 18). Since it is the adaptation of $Z$ that constitutes a customer order unique solution, the demand adapts lead-time $A_D = 2$ periods and $Y$ is customer unique. From an engineering and production perspective it would be possible to also make item $U$ customer order unique which means that the supply adapt lead-time $A_{S,U} = 9$ periods. The items $V$ and $Q$, finally, are purchased in a traditional way from suppliers. These two items represent the end of two branches with an external lead-time ($E$). An internal lead-time ($I$) can then be calculated for each branch of the bill-of-material. The set Leafs is assumed to contain all $N$ items that are at the lowest level of each branch and would in many cases correspond to purchase items. In the example of Figure 10 it would mean that $N = 2$ and Leaf $= \{V, Q\}$. For the branch of item $V$ the cumulative lead-time is $S_V = E_V + I_V = 2 + 8$ and for item $Q$ the cumulative lead-time is $S_Q = E_Q + I_Q = 3 + 9$. The branch with the longest cumulative lead-time is also equal to the supply lead-time $S = \max_{n \in \text{Leaf}} \{S_n = E_n + I_n\}$. A more elaborate analysis can also be performed where a $S$ is defined for each item in the bill-of-material (Bäckstrand & Wikner, 2013). This approach provides the opportunity to identify for each item if it is within $D$ and can be customer-order-driven, or if it is longer than $D$ and must be forecast-driven.

### 5.3.   Lead-time context

If only one branch of the product structure in Figure 10 is considered, a network without different branches is obtained and this can be labelled as a 'linear' chain. Using the terminology introduced above the chain may be referred to as consisting of two logical entities, one focal entity ('focal' refers to the unit that is in focus for the analysis) and one supplying entity, i.e. two logical entities in sequence and with an implicit entity represented by a customer. The context of this analysis is therefore a triad consisting of a customer entity and two logical entities representing the supply. Each one of the logical entities can be illustrated with the entity's strategic lead-times SEIAD resulting in

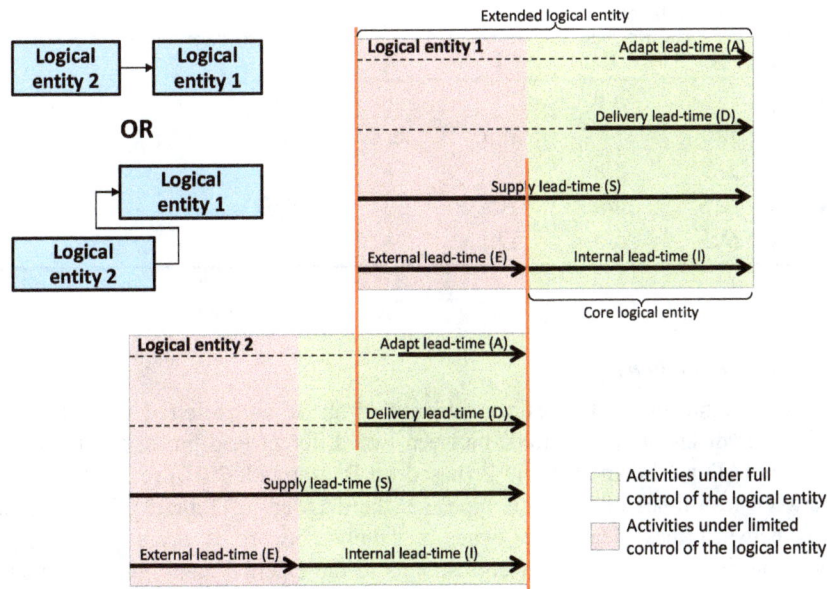

Figure 11.   Dyad of two logical entities with no integration.

Figure 11. Logical entity 1 would then correspond to all activities required to supply with item $Z$ in the example of Figure 10, i.e. the focal entity. Logical entity 2 is related to the supply of item $V$ or $Q$ (since $E + I = S$ in logical entity 1). These items may in turn consist of other items but that is not shown in the example but covered by other structures within the supplying logical entity (logical entity 2). The green part on the right of each logical entity represents the part of the flow that is controllable for the logical entity and the red part on the left represents the part that is not controllable for the logical entity. As shown in Figure 11 there is an overlap between the entities but in the overlap only one logical entity has the ability to control the activities (logical entity 2 in the overlap in Figure 11). This assumption is relaxed later in section 6.3.2. This type of overlap is illustrated in Figure 11 on the top left of the figure where two logical entities are in sequence with some overlap. The top entity-dyad illustrates in this way the core logical entities (green segment) while the lower entity-dyad illustrates the extended logical entity (which corresponds what is referred as a logical entity in the text unless otherwise stated).

The external lead-time ($E$) for logical entity 1 corresponds to the lead-time for a purchase order, e.g. for item V or item Q in Figure 10. From the supplying entity's perspective (logical entity 2) it corresponds to the delivery lead-time ($D$) for a customer order. Hence, this is also an illustration of how a purchase order of the focal company is connected to a customer order at the supplier.

The strategic lead-times thus represent a link between different logical entities as exemplified by $E$ for logical entity 1 and $D$ for logical entity 2. Within each logical entity a number of important lead-time relations can be identified. In particular three lead-time relations are of key important to identify decision categories for COBM and related strategic choices in terms of positioning of decoupling points and decoupling zones.

Table 2.   Lead-time relations.

|     |   S   |   E   |   I   |   A   |   D   |
|-----|-------|-------|-------|-------|-------|
| S   |   -   |  n/a  |   ×   |   ×   |   ×   |
| E   |  n/a  |   -   |  n/a  |  n/a  |  n/a  |
| I   | *I:S* |  n/a  |   -   |   ×   |   ×   |
| A   |  A:S  |  n/a  |  A:I  |   -   | *A:D* |
| D   | *D:S* |  n/a  |  D:I  |   ×   |   -   |

## 5.4.   *Lead-time relations*

Five different strategic lead-times were identified above and referred to as SEIAD. In total 20 different lead-time relations, between two different lead-times, can be identified based on SEIAD as shown in Table 2 (based on Bäckstrand, 2012) (excluding the relation between lead-times of the same type, i.e. the diagonal of Table 2). There are some overlaps between these lead-times when a supply network is investigated since the delivery lead-time $D$ for a supplier corresponds to the external lead-time $E$ for that item at the customer. Note, however, that the analysis performed here is based on one logical entity. Since the focus here is on COBM the emphasis is on customer facing lead-times (see Figure 11), i.e. $A$, $D$, $S$, and $I$ (lead-time relations related to $E$ are therefore indicated with *n/a* in Table 2 and since $S = E + I$, the $E$-based relations may be derived based on $I$-based relations if necessary). The lead-time relations above the diagonal are the inverse of the lead-time relations below the diagonal and only one relation of each 'pair' is included. As a general rule the relation where we normally would expect the smaller value in the numerator is included and the other is indicated by a '×' in Table 2. In total there are hence six potential lead-time relations left to investigate: $D:S$, $A:D$, $I:S$, $A:I$, $A:S$, and $D:I$.

In a lead-time relation one lead-time is used as a point of reference and the relation then shows how long the other lead-time is in relation to the reference lead-time. The lead-time relations further investigated are mainly related to $D$ since the delivery lead-time $D$ is the fundamental lead-time from customers' perspective. Note that the $I:S$ relation and the $I:A$-relation are multiple relations since there are $N$ pieces of $I$ lead-times (one for each branch of the bill-of-material) and hence also $N$ pieces of $I:S$ relations. In case of the $A:I$ relations it is even more complex since there may also be multiple $A$ lead-times, and in addition $A$ may be supply based or demand based. The $A:I$ relation could be of some interest as it can be used to analyse if customization affects more than the focal entity. This information is, however, implicitly provided by the $I:S$-relation in combination with the $A:D$-relation when the compounded decision domains are introduced. In a similar fashion the $A:S$ relation represents how customization affects supply but it is less exact compared to the $A:I$-relation and is hence not included. The $D:I$ relation, finally, could provide some interesting information on if customer-order-driven flow impacts suppliers but this information can also be obtained from compounded decision domains which are described later. The three remaining lead-time relations are of key importance to identify important decision categories for COBM and they are highlighted in Table 2 as bold and italic. As a result there are three lead-time relations that connects the two aspects of supply ($I$ and $S$), the two aspects of demand ($A$ and $D$, in particular if $A_D$ is assumed), and finally the key link between supply and demand ($D$ and $S$).

### 5.4.1. D:S-relation

The fundamental lead-time relation is based on the delivery lead-time ($D$) in relation to the supply lead-time ($S$). If the customers can accept to wait longer than it takes the supplier to provide the product it becomes possible to perform all provisioning activities at the request of a customer. If, on the other hand, the customer cannot accept to wait the time it takes to perform the activities at least some of these activities must be performed on speculation about future customer orders. This lead-time relation has received relatively large attention in both practical application and literature (Shingō, 1989) (originally published in Japanese in 1981) and is usually seen as the earliest reference on this even if Shingō used the denotation 'relation of $D$:$P$'. This relation was introduced to a broader audience by Mather (1984) who observed the potential of this concept. Mather did however note that $D$:$P$ could be mixed up with DP, which at that time was widely used as an acronym for 'data processing'. To reduce the risk for this confusion he suggested that it would be better to call it the $P$:$D$ relation (or $P$:$D$ ratio as Mather called it). Unfortunately the numerical result of the $P$:$D$ relation does not provide an intuitive sense of how large part of $P$ that is customer-order-driven. Due to this the original relation $D$:$P$, as defined by Shingō, is used here as a point of departure.

Initially (Shingō, 1989) referred to $P$ as the product lead-time but it has also been associated with production lead-time. Today supply is used as a terminology for all types of activities related to the provisioning of goods and hence $S$ is here used instead of $P$. Both Shingō (1989) and Mather (1984) used $D$ as an abbreviation for delivery lead-time and this is also used here. Note that $D$ implies that activities within $D$ may be performed to customer order. It is, however, not a requirement but should be seen as an 'option' to provide to customer order. If the products are standardized and delivered frequently the activities may also be performed on speculation with reasonable risk taking. This option is sometimes not fully used by for instance production engineering reasons where it is suitable to produce a quantity that deviates from an individual customer order. Some standard examples of the $D$:$S$-relation are:

- $D \ll S$: Basically all activities must be performed on speculation (corresponds to the strategy Make-to-Stock, MTS).
- $D < S$: Some activities are performed on speculation (corresponds to the strategy Assemble-to-Order, ATO).
- $D \approx S$: All activities can be performed on commitment (corresponds to the strategy Make-to-Order, MTO, Purchase-and-Make-to-Order, PMTO, or Engineer-to-Order, ETO, depending on if engineering activities are included or not).

### 5.4.2. A:D-relation

The relation between adapt lead-time ($A$) and delivery lead-time ($D$) is relating two aspects of demand to each other. The distinction made here is based on how large part of $D$ that is actually related to a part of the lead-time that is customer order unique, $A$. This relation highlights that even if the flow is customer-order-driven it is not necessarily also tailor made for that particular customer. The $D$:$S$ relation shows what is customer-order-driven and therefore also that there is an 'option' for providing something customer order unique. A more detailed approach could also, as mentioned earlier, differentiate if $A$ is what is actually possible from a supply perspective ($A_S$) or if $A$ is related to what is required by the customer ($A_D$) (Bäckstrand & Wikner, 2013). From a supply perspective the different branches of the bill-of-material may provide different opportunities for customization and hence $A_s$ should also be defined for each and every branch of the bill-of-material. In summary $A_D$ represent the market requirements and $A_s$ the opportunities available from a supply perspective. Hence the difference corresponds to the options available for responding to new market requirements due to e.g. increasing competition. In this context $A$ is however assumed to represent the customers' requirements. The $A{:}D$ relation shows to what extent the provider has decided to exercise the 'option' to provide something customer unique:

- $A \ll D$: Basically all activities are standardized.
- $A = D$: Customer-order-driven activities are also customized.
- $A > D$: Some forecast-driven activities are customer order adapted (a high risk strategy that usually should be avoided).

The connection between customer adaptation and customer-order-driven flow has many facets covering e.g. that it is not necessary for a product to first be completely engineered and then produced. It may also involve, for example, engineering adaptations where some semi-finished goods are in stock when engineering adaptations are made. This can be described as a two dimensional problem where engineering and production activities can be combined in different ways (Wikner & Rudberg, 2005a).

### 5.4.3. I:S-relation

In contrast to the $D{:}S$ and $A{:}D$ relation, which are based on demand, the $I{:}S$ relation is completely based on the supply side. The internal lead-time ($I$) represents the lead-time performed within control of the focal logical entity and the supply lead-time ($S$) represents the cumulative lead-time of the system in question. In the case of traditional relations to the suppliers, where integration is low and purchase order and customer order are the foundation for cooperation, this means that if $I$ is much smaller than $S$, basically all activities within $S$ are performed by suppliers upstream in the flow. But, the supplier plays a less important role in the supply chain in case the total lead-time $S$ is short and $I$ therefore relatively long. Also here three cases are of particular interest and note that in this case there is a lead-time relation for each 'branch', cf. Figure 10:

- $I_k = S_k$: Basically all activities of the $k$-branch are performed by the core logical system.
- $I_k < S_k$: Some activities of the $k$-branch are outsourced to a supplier.
- $I_k \ll S_k$: Basically all activities of the $k$-branch are outsourced to a supplier (the extended logical entity).

Based on the three lead-time relations defined above it is possible to define and position three decoupling points and three process-based decoupling zones that constitute three decision categories and thus also the core of the flow based framework for COBM.

### 5.5. Example: Lead-time relations

Going back to the example of Figure 10 a set of lead-time relations can be calculated and used to classify both the whole product and the individual items. In this case the

Table 3.    Example: Item-based $D{:}S$ relations.

| Item ($k$) | Make/buy | $L_k$ | $S_k$ | $D{:}S_k$ |
|---|---|---|---|---|
| $Z$ | Make | 2 | 2 | $6/2 = 3$ |
| $Y$ | Make | 3 | 5 | 1.2 |
| $X$ | Make | 1 | 6 | 1 |
| $W$ | Make | 2 | 8 | 0.75 |
| $V$ | Buy | 2 | 10 | 0.6 |
| $U$ | Make | 3 | 9 | 0.67 |
| $Q$ | Buy | 4 | 12 | 0.5 |

Table 4.    Example: classification using item-based $D{:}S_k$ relations

| Category | Item |
|---|---|
| Requires speculation | $W, V, U, Q$ |
| Possible to speculate on | $X$ |
| Not suitable for speculation | $Y, Z$ |

$D{:}S$ relation is analysed and in particular the item based $D{:}S$ relations are analysed. The indented bill-of-material of Table 3, based on Figure 10, includes the item based $D{:}S$ relations in the rightmost column. $S_k$ is calculated for each item based on the cumulative lead-time for that particular item. For item X the cumulative lead-time $S_X = 2 + 3 + 1 = 6$ periods which is the same as the delivery lead-time according to Figure 10 (D=6). Consequently the item based $D{:}S_k$ relation is one which means that item X can be produced based on customer order.

According to the introduction of the example, item $Z$ is customer order unique and $Y$ is customer unique. Both of these items are within the delivery lead-time which means that they can be produced based on customer order. Since they are unique for the customer they are not suitable for speculation even if recurring customer orders would make it possible to speculate on $Y$ since it is unique for the customer and is hence the same for different customer orders from the customer, The items $W$, $V$, $U$, and $Q$ all have a $D{:}S_k$ relation $< 1$ and must be made to forecast and thus requires speculation. Item $X$ finally is interesting since it has a $D{:}S_k$ relations $= 1$ indicating that is should be customer order driven but at the same time it is a standard item used for all types of customer. In this case the $D{:}S_k$ relation should be interpreted as a standard item with an option to produce either to customer order or to forecast, i.e. on speculation. This discussion is summarized in Table 4.

## 6.    Customer-order based management

The different lead-time relations introduced above each represents a way of dividing the flow into two parts with different preconditions. The interface between these two types of flow involves a discontinuity along the flow in the sense that the preconditions change significantly when the flow crosses that point, which also corresponds to the interface between two decision domains. Each one of the lead-time relations identified as important in Table 2 corresponds to such an interface and is of critical importance as decision support for COBM. The three decision categories covered by this analysis are flow driving, flow differentiation and flow delimitation. In addition the DC flow transparency, with a slightly different approach, is introduced below.

## 6.1.   Decision category: flow driving

Flow driving refers to what it is that decides if a process should be initiated or not. In the case of *flow driving*, it is assumed that a high *level of certainty* of the driver is defined from a customer perspective. The decoupling point that separates certain from uncertain flow driving is referred to as the CODP (see e.g. Giesberts & van der Tang, 1992) and based on the *D:S*-relation, see Table 2. Downstream from the CODP demand is certain in terms of a 'perfect' customer order where what (in terms of form/place) the customer wants and when the customer wants delivery is determined. The hybrid decision domain here represents that there is some knowledge of what will be demanded but that some information is missing for complete certainty. This domain corresponds to the concept customer order decoupling zone (CODZ), introduced by Wikner & Rudberg (2005b). The starting point of the flow in Figure 12 is Source, which represents what is beyond the extended logical entity, and Sink, which is the end point (usually also corresponding to a customer) of the logical entity.

### 6.1.1.   Customer order decoupling point

The interest in the CODP gained ground when different so called hybrid strategies (see e.g. Sun, Ji, Sun, & Wang, 2008) came into focus. The pure make-to-stock (MTS) and make-to-order (MTO) strategies were employed in different companies, but as customer requested short delivery lead-times in combination with customization the assemble-to-order (ATO) strategy was highlighted (Wemmerlöv, 1984). The ATO strategy can be described as a combination of MTS with MTO where MTS is applied upstream from the CODP and MTO downstream. In this context, the CODP is usually seen as identical to a physical stock point. This material-based approach to decoupling points has a long tradition in the literature. There are a number of 'local' publications in e.g. industry magazines in different countries but the two publications usually quoted as the origins of this concept are Hoekstra & Romme (1992), originally published in 1985 in Dutch and Sharman (1984). Sharman referred to the concept as order penetration point (OPP) since it is a measure of how far deep into the supply process (from a lead-time perspective) that the actual customer order penetrates. The CODP has been used in many contexts and is also known as e.g. decoupling point (Hoekstra & Romme, 1992), customer order point (COP) (Mason-Jones & Towill, 1999; Olhager & Östlund, 1990), OPP (Sharman, 1984), supply stream decision point (Hines & Rich, 1997), material

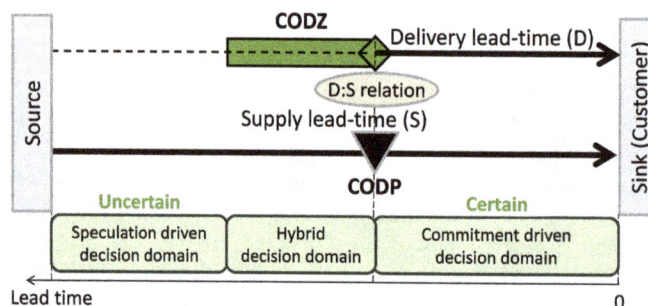

Figure 12.   DC for customer-order based management – flow driving.

decoupling point (Mason-Jones & Towill, 1999) and push-pull boundary (Chopra & Meindl, 2004). In the postponement literature, the CODP corresponds to time postponement relating to the time when the customer order actually drives the activities (Pagh & Cooper, 1998). This explicit focus on the physical stock point (see e.g. Banerjee et al., 2011; Sun et al., 2008) as the CODP is the dominant perspective in the literature but here the CODP is first and foremost seen as something related to decision-making in line with Wikner and Rudberg (2005b) and the CODP is therefore defined as:

> The *customer order decoupling point* (*CODP*) separates decisions about initiating flow based on speculation on future customer orders from commitment against actual customer orders.

The CODP is based on the *D:S*-relation, where *D* indicates the part of the flow where the (future) customer's demand is known. In Figure 12, the decision-based aspect of the CODP is indicated with a diamond (quadrangle). The stock point (black triangle in Figure 12) corresponding to the CODP constitutes a buffer between the speculation-driven decision domain and the commitment-driven decision domain. The CODP buffer is thus a safety stock for uncertainty in demand. The material in the CODP buffer is replenished based on speculation (forecast), i.e. an estimation of future demand for materials in the stock point has been done. The commitment-driven withdrawals represent real demand and the dimensioning of the CODP buffer is therefore based on the deviation between forecasted withdrawals and actual withdrawals which is in line with how safety stocks usually are calculated. It is however important to observe that the 'real' demand used in dimensioning should reflect the demand that the commitment-driven process can deliver. If there is capacity to handle variations in demand downstream, the pressure on the material availability in the CODP buffer increases, but if the capacity downstream is not that flexible it is also a constraint for how much the withdrawals can vary. The CODP hence play a critical role in order promising and this is further discussed in e.g. Fleischmann and Meyr (2004); Rudberg and Wikner (2004). Dimensioning of the CODP buffer should therefore be based on the flexibility that the flow downstream from the CODP can offer. Product mix variations can, however, increase demand on the buffer even if the volume is stable. Another interpretation of this scenario is that flow upstream from the CODP is goods based but downstream also can involve services (Fließ & Kleinaltenkamp, 2004). In this context, the CODP has also been called the service decoupling point (Wikner, 2012b).

### 6.1.2. Customer order decoupling zone

The CODZ involves a mix of uncertainty and certainty about customer orders' timing. Downstream from CODZ the flow is driven by customer orders and upstream from CODZ the flow is driven by a market-based forecast. Within the CODZ, the flow driver is uncertain to some extent depending on either partial customer order information (process-based CODZ) or aggregate capacity requirements for resources where some of the materials are forecast-driven and some are customer-order-driven (resource-based CODZ).

> The *customer order decoupling zone* (*CODZ*) covers decisions about initiating flow based on a compromise of speculation and commitment on customer orders.

For process-based CODZ, the certainty increases along the flow (Wikner & Rudberg, 2005b) and the level of certainty can then be defined by two dimensions based on igure 6:

- Spatial dimension (what?): On a scale from uncertain to certain about what the customer wants (in terms of form and place).
- Temporal dimension (when?): On a scale from uncertainty to certainty of when the customer wants delivery.

Resource-based CODZ is based on that the level of certainty can vary depending on to what extent resources are involved in a combination of speculation-driven and commitment-driven decision domains. This type of CODZ has been noticed in different contexts such as the assortment hybridity (Giesberts & Tang, 1992). The mixed load on resources can be created both as a consequence of that customers with different demand in terms of $D$ purchase the same product and that different products are loading the same resource but with different flow drivers. If the customers request different delivery lead-times, a backlog profile can be created where the resource-based CODZ represents the time horizon where some customer orders are known, but not all, and hence the resources performing the corresponding activities operate under mixed load. This could also be referred to as customer-based CODZ. Correspondingly, different products may create a mixed load on resources which means that the aggregate could be referred to as product-based CODZ. But, for simplicity, both these types of mixed load are associated with the concept resource-based CODZ. In both these cases, there is a challenge in terms of how to schedule the flow on the resources. For example, the forecast-driven flow can generate a levelled first loading and then being complemented with customer-order-driven load, or the customer-order-driven flow may be prioritized and first loaded on the resources and thereafter the forecast-driven flow can be used to level out the load on the resources.

## 6.2.  *Decision category: flow differentiation*

Flow driving concerns the driver of the process but provides no information about the level of customization involved, which is a separate issue (see e.g. García-Dastugue & Lambert, 2007; Hoekstra & Romme, 1992; Mason-Jones, Naylor, & Towill, 2000; Olhager & Östlund, 1990; Wikner & Wong, 2007). For decisions about customization, the point of departure is *flow differentiation*, which means the *level of uniqueness* in terms of form and place. A completely unique process is adapted for a specific customer order and is limited, from a flow perspective, by the customer adaptation decoupling point (CADP) related to the *A:D*-relation. Generic processes are upstream from CADP and provide standard products and therefore lack connection to specific customers in terms of form and place at this stage. The customer adaptation decoupling zone (CADZ) corresponds to some degree of adaptation, such as when the flow is customer unique, i.e. the product is customer unique, but not customer order unique, in terms of design (Wikner & Bäckstrand, 2012). The CADZ may hence extend upstream from the CODZ since speculation may be performed even on customer unique materials if the customer orders are recurring. A more comprehensive discussion on these issues, from a customization aspect, can be found in Graca et al. (1999). These concepts are illustrated in Figure 13.

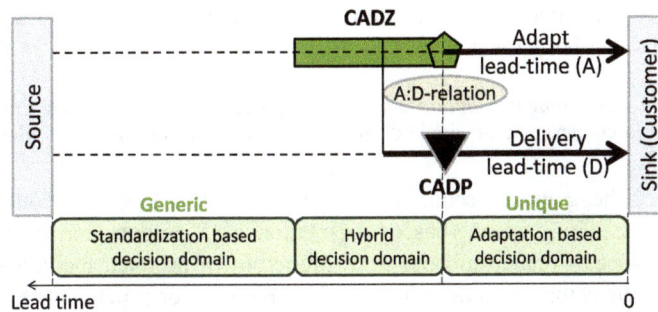

Figure 13.    DC for customer-order based management – flow differentiation.

### 6.2.1.   Customer adaptation decoupling point

The CADP is based on the *A:D* relation (Bäckstrand & Wikner, 2013; Wikner, 2012a) and represents the interface between flow of standard products and customer order adapted products (indicated with a pentagon in Figure 13). As mentioned above, the *A:D*-relation can be defined from a demand perspective or a supply perspective. When the decoupling point and decoupling zone for customer adaptation are defined, the demand perspective is assumed but in most cases the discussion is equally valid for the supply perspective.

> The *customer adaptation decoupling point* (*CADP*) separates decisions about differentiating flow based on standardization for a market of different customers from adaptation against actual customer orders.

The requirement for a stock point positioned at the CADP is in this case not as obvious as in the case of the CODP. It is, however, important to note that upstream from CADP there is recurring need for the same product which enables some speculation (it is here assumed that CADP is positioned downstream from CODP since it otherwise would result in speculation on customer order unique products, which is not recommended). If it is possible to perform standard activities between the CODP and the CADP to customer order there might be a lot-sizing buffer of standard products positioned at the CADP as a stock point. This can be useful when it is preferable to supply with larger quantities (due to e.g. long set-up time) than the customer order that initiated the batch requires (a lot-sizing method other than lot-for-lot is used). This type of lot-sizing inventory replenishement can also be created in other places upstream from CADP but CADP indicates the last point in the flow where this type of inventory should occur.

### 6.2.2.   Customer adaptation decoupling zone

The CADZ (Wikner & Bäckstrand, 2012) is a hybrid between what is generic from a customer perspective and what is unique from a customer order perspective. This compromise means that the flow is customer unique but not unique for a single customer order. This distinction between what is customer unique and what is customer order unique is in many cases of decisive importance when it is combined with the decision to speculate since it can be a reasonable decision for customer unique products but basically not for customer order unique products. In general, the CADZ is based on the *A:D*-relation and can be defined as:

> The *customer adaptation decoupling zone* (CADZ) covers decisions about differentiating flow based on a compromise of standardization and adaptation for customer orders.

As indicated in the general decoupling theory, the decoupling zones can be of two types: process-based or resource-based. Adaptation of a product can be performed in many different aspects related to both form properties of product, including the service performed, and the actual location, place, of the product. For process-based CADZ, two key dimensions for flow differentiation can hence be identified in line with Figure 6:

- Form dimension (product): On a scale from generic form (standard product) to unique form (customer order unique product).
- Place dimension (location): On a scale from generic place (e.g. central distribution site) to a unique place (e.g. the customer's site or a specific delivery point/site).

Within the process-based CADZ the product is not completely standard but has some characteristics that only is of interest to some customers which means that the product can be adapted in the meaning customer unique but not customer order unique (Wikner & Bäckstrand, 2012). The CADZ hence provides a more detailed view of customization where one level of customization is product related in the sense that it is unique for a particular customer, whereas customization for a particular customer order is unique for that particular customer order. In some cases, only the form dimension is applied since it could be argued that place is only unique once the product is in the hands of the customer. Unless form and place uniqueness are handled separately, the form uniqueness may then be dominated by the place uniqueness.

For resource-based CADZ, it means that one resource is involved in multiple processes that have different characters in terms of differentiation. This can create a number of challenges from a capability perspective since standard products usually have cost as an order winner, and customer order adapted products usually have order winners related to delivery speed or delivery precision. This of course also affects the loading challenge indirectly since products with focus on delivery precision have other requirements on the process compared to products of standard character which can be kept in inventory.

### 6.2.3.  Product differentiation

Customer adaptation (or customization) as a concept is part of the wider concept of product differentiation since the latter by definition is not explicitly connected to customer demand but covers products in general, including customer generic products. Also for the flow upstream from the CODZ, a number of important decisions may be made related to what in the TOC (see e.g. Blackstone, 2008) are known as control points related to diverging flows which means that one particular material or component may be used for different purposes due to supply-based differentiation. A set of standard components can be assembled to a wide range of products based on forecast and available inventory for each product. Hence, the decision to use the components for a particular product is in this case also resulting in differentiation since different variants could be created by the standard components (see e.g. Hoekstra & Romme, 1992). Note however that the differentiation in this case is based on supply aspects and not related to a particular customer. The subject can therefore be divided into demand-based and supply-based differentiation:

*Demand-based differentiation* is based on customer requirements as above, which means that the product (the result from a process) in some sense is customer adapted. The customer unique flow is related to the CADZ in Figure 13 and the customer order unique flow is related to the flow downstream from the CADP.

- Customer adaptation type A: Customer order unique product: Properties are related to requirements that are unique for a specific customer order of one-time character, wherefore the product is difficult, and maybe even impossible, to sell at a later time.
- Customer adaptation type B: Customer unique product: Properties that are related to requirements that are unique for a particular customer but covered by recurring customer orders. This type of products provide some opportunity for speculation since it is a recurring customer.

*Supply-based differentiation* is independent of specific customer requirements and is therefore only related to customer generic product (standard products). In Figure 13, this corresponds to the flow upstream from the CADZ but may in some instances also be an additional cause of differentiation in the CADZ due to the design of the product in terms of the process-based CADZ or due to mixing different flows in terms of resource-based CADZ. Note that $CADP_S$ represents potential customer-based differentiation, whereas supply-based differentiation is independent of customer requirements. Supply-based differentiation is therefore an important concept, but not in focus for COBM as outlined here.

- Standard products can be sold to a market consisting of multiple customers. Product differentiation in this case is related to that material changes in terms of form or place to be part of different types of customer generic products.

### 6.3. Decision category: flow delimitation

The third flow-based DC complements the other two in the sense that flow driving is related to why and when a flow is initiated and flow differentiation is related to decisions about the uniqueness of the flow. The remaining issue, according to the analysis of Table 2, is how the flow is managed and this DC is here referred to as *flow delimitation*. This DC is based on the extension of the logical entities and is related to the *I:S*-relation and the *level of finiteness* that can be applied in managing the flow. The decoupling point is defined from a control perspective but is despite this referred to as the purchase order decoupling point (PODP), (Wikner et al., 2009). Since the PODP is related to the *I:S*-relation it is important to remember that since there might be multiple *I:S*-relations, there might correspondingly also be multiple PODPs related to one product (see e.g. Figure 10 with $I_V$ and $I_Q$). The background to this name is that what is referred to as 'external' and 'uncontrollable' frequently is also related to suppliers and hence related to purchase orders (note that this is however not necessary). The analogy with CODP and customer order also supports the naming of the decoupling point in the framework. Somewhat simplified, processes performed by the focal company (same as focal logical entity in this case) can be considered as controllable as only own resources are involved. Processes performed by suppliers can in a corresponding fashion be considered as uncontrollable. In some cases, the customer can control parts of the supplier's resources (for example, when capacity is 'purchased') or depending on divided responsibility so that all the own resources can be controlled as one entity. Correspondingly, a supplier may have influence on some of the customers flow through vendor-managed inventory (VMI), which would move the PODP downstream compared to the situation before VMI was implemented. Wikner and Bäckstrand (2011) provide an overview of different configurations. Since controllability cannot unambiguously be related to customer/supplier interface the controllable decision domain is referred to as 'Internal' and the uncontrollable decision domain as 'External'.

Also in this case, a hybrid decision domain can be identified. Supply of material and capacity can vary between finite and infinite. If material and capacity can be analysed as finite resources there is full controllability but this is not always possible to achieve. The hybrid decision domain, where resources to some extent can be managed as finite, are in Figure 14 represented by the purchase order decoupling zone (PODZ).

### 6.3.1. Purchase order decoupling point

The PODP is indicated with a hexagon in Figure 14 and is based on the *I:S*-relation. The interface between the decision domains is in this case based on the controllability of the process which means that the PODP can be defined in line with CODP and CADP:

> The *purchase order decoupling point* (*PODP*) separates decisions about delimiting flow based on what is external to the logical entity from what is internal and hence controllable.

The PODP is based on the *I:S* relation and the interface is in this case based on the flow delimitation and therefore also the controllability. Upstream from PODP is another logical entity that 'owns' the controllability but downstream the flow is controllable from the focal logical entity's perspective. Since the flow upstream from the PODP is not controllable for the focal entity, it can also involve some uncertainty considering the supply to the stock point indicated in Figure 14 at the PODP. To handle this uncertainty, either safety lead-time is used, which results in some stock since materials are delivered

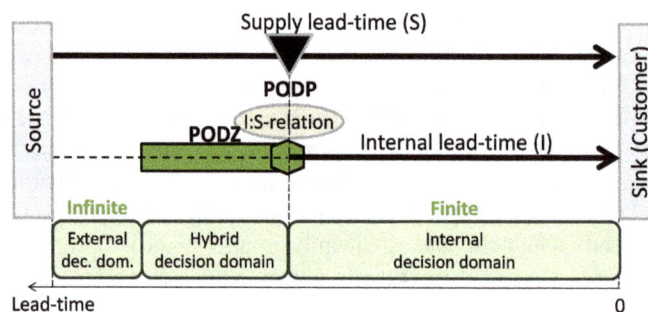

Figure 14.   DC for customer-order based management – flow delimitation.

before there is a requirement, or a traditional safety stock is used with a physical quantity. In combination with CODP, the PODP creates different preconditions for supplier interaction (Wikner & Bäckstrand, 2011).

### 6.3.2. Purchase order decoupling zone

Planning and control of materials flow has traditionally been divided into planning and control for manufacturing and for materials supply (basically purchased materials). This delimitation along the flow is becoming less significant and focus is more and more on creating a holistic and integrated management approach to the whole flow independent of, if it covers one department or multiple integrated actors in a supply network. In this context, the logical entities play a crucial role since they are defined from a management and control perspective. The activities that are included in the core logical entity, see Figure 11, are controllable and finite principles considering resource constraints can be applied. The upstream activities are considered as non-controllable which means that infinite control principles can be applied where a lead-time is given but the availability of resources in terms of materials and capacity is unknown to the core logical entity. The delimitation between these domains is not always that obvious but different degrees of finite can exist, which is also what defines the PODZ.

> The *purchase order decoupling zone* (*PODZ*) covers decisions about delimiting flow based on a compromise on what is external and what is internal to the logical entity.

Two key dimensions for flow control can be identified and they are applied in many contexts related to planning and control. For example, in MRPII, materials are planned using a time-phased approach and then a capacity evaluation is performed based on these plans (Wight, 1984). In rate-based approaches, capacity is usually planned first through e.g. line balancing and then the product mix and required materials are scheduled as needed and replenished using pull systems (see e.g. Duggan, 2012). At an overall level, most techniques for planning and control are emphasizing these two dimensions:

- Material dimension (transformed resources): On a scale from infinite to finite materials and with focus on availability of materials at stock points.
- Capacity dimension (transforming resources): On a scale from infinite to finite capacity and with focus on availability of capacity at resources.

For CODZ and CADZ, it is obvious that there is a process-based and a resource-based variant of the decoupling zone. For PODZ, it is however not as intuitive to identify these two perspectives since resources play such a key part in this DC on controllability. But, also in this case the two types of decupling zones can be identified.

Process-based PODZ represents a scenario where the possibility to apply finite techniques gradually increases along the flow. In general, it is assumed that the more downstream the more information about the resources is available to the logical entity. It is, for example, possible to apply finite scheduling in the own production (at least in theory, in practice all information might not be available due to e.g. technical reasons) but in the parts that involve suppliers, the information about availability might be limited and the only given information might be the planned lead-times. For the receiving logical entity, it means that the supplier is upstream from PODP and hence classified as infinite. It could however be that the availability of some of the capacity at the supplier is known and hence it would be possible to consider as finite, which would correspond to being inside a process-based PODZ, where some of the capacity can be finite scheduled or that the capacity can be finite scheduled to some degree.

The resource-based PODZ is an aggregate of multiple flows that consumes the same resource. In most cases, all load would be treated in the same way meaning for instance that if a resource is scheduled using finite capacity all load would have to compete for this capacity using some kind of priority as a base line but when the maximum capacity is reached more capacity is only added as an exception. An example of this scenario is when different customers have different priority. Normally, the resources are finite scheduled but in some cases fast delivery is highest priority and by using, e.g. overtime the extra capacity can be provided to secure delivery reliability. In this way, the resource is managed as finite for some flows/customers and as infinite for other flows/customers.

### 6.4. Decision domains for customer-order-based management

With flow-based decision-making as a point of departure, three different decision categories have been identified above with support from analysis of lead-time relations. Three different sets of decision domains have been identified together with three decoupling points and related decoupling zones. The four strategic lead-times $S$, $I$, $A$, and $D$ have been used to position the decoupling points and therefore also decoupling zones from a flow perspective. To be able to decide the starting point of the zones require further information which either can be based on the first decision domain in the flow or simply the length of the decoupling zones, which is included in Figure 15 as CODZ lead-time, CADZ lead-time or PODZ lead-time. From a planning and control perspective, it can also be observed that the supply lead-time $S$ corresponds to the planning time fence (PTF) if master scheduling (Blackstone, 2008) is applied at the level of the delivery object and defines how far into the future that flow is a constraint according to planned lead-times. Note however that sometimes a master schedule is used at lower levels in the flow such as at the CODP, and in this case the PTF would cover the time

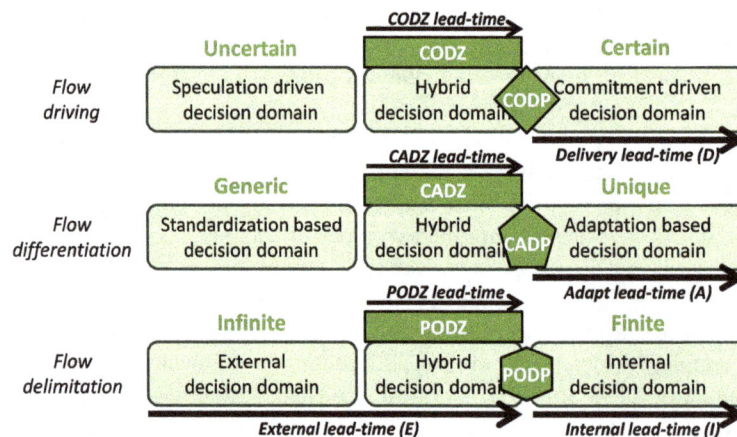

Figure 15.   Decision domains for three decoupling points and three decoupling zones.

span from the CODP to the beginning of the flow, i.e. the PTF would at the CODP be set to PTF = $S$–$D$. If there are several branches in the product structure, such as in the example of Figure 10, PTF is related to the longest branch in terms of lead-time. PTF is also referred to as frozen time fence, since firm planned order is used within PTF. Correspondingly, the demand time fence (DTF) that also can be used for planning and control of delivery objects (Blackstone, 2008) is associated with the CODP since DTF indicates the part of the flow that, from a lead-time perspective, is customer-order-driven with no respect to forecast. DTF is also used in master scheduling. All deliveries within DTF must be related to customer order to make it possible for all customer-order-driven activities to be performed within the delivery lead-time. In case master scheduling is applied at the CODP, the DTF would be positioned at time zero since the master schedule then would be applied in a MTS-scenario.

The three decoupling points and the three decoupling zones can therefore be summarized as:

- CODP: $D$:$S$ relation, safety stock for uncertain demand, DTF.
- CODZ: Process-based: Different levels of certainty between uncertain and certain for the attributes what and when. Resource-based: Mixed load for resources based on forecast-driven and customer-order-driven flows.
- CADP: $A$:$D$ relation, last point in flow with lot-sizing inventory.
- CADZ: Process-based: Different levels of uniqueness between customer generic and customer order unique for the attributes form and place. Resource-based: The same resource is involved with different levels of customization.
- PODP: $I$:$S$ relation, safety stock for uncertain supply.
- PODZ: Process-based: Different levels of controllability between infinite and finite for the attributes material and capacity. Resource-based: Different levels of finiteness are applied to different load-generating activities for the same resource.

Decoupling points thus mean that the preconditions for decision-making can vary depending on which part of the logical entity that is considered. This more sophisticated description of the decision problem is modelled in Figure 15 in terms of three decision categories with three decision domains each. The resource-based decoupling zones are however not explicitly illustrated in Figure 15.

So far the three decision categories have been analysed individually, but in practice a combination of multiple decision categories are usually relevant and the compounded decision domains are suitable to use for this purpose in COBM.

### 6.5. *Compounded decision domains for customer-order-based management*

The three decision perspectives related to flow driving, flow differentiation and flow delimitation are important but do not exist in isolation from each other. In some decision situations, one of these could dominate the others which may then be less important and therefore neglected. From a more general perspective, the context of the decision in a specific situation is based on a combination of different decision domains. A flow decision may, for example, be based on a certain requirement (flow driver is a customer order) for a standard product (flow uniqueness is standard in terms of standard product from stock) and own resources are used (flow control can be performed in terms of finite material and finite capacity for delivery from finished goods inventory).

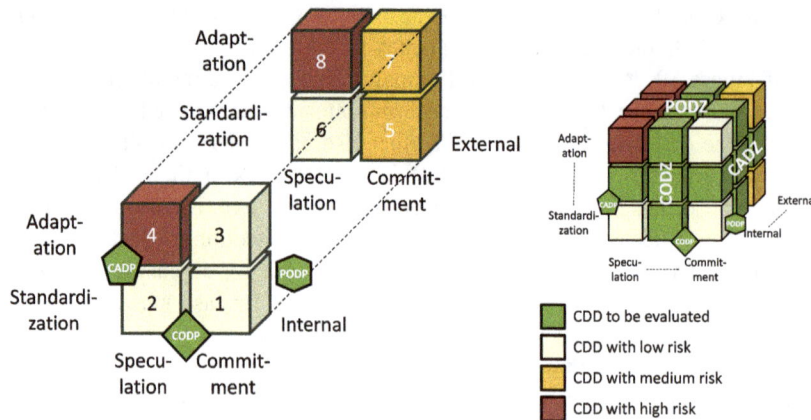

Figure 16. Compounded decision domains.

Since the three different types of decision domains, based on lead-time relations, are not mutually dependent, in terms of what they represent, they can be considered as three separate dimensions as shown in Figure 15. By combining these three dimensions of decision domains, eight compounded decision domains (CDD) can be identified if only decoupling points are included (on the left of Figure 16) or a total of 27 CDD if also decoupling zones are included (on the right of Figure 16). This approach is based on Figure 9. A CDD can be used to identify risk levels for different process configurations. In the case with eight CDD, the CDDs 1, 2, 3 and 6 (as indicated in Figure 16), represent scenarios that are well-established scenarios with relatively low risk. CDDs 5 and 7 represent a slightly higher level of risk when uncontrollable parts of the flow (e.g. a traditional supplier) are involved in the customer-order-driven part of the flow (which also can be customer order unique). CDDs 4 and 8, finally, have high risk since a customer order unique flow is performed on speculation.

To make it easier to reference to individual CDDs in the more complex model with 27 CDDs, to the right in Figure 16, a typology is introduced to be able to refer to one individual CDD where the first dimension ($x$-axis, to the right) represents flow driving, the second dimension ($y$-axis, upwards) flow differentiation and the third dimension ($z$-axis, outwards) flow delimitation. The example above with a certain demand for standard products supplied by own resources would then be represented by CDD (3,1,3), i.e. the cube to the right, furthest down and at the front, which is a CDD with low risk.

The three light yellow CDDs correspond to the standard cases with own production of standard products to forecast (1,1,3) and customer order (3,1,3), respectively, and to make something customized to customer order (3,3,3). The slightly darker CDDs at the back on the right (3,1,1) and (3,3,1) are CDDs with medium risk in the sense that this is about standard products but the supply is performed by resources that are outside the control of the logical entity (e.g. being performed by a supplier). CDDs (1,3,1), (1,3,2) and (1,3,3) are CDDs with high risk since they involve customer order unique production under speculation, which obviously should be avoided. The remaining CDDs include one or more zones and require a more thorough analysis to be evaluated. The present knowledge about these CDDs is simply not so developed in neither the literature nor in practical application. A further complicating factor is that the decoupling zones can be of two types, i.e. process-based or resource-based. Above the decoupling zones included were process-based. As discussed in the section on CDDs and RB-DZs, the complexity would increase even further if RB-DZs were included since intersections of the 27 different CDDs would then be involved.

### 6.6. Decision category: flow transparency

The lead-time-based decoupling points define different decision categories in terms of decision domains as outlined above. A complementary type of DC is based on the sharing of information. This type of decoupling point does not define the decision situation or how the information may be used but indicates if a specific type of information is available as decision support. If there is a need for information that the owner of the information cannot, or does not want to, share this is indicated in the flow by an information-based decoupling point. From a flow perspective, there are in particular two types of information that are in focus: information about demand and information about supply.

Sharing of demand information upstream in the flow is sometimes associated with sharing sales information such as point-of-sales data. From a more general perspective, all types of sales information are involved related to, e.g. customer orders, direct sales, delivery schedules and call-offs. The sharing of demand information upstream is limited by the demand information decoupling point (DIDP), which corresponds to information decoupling point according to Mason-Jones and Towill (1999). The DIDP should be positioned upstream from the CODZ, or upstream from the CODP if no CODZ is present (Olhager, Selldin, & Wikner, 2006), as shown in Figure 17.

> The *demand information decoupling point* (*DIDP*) constrains the transparency upstream of demand information.

If DIDP was positioned downstream from the CODP, it would mean that parts of the customer-order-driven flow would not have access to information about the customer order which would obviously be problematic. This could, however, occur if the CODP is positioned at a supplier who is not allowed to take part of the customer order information directly but only after it has been 'filtered' by the logical entity in direct contact with the customer. Considering also the CODZ, the DIDP should be positioned upstream of CODZ, if it is included in the analysis, since the CODZ indicate that some

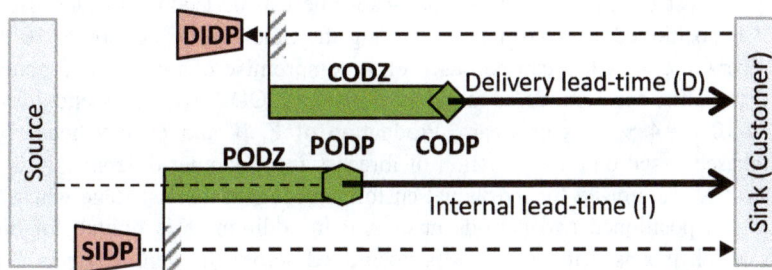

Figure 17.    DC for customer-order-based management – flow transparency.

of the process driving is based on customer order. DIDP therefore highlights the interface between the part of the flow where all requested demand information is available and the part where different types of filter have been applied to the demand information.

Sharing of supply information downstream, i.e. downstream units having access to information of an upstream state, in the flow, enables for decision-makers to have information available regarding supply resources, such as available capacity and/or the usage of capacity. The will to share information may also in this case be limited. The user requesting the information is sometimes not allowed to take part of it and this constraint in information sharing is indicated by the supply information decoupling point (SIDP), see Figure 17.

> The *supply information decoupling point* (*SIDP*) constrains the transparency downstream of supply information

Since PODP marks a limit in the flow what concerns the possibility to control the flow, it also means that SIDP cannot be positioned downstream from PODP as indicated in Figure 17. If PODZ is also considered, the SIDP should be positioned upstream of PODZ since the PODZ indicates that some information on availability is required. Also note that since the general case can consist of multiple PODPs there would also be corresponding multiple SIDPs.

In both the case with demand information and the case with supply information, it is about the availability of information. The actual use of information is instead related to the decision domains or CDDs in combination with resource-based decoupling zones.

### 6.7. Example: decision categories

A fictitious product Z was introduced in Figure 10 in terms of a time-phased bill-of-material. Strategic lead-times were identified for the product and later some of the lead-time relations were analysed, see e.g. Table 3. Continuing on this example, some aspects of COBM can be illustrated. As a first step, a representative order book is assembled. In this example, the representative order book consists of 15 customer orders. The order book is sorted according to the requested delivery lead-time and illustrated as a time-phased order book in Figure 18. An alternative would be to use the promised delivery lead-time but to actually reflect the market requirements the requested delivery lead-time is preferable. The customer orders are positioned horizontally, where the requested delivery lead-time corresponds to the orange part at the right of each bar. The delivery lead-times vary between 6 periods and 10 periods. Consequently, all activities during the last 6 periods of the supply lead-time can be customer-order-driven and the CODP is positioned at $D = 6$ periods. During the time from 6 periods to 10 periods before delivery, the activities can be based on a compromise of forecast and actual customer orders. This corresponds to a resource-based CODZ which extends from the CODP and $10-6 = 4$ periods upstream. Production of $V$, $W$ and $U$ may hence be produced/purchased based on a combination of forecast and customer orders.

Product Z is customized for individual customer orders at the last stage which means that $CADP_D$ is positioned two periods upstream. In addition, $Y$ is unique for different customers which means that a $CADZ_D$ is positioned across the item $Y$. Items $V$ and $Q$ are purchased items, which here corresponds to the respective PODP since the product is assumed to be produced in one logical entity. The information about the activities

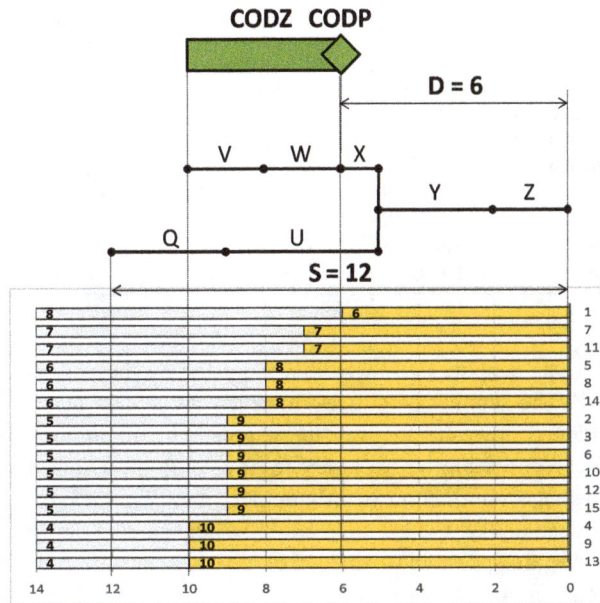

Figure 18. Example: the time-phased bill-of-material in relation to the time-phased order book.

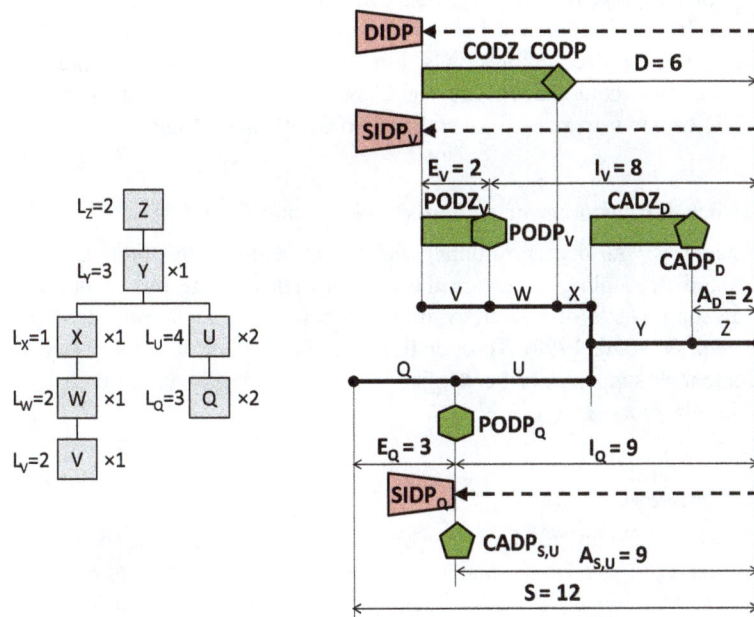

Figure 19. Example: the time-phased bill-of-material with decoupling points and decoupling zones.

and resource load at the supplier of $Q$ is limited, wherefore the $SIDP_Q$ is positioned at the $PODP_Q$. The supplier of $V$ provides load information but still is independent, wherefore this supplier is positioned within $PODZ_V$. Demand information (i.e. point of sales

Table 5.   Example: CDD classification of the items (columns ordered in line with Figure 20).

| Item ($i$) | Driving | Differentiat. | Delimitation | CDD | SIDP | DIDP |
|---|---|---|---|---|---|---|
| Z | Customer order | Customer order | Internal | (3,3,3) | DS | DS |
| Y | Customer order | CADZ | Internal | (3,2,3) | DS | DS |
| X | Customer order | Standard | Internal | (3,1,3) | DS | DS |
| W | CODZ | Standard | Internal | (2,1,3) | DS | DS |
| V | CODZ | Standard | PODZ | (2,1,2) | DS | DS |
| U | CODZ | Standard | Internal | (2,1,3) | DS | DS |
| Q | Forecast | Standard | External | (1,1,1) | US | US |

data) is provided to the supplier of V which is indicated by the position of DIDP which is upstream from $V$. The supplier of $Q$ does however not receive demand information as the DIDP is positioned downstream from the supplier. In this case, only one DIDP is used but in some cases it might be necessary to have multiple DIDP depending on how the provision of demand information is differentiated between different suppliers. Using the notation introduced above, the example could be illustrated as in Figure 19.

At this stage, it is also possible to categorize all items according to CDDs. Based on Figure 19, each item can be categorized in relation to the decoupling points as shown in Table 5. Item $Y$ for instance is positioned downstream from the CODP and is hence customer order driven. It is also customer unique which corresponds to a position within the CADZ and finally it is manufactured and downstream of the PODPs which correspond to internal supply. In summary, this corresponds to CDD (3,2,3). It is important to note that each item is classified based on the most upstream position of the item and how it is related to the decoupling points. Finally, the information decoupling points are included in the two columns on the right. Each item is then categorized as being upstream (US) or downstream (DS) in relation to SIDPs and DIDP.

## 7.   Framework for customer-order-based management

Decision categories with decision domains related to lead-time relations and also decoupling points and decoupling zones are the core elements of the framework for COBM. This type of components of a strategy framework is usually referred to as the *content* (Leong, Snyder, & Ward, 1990). To operationalize the content, it is necessary to identify how the content is supposed to be applied and this is usually referred to as the work process or simply *process*.

### 7.1.   COBM framework: content

The framework's different decision categories, i.e. its content, have been defined and exemplified above. Figure 20 summarizes the most important elements of the four decision categories. Three of the decision categories in Figure 20 are decision based and of similar structure. The DC Flow transparency is of information character and has in this way a more supportive function, but nevertheless it is covering decisions about level of transparency and hence also about positioning of two decoupling points. Flow transparency is also absolute in the sense that it is based on lead-times, whereas the other decision categories are relative since they are based on lead-time relations.

The four decision categories of Figure 20 represent the core aspects of the COBM framework as it is defined here. The purpose of the approach is to highlight that by

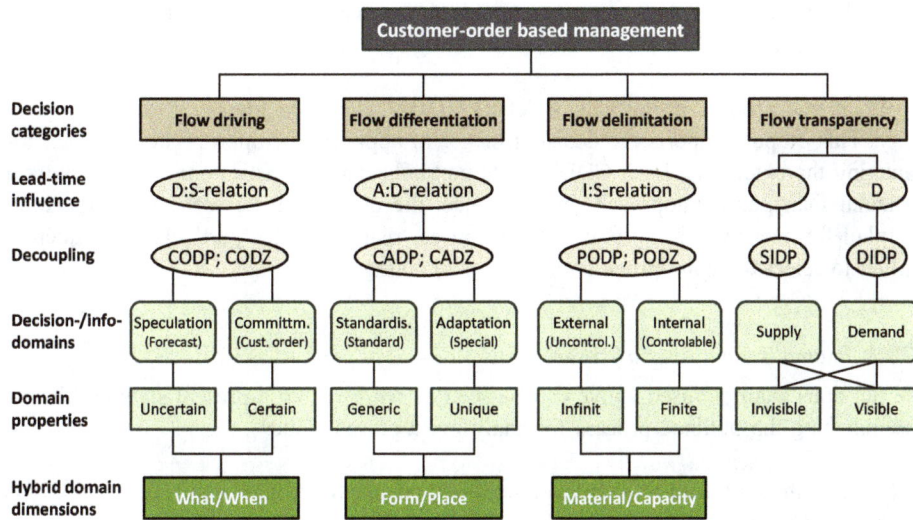

Figure 20.    Key content in the framework for customer-order-based management.

identifying the different decision domains, positioning decoupling points and decoupling zones, and consequently identifying the CDDs, an appropriate approach to flow management can be outlined. The procedure for applying this type of approach is then referred to as the process of the framework.

### 7.2.    COBM framework: process

A complete process for analysing COBM is relatively comprehensive with a large number of alternative activities depending on the specific circumstance in the particular case. The processes below are only outlined to provide the general idea and are based on Wikner (2012a). A more comprehensive process for supplier interaction applied in industrial cases is presented in (Bäckstrand, 2012; Bäckstrand et al., 2013). There are mainly two scenarios that are of interest and they are based on the analysis of a present state and the design of a future state, respectively. In both cases, the point of departure is usually the analysis of one product family, i.e. one value stream, at a time.

#### 7.2.1.    Analysing a present state

When analysing a present state, the objective is usually to investigate if the used control model is suitable for the present preconditions. The process can be summarized in seven steps:

(1)  Identify the strategic lead-times $S$, $A_s$, $D$ and the relevant $I_s$.
(2)  Analyse lead-time relations.
(3)  Position the strategic decoupling points and use decoupling zones when suitable.
(4)  Analyse the positioning of the information decoupling points SIDP and DIDP.
(5)  Identify which CDDs that are applicable.
(6)  Evaluate the concerned supplier/customer relations from a CDD perspective.

(7) Decide/evaluate the appropriate approach for managing the flow based on identified CDDs and resource-based decoupling zones.

When this process is finished, a set of CDDs have been identified for the logical
entity. This output supports the decision-maker in applying the appropriate approach for
managing the enterprise. Each CDD has its particular characteristics and hence require a
particular management approach. In this sense, the output can be seen as a foundation
for 'plan for every CDD' in a similar fashion as 'plan for every part' that is frequently
applied to select the appropriate approach for planning and control.

### 7.2.2.  Example: analysing a present state

The fictitious example used above can be seen as representing a present state and hence
also matching the outlined process for analysing a present state.

(1) The strategic lead-times were identified and illustrated in Figure 10.
(2) Lead-time relations were analysed in Tables 3 and 4.
(3) Position of the strategic decoupling points and decoupling zones was illustrated
in Figures 18 and 19.
(4) SIDPs and DIDP were positioned in Figure 19.
(5) CDDs were identified in Table 5.
(6) The relation with the two suppliers were briefly discussed based on Figure 19
and Table 5.
(7) The CDDs were identified in Table 5 and provides the foundation for managing
flow since items are categorized according to driving, differentiation and delimitation of flow. The details of flow management are however not within the scope
of this paper.

### 7.2.3.  Designing a future state

Design of a future state involves a completely different challenge compared to analysis
of a present state. Instead of focusing on a present state this process is about innova-
tively designing a future flow in the best possible way. This can of course involve many
types of scenarios, such as designing new flows when new products are introduced,
which involves the largest degrees of freedom, to designing a desired future state for an
existing value stream with existing products and resources. At an overall level, the process has similar properties for most alternative scenarios. Also, in this case, the process
can be summarized in seven steps:

(1) Identify customer value and requirements on the delivery lead-time $D$ and cust-
omizations (related to $A_D$).
(2) Position the CODP and decide the length of CODZ to handle variations in the
requirement(s) for $D$.
(3) Position CADPs and decide the length of CADZs to highlight potential differences between standard, customer unique and customer order unique.
(4) Position the PODPs that are required to identify where the control possibilities
are limited, e.g. supplier interfaces.
(5) When required, decide on the length of the respective PODZs.

(6) Position the information decoupling points SIDP and DIDP.

(7) Decide on the appropriate approach for managing the flow based on identified CDDs and resource-based decoupling zones.

Designing a future state obviously provides more degrees of freedom. It is a great opportunity to balance efficiency and effectiveness and consequently to establish competitive advantage by carefully positioning decoupling points and decoupling zones.

## 8. Discussion

Producing standard products to a mass market is still the main business for many companies. In these cases, the main competitive advantage is through cost efficiency and this has accelerated the trend for outsourcing based on finding the supplier that can provide the lowest possible cost. An alternative avenue is to differentiate the offering to the market. Differentiation can be based on increasing the variety of standard products but unless the variety is large the possibility of fulfilling individual customer needs is limited. Instead, different approaches for customization and servitization have developed over time to support this type of agility. Some activities are performed in the time span between the customer order is received and the delivery is performed and this is the target scenario for the framework developed here. Strategic decoupling points are an important concept for understanding the implications of these trends but so far the different perspectives have mainly been covered separately in the literature. In most cases the focus has been on the process driver which also can be seen as the origin of decoupling point analysis. The interest in a more integrated approach has however increased and in particular 'multiple decoupling points' (see e.g. Banerjee et al., 2011; Sun et al., 2008) has been used as a keyword. In addition, a number of CODP-based extensions have evolved over time. The most extensively explored is the different aspects of customization. Even in the early literature on CODP, such as Hoekstra and Romme (1992), this aspect was thoroughly investigated and many have followed. In parallel, the postponement literature provided an integrated approach to these two perspectives (see e.g. Pagh & Cooper, 1998). The CODP itself has been extended with the introduction of CODZ (Wikner & Rudberg, 2005b). Engineering has mainly been seen as occurring upstream from manufacturing but later these were treated as two separate dimensions (Giesberts & Tang, 1992; Wikner & Rudberg, 2005a). In addition, purchasing has been suggested as an additional dimension (Wallin, Rungtusanatham, & Rabinovich, 2006; Wikner & Bäckstrand, 2011). The resources and planning & control and order promising perspective has also been used in some cases (Giesberts & Tang, 1992; Rudberg & Wikner, 2004), and also implications for reverse logistics (Wikner & Tang, 2008) and services (Fließ & Kleinaltenkamp, 2004) have been included. The list above is not all-inclusive but still all these aspects are introduced more or less separately and not based on an integrative approach and on a common platform.

The frameworks outlined here provide both a common platform, in terms of the framework for flow-based decision-making, which is a general framework that can be used for many purposes, and the COBM framework, which provides an integrated approach to all these aspects of extending the CODP in terms of, e.g., multiple decoupling points. The generic framework for flow-based decision-making is related to the first research objective and outlines the fundamental aspects of COBM in terms of 10 key concepts. The concepts provide the building blocks of the decision categories for COBM. The second research objective resulted in the identification of SEIAD, the five

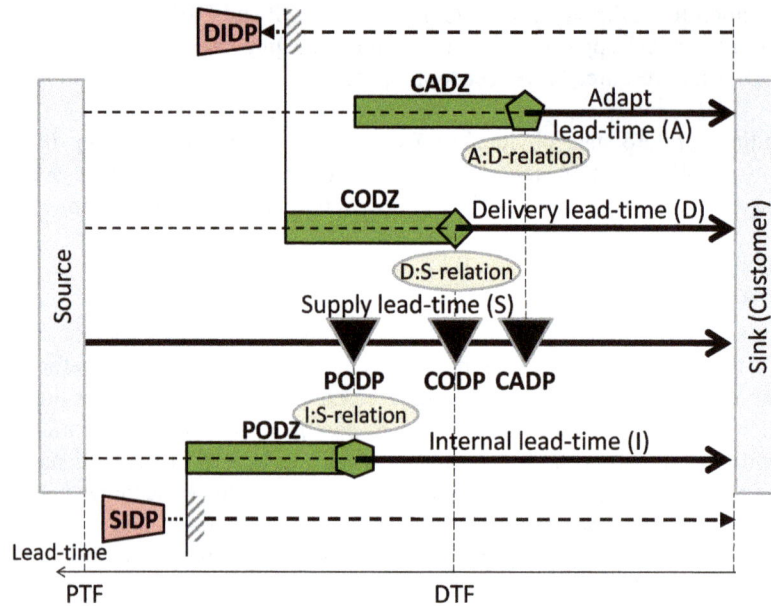

Figure 21.   Lead-time-based summary of content for customer-order-based management.

strategic lead-times. Three of the decision categories are defined based on these strategic lead-times and each involves a lead-time relation, a decoupling point, decoupling zones and a strategic stock point. In addition, a DC related to transparency is included which resulted in a total of four decision categories which correspond to the third research objective. Figure 21 provides a lead-time-based overview of these four decision categories. Note however that the compounded decision domains and the resource-based decoupling zones are not explicitly represented in Figure 21. These two aspects are related to the intersection between multiple flows and different decision categories, respectively.

In addition to supporting the design of the flow, the COBM framework has also proved itself to be very useful for communication between different functions within a company and for communication with suppliers. Financial information is readily expressed using cost accounting terminology, engineering information in terms of product design and material selection, quality information in six sigma-based terminology, etc. Logistics have however had more difficulties in providing a comprehensive picture of the implications of different types of decisions on logistics-related issues (also including manufacturing). The COBM framework highlights strategic lead-times and strategic decoupling points and decoupling zones based on a time-phased bill-of-material and a time-phased order book. In combination they provide a comprehensive picture that highlights implications of lead-times. For example, a lead-time reduction that moves the CODP upstream in the bill-of-material has more important implications than a lead-time reduction that does not affect the position of the CODP. In a similar way, the procurement of focusing on low cost might extend the lead-time and thus changes a purchase-to-order to a purchase-to-forecast scenario. If this is also within the CADZ, it will be even more critical since customer unique material should preferably not be purchased to forecast.

The COBM framework rests on a generic foundation independent of, e.g., industry and product portfolio. Some level of customer order influence on the flow obviously makes the approach more relevant. The application can, however, be demanding during some circumstances. An important challenge is if the context is very dynamic, the strategic lead-times and consequently the position of decoupling points and decoupling zones may be dynamic, i.e. their position may change over time. In this type of situation, it would be necessary to either re-evaluate the positioning according to the decision categories, or to prepare for a number of scenarios with predefined configuration of COBM for each scenario. Another challenge is when the demand is not easily classified as customer order or forecast. In some cases, the contracts may stipulate some flexibility for the customer depending on the time horizon. The PB-CODZ might be suitable to handle this but still it can be difficult to actually position the CODP. This is even more accentuated in rate-based environments where the rate is set for an extensive time period.

## 9. Conclusions

The COBM framework has been derived using a three-layer approach, where the generic framework consisting of 10 key concepts provides the foundational constructs. The first level of instantiation is represented by the COBM framework consisting of content (see Figure 20) and process. The second level of instantiation is represented by the example used throughout the text, which is an instantiation of the COBM framework. The example highlights some detailed aspects of the framework but the implications of applying the framework are important also at a more general management perspective. Enterprise development with emphasis on operational efficiency has played a central role for many years in improving industrial productivity. Lean production has been a key enabler in this context and also introduced more emphasis on the customer. A cost efficient internal flow is however only the first step towards improved competitiveness. To reach a higher level of competitiveness, it is important to also develop the effectiveness, which is here referred to as the strategic efficiency, to provide the right preconditions for the operational flow through improved interaction. The flow-based COBM framework introduced here can hopefully provide increased understanding of, and support for, how strategic efficiency may be established.

The suggested COBM framework provides numerous opportunities for further research related to theory development, as well as industrial applications. In particular, three areas are being developed further:

- *Strategic decoupling and supply chain strategies* connects the logical entities to the physical entities and focus on how common supply chain strategies can be interpreted in terms of the COBM framework.
- *Strategic decoupling points and supplier/customer interactions* connects the logical entities to the legal entities and focuses on how supplier/customer interactions can be interpreted in terms of the COBM framework.
- *Strategic decoupling points and implementation* establishes a method for how to implement competitive COBM and thus also how to best position decoupling points and decoupling zones.

## Acknowledgements

The development of the COBM framework has been partly supported by six companies based on the project KOPeration, covering the alignment of key aspects of purchasing strategy with operations strategy, and the project KOPtimera, focusing on the evaluation of positioning of decoupling points and decoupling zones. The projects are funded by the Swedish Knowledge foundation (KKS) and by the participating companies.

## References

Bäckstrand, J. (2012). *A method for customer-driven purchasing: Aligning supplier interaction and customer-driven manufacturing.* Jönköping: School of Engineering, Jönköping University.

Bäckstrand, J., Johansson, E., Wikner, J., Andersson, R., Carlsson, B., Kornebäck, F., ... Spaak, B. (2013). *A method for customer-driven purchasing.* Paper presented at the 20th EurOMA Conference, Dublin, Ireland.

Bäckstrand, J., & Wikner, J. (2013). *Time-phasing and decoupling points as analytical tools for purchasing.* Paper presented at the IEEE International Conference on Industrial Engineering and Engineering Management, Bangkok, Thailand.

Banerjee, A., Sarkar, B., & Mukhopadhyay, S. K. (2011). Multiple decoupling point paradigms in a global supply chain syndrome: A relational analysis. *International Journal of Production Research, 50,* 3051–3065.

Berry, W. L., Hill, T. J., & Klompmaker, J. E. (1995). Customer-driven manufacturing. *International Journal of Operations & Production Management, 15,* 4–15.

Bertrand, J. W. M., Wortmann, J. C., & Wijngaard, J. (1990). *Production control: A structural and design oriented approach.* Amsterdam: Elsevier.

Blackstone, J. H. (Ed.). (2008). *APICS dictionary* (12th ed.). Athens, GA: APICS.

Borgström, B., & Hertz, S. (2011). Supply chain strategies: Changes in customer order-based production. *Journal of Business Logistics, 32,* 361–373.

Bucklin, L. P. (1965). Postponement, speculation and the structure of distribution channels. *Journal of Marketing Research, 2,* 26–31.

Chandrashekar, A., & Schary, P. B. (1999). Toward the virtual supply chain: The convergence of IT and organization. *International Journal of Logistics Management, 10,* 27–39.

Childerhouse, P., & Towill, D. R. (2003). Simplified material flow holds the key to supply chain integration. *Omega, 31,* 17–27.

Chopra, S., & Meindl, P. (2004). *Supply chain management: Strategy, planning, and operation* (2nd ed.). Upper Saddle River, NJ: Prentice Hall.

Clark, J. T. (1979). *Capacity management.* Proceedings of the 22nd APICS Annual International Conference.

Clark, A. J., & Scarf, H. (1960). Optimal policies for a multi-echelon inventory problem. *Management Science, 6,* 475–490.

Duggan, K. J. (2012). *Creating mixed model value streams: Practical lean techniques for building to demand.* New York, NY: CRC Press.

Fleischmann, B., & Meyr, H. (2004). Customer orientation in advanced planning. In H. Dyckhoff, R. Lackes, J. Reese, & G. Fandel (Eds.), *Supply chain management and reverse logistics* (pp. 297–321). Berlin: Springer.

Fließ, S., & Kleinaltenkamp, M. (2004). Blueprinting the service company: Managing service processes efficiently. *Journal of Business Research, 57,* 392–404.

Ford, H., & Crowther, S. (1988). *Today and tomorrow.* Cambridge, MA: Productivity Press.

Forrester, J. W. (1958). Industrial dynamics. *Harvard Business Review, 36,* 37–66.

García-Dastugue, S. J., & Lambert, D. M. (2007). Interorganizational time-based postponement in the supply chain. *Journal of Business Logistics, 28,* 57–81.

Giesberts, P. M. J., & van der Tang, L. (1992). Dynamics of the customer order decoupling point: Impact on information systems for production control. *Production Planning & Control, 3,* 300–313.

Goldratt, E. M. (1990). *What is this thing called theory of constraints and how should it be implemented?* Croton-on-Hudson, NY: North River Press.

Goldratt, E. M., & Cox, J. (1984). *The goal: Excellence in manufacturing.* Croton-on-Hudson, NY: North River Press.

Graca, A., Hendry, L., & Kingsman, B. (1999). Competitive advantage, customisation and a new taxonomy for non make-to-stock companies. *International Journal of Operations & Production Management, 19*, 349–371.

Hammer, M., & Champy, J. (1993). *Reengineering the corporation: A manifesto for business revolution*. New York, NY: HarperBusiness.

Harland, C. M. (1996). Supply chain management: Relationships, chains and networks. *British Journal of Management, 7*, S63–S80.

Hayes, R. H., & Wheelwright, S. C. (1984). *Restoring our competitive edge: Competing through manufacturing*. New York, NY: Wiley.

Hines, P., & Rich, N. (1997). The seven value stream mapping tools. *International Journal of Operations & Production Management, 17*, 46–64.

Hoekstra, S., & Romme, J. (Eds.) (1992). *Integral logistic structures: Developing customer-oriented goods flow*. New York, NY: Industrial Press.

Jones, D. T. (2001). The lean transformation. In D. H. Taylor & D. Brunt (Eds.), *Manufacturing operations and supply chain management: The Lean approach* (pp. 17–24). London: Thomson Learning.

Leong, G. K., Snyder, D. L., & Ward, P. T. (1990). Research in the process and content of manufacturing strategy. *Omega, 18*, 109–122.

Mason-Jones, R., Naylor, B., & Towill, D. R. (2000). Engineering the leagile supply chain. *International Journal of Agile Management Systems, 2*, 54–61.

Mason-Jones, R., & Towill, D. R. (1999). Using the information decoupling point to improve supply chain performance. *International Journal of Logistics Management, 10*, 13–26.

Mather, H. (1984). *Attack your P:D ratio*. Paper presented at the APICS Conference.

Naylor, B. J., Naim, M. M., & Berry, D. (1999). Leagility: Integrating the lean and agile manufacturing paradigms in the total supply chain. *International Journal of Production Economics, 62*, 107–118.

Naim, M. M., & Gosling, J. (2011). On leanness, agility and leagile supply chains. *International Journal of Production Economics, 131*, 342–354.

Olhager, J., & Östlund, B. (1990). An integrated push-pull manufacturing strategy. *European Journal of Operational Research, 45*, 135–142.

Olhager, J., Selldin, E., & Wikner, J. (2006). Decoupling the value chain. *International Journal of Value Chain Management, 1*, 19–32.

Olhager, J., & Wikner, J. (1998). A framework for integrated material and capacity based master scheduling. In D. Andreas & A. Kimms (Eds.), *Beyond manufacturing resource planning (MRP II): Advanced models and methods for production planning* (pp. 3–20). Berlin: Springer-Verlag.

Ōno, T. (1988). *Toyota production system: Beyond large-scale production*. Cambridge, MA: Productivity Press.

Pagh, J. D., & Cooper, M. C. (1998). Supply chain postponement and speculation strategies: How to choose the right strategy. *Journal of Business Logistics, 19*, 13–33.

Plossl, G. W. (1985). *Production and inventory control: Principles and techniques* (2nd ed.). Englewood Cliffs, NJ: Prentice-Hall.

Rother, M., & Shook, J. (1998). *Learning to see: Value stream mapping to add value and eliminate muda*. Brookline, MA: Lean Enterprise Institute.

Rudberg, M., & Wikner, J. (2004). Mass customization in terms of the customer order decoupling point. *Production Planning & Control, 15*, 445–458.

Schmenner, R. W. (2001). History of technology, manufacturing, and the industrial revolution: A rejoinder. *Production and Operations Management, 10*, 103–106.

Schmenner, R. W., & Swink, M. L. (1998). On theory in operations management. *Journal of Operations Management, 17*, 97–113.

Sharman, G. (1984). The rediscovery of logistics. *Harvard Business Review, 62*, 71–79.

Shingō, S.(1989). *A study of the Toyota production system from an industrial engineering viewpoint* (rev. ed.). Cambridge, MA: Productivity Press.

Smith, S. B. (1989). *Computer-based production and inventory control*. Englewood Cliffs, NJ: Prentice-Hall.

Stalk, G., & Hout, T. M. (1990). *Competing against time: How time-based competition is reshaping global markets*. New York, NY: Free Press.

Stein, R. E. (1996). *Re-engineering the manufacturing system: Applying the theory of constraints.* New York, NY: M. Dekker.

Sun, X. Y., Ji, P., Sun, L. Y., & Wang, Y. L. (2008). Positioning multiple decoupling points in a supply network. *International Journal of Production Economics, 113,* 943–956.

Vargo, S. L., & Lusch, R. F. (2004). The four service marketing myths. *Journal of Service Research, 6,* 324–335.

van der Veeken, D. J. M, & Rutten, W. G. M. M. (1998). Logistics service management: Opportunities for differentiation. *International Journal of Logistics Management, 9,* 91–98.

Waddell, W., & Bodek, N. (2005). *Rebirth of American industry: A study of lean management.* Vancouver, WA: PCS Press.

Wallace, T. F. (1992). *Customer-driven strategy: Winning through operational excellence.* Essex Junction, VT: Oliver Wight.

Wallin, C., Rungtusanatham, M. J., & Rabinovich, E. (2006). What is the "right" inventory management approach for a purchased item? *International Journal of Operations & Production Management, 26,* 50–68.

van Hoek, R. I. (2001). The rediscovery of postponement a literature review and directions for research. *Journal of Operations Management, 19,* 161–184.

Wemmerlöv, U. (1984). Assemble-to-order manufacturing: Implications for materials management. *Journal of Operations Management, 4,* 347–368.

Wheelwright, S. C. (1984). Manufacturing strategy: Defining the missing link. *Strategic Management Journal, 5,* 77–91.

Wight, O. W. (1984). *Manufacturing resource plannig: MRP II* (rev. ed.). Essex Junction, VT: Wiley.

Wikner, J. (2011). *Ledtidsbaserat ramverk för kundorderstyrning* [in Swedish] [Lead-time based framework for customer-order based management]. Paper presented at the 13th Plan Conference on research and application, Norrköping, Sweden.

Wikner, J. (2012a). Grad av kundorderstyrning [Level of customer-order based management]. In G. Hultén (Ed.), *Inköp & Logistik* (pp. 5.1/1–5.1/25). Stockholm: Bonnier.

Wikner, J. (2012b). A service decoupling point framework for logistics, manufacturing, and service operations. *International Journal of Services Sciences, 4,* 330–357.

Wikner, J., & Bäckstrand, J. (2011). *Aligning operations strategy and purchasing strategy.* Paper presented at the EurOMA Conference, Cambridge, UK. Retrieved from: http://urn.kb.se/resolve?urn=urn:nbn:se:hj:diva-17224

Wikner, J., & Bäckstrand, J. (2012). *Decoupling points and product uniqueness impact on supplier relations.* Paper presented at the 4th World Conference P&OM, Amsterdam.

Wikner, J., Johansson, E., & Persson, T. (2009). *Process based inventory classification.* Paper presented at the The 21st Annual Nofoma Conference, Jönköping, Sweden.

Wikner, J., Naim, M. M., & Towill, D. R. (1992). The system simplification approach in understanding the dynamic behaviour of a manufacturing supply chain. *Journal of Systems Engineering, 2,* 164–178.

Wikner, J., & Rudberg, M. (2005a). Integrating production and engineering perspectives on the customer order decoupling point. *International Journal of Operations & Production Management, 25,* 623–641.

Wikner, J., & Rudberg, M. (2005b). Introducing a customer order decoupling zone in logistics decision-making. *International Journal of Logistics Research and Applications, 8,* 211–224.

Wikner, J., & Tang, O. (2008). A structural framework for closed-loop supply chains. *International Journal of Logistics Management, 19,* 344–366.

Wikner, J., & Wong, H. (2007, September 17–19). Postponement based on the positioning of the differentiation and decoupling points. In J. Olhager & P. Fredrik (Eds.), *International IFIP TC 5, WG 5.7 Conference on Advances in Production Management Systems (APMS 2007)* (Vol. 246, pp. 143–150). Linköping: Springer-Verlag.

Womack, J. P., & Jones, D. T. (1996). *Lean thinking: Banish waste and create wealth in your corporation.* New York, NY: Simon & Schuster.

Wortmann, J. C., Muntslag, D. R., & Timmermans, P. J. M. (1997). *Customer-driven manufacturing* (1st ed.). London: Chapman & Hall.

# 11

# Advanced optimal tolerance design of machine elements using teaching-learning-based optimization algorithm

R.V. Rao* and K.C. More

*Department of Mechanical Engineering, S.V. National Institute of Technology, Ichchanath, Surat, Gujarat 395 007, India*

Tolerance design has become a very key issue in product and process development because of an informal compromise between functionality, quality, and manufacturing cost. The problem formulation becomes complex with simultaneous selection of design and manufacturing tolerances and the optimization problem is difficult to solve with the traditional optimization techniques. In this paper, a recently developed optimization algorithm called teaching-learning-based optimization (TLBO) is used for optimal selection of design and manufacturing tolerances with an alternative manufacturing process to obtain the optimal solution which is nearer to global optimal solution. Three problems are considered and these are: overrunning clutch assembly, knuckle joint assembly with three arms, and a helical spring. Out of these three problems, the problems of overrunning clutch assembly and knuckle joint assembly with three arms are multi-objective optimization problems and the helical spring problem is a single-objective problem. The comparison of the proposed algorithm is made with the genetic algorithm (GA), Non-dominated sorting genetic algorithm-II (NSGA-II), and multi-objective particle swarm optimization algorithm (MOPSO). It is found that the TLBO algorithm has produced better results when compared to those obtained by using GA, NSGA-II, and MOPSO algorithms.

**Keywords:** tolerance design; multi-objective optimization; machine elements; teaching-learning-based optimization algorithm

## 1. Introduction

Assembly tolerance in an acceptable functioning of a product assembly is ensured by making the assembly response to be included in a specified range. Allocation of this assembly tolerance among the ingredient dimensions while gathering the functionality requirements is known as tolerance design. Design engineers normally tend to specify rigid tolerances for ensuring the performance and functionality of the product assembly; while manufacturing engineers do prefer slack tolerances to produce parts economically. Therefore, design of tolerance plays an essential role in bonding the between designs and manufacturing phases. Selection of optimal tolerances for minimum manufacturing cost and meeting the functional requirements of the product assembly has been an appealing topic of research for some decades (Singh, Jain, & Jain, 2005).

The tolerance design problem becomes more intricate for manufacturing every dimension in the existence of different processes or machines, as shown in Figure 1 (Prabhaharan, Asokan, & Rajendran, 2005). Due to this the manufacturing cost–tolerance

*Corresponding author. Emails: ravipudirao@gmail.com, rvr@med.svnit.ac.in

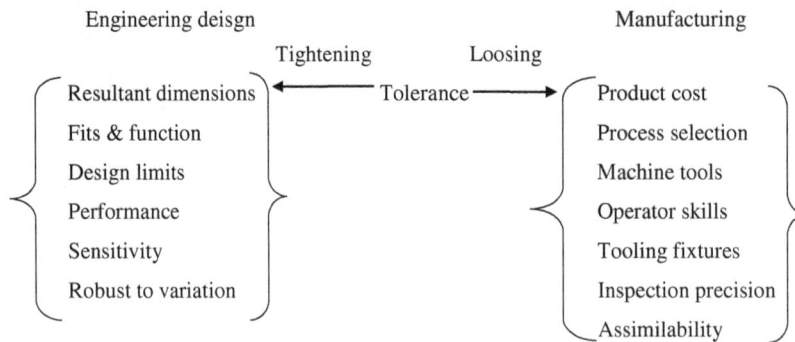

Figure 1.   Critical link between engineering design and manufacturing (Prabhaharan et al., 2005).

features change from machine to machine and from process to process. The total costs incurred throughout a product's life cycle can be divided in two main categories: manufacturing cost, which occurs before the product reaches the customer; and quality loss, which occurs after the product is sold. A rigid tolerance (i.e. high manufacturing cost) indicates that the variety of product quality characteristics will be low (i.e. little quality loss). On the other hand, a slack tolerance (i.e. low manufacturing cost) indicates that the variability of product quality characteristics will be high (i.e. high quality loss). Therefore, there is a need to adjust the design tolerance between the quality loss and the manufacturing cost and to reach an economic balance during product tolerance design.

During the last few decades, researchers have focused their attention on obtaining the best tolerance allocation in such a way that the product design not only meets the efficient needs but also minimizes manufacturing cost. In order to solve the tolerance allocation problem, various numerical methods were employed to deal with complicated computations associated with tolerance design models. Forouraghi (2002) introduced an approach to worst-case tolerance design using a genetic algorithm (GA) for the optimal design of a clutch assembly without considering alternative manufacturing process selection. Ye and Salustri (2003) introduced a new method of synthesizing tolerances simultaneously for both manufacturing cost and quality. The authors had constructed a non-linear optimization model and had minimized the quality loss and the manufacturing cost simultaneously in a single-objective function by setting both the process tolerances and design tolerances. Singh, Jain, and Jain (2003, 2004) obtained an optimal solution to a set of tolerance design problems with simple dimension chain involving sets of alternative processes using GAs.

Prabhaharan et al. (2005) used ant colony algorithm as an optimization tool for minimizing the critical dimension deviation and for allocating the cost-based optimal tolerances. Krishna and Rao (2006) used a scatter search method to simultaneously allocate both the design and manufacturing tolerances based on minimum total manufacturing cost. Huang and Shiau (2006) obtained tolerance allocation of a sliding vane rotary compressor and the required reliability, minimum cost, and quality loss were optimized. Huang and Zhong (2007) established the sequential linear optimization models to maximize the 2D-sized, angular, and directional working tolerances of a 3D-machined part based on the process capabilities. This approach releases the working tolerances, reduces manufacturing costs, and enhances the acceptance rate of machined parts.

Singh, Jain, and Jain (2008) introduced GA to obtain an optimal solution to the advanced tolerance synthesis problem by considering a continuous cost function and this method was used for both single and multiple tolerance stack-ups. Huang and Zhong (2008) implemented a linear programming approach to obtain process tolerances concurrence from an assembly by using the information of process planning. Sivakumar, Kannan, and Jayabalan (2009) used a hybrid algorithm for optimum tolerance allocation in complex assemblies with alternative process selection. For hybridization, the authors had used a tabu search and heuristic algorithm.

Gonzalez and Sanchez (2009) developed a methodology to allocate optimal statistical tolerances to dependent variables and the manufacturing costs were minimized using a statistical approach. Forouraghi (2009) developed a methodology to allow an automatic tolerance allocation capable of minimizing manufacturing costs based on particle swarm optimization (PSO) algorithm. Sivakumar and Stalin (2009) used a Lagrange multiplier method for selection of alternative processes for tolerance allocation and total manufacturing cost minimization. Wu, Jean-Yves, Alain, Ali, and Patrick (2009) adopted Monte Carlo simulation and GA methodology for tolerance allocation of minimizing both manufacturing cost and quality loss.

Muthu, Dhanalakshmi, and Sankaranarayanasamy (2009) used metaheuristic techniques such as GA and PSO to the design and process tolerances simultaneously for minimum combined manufacturing cost and quality loss over the life of the product. Sivakumar et al. (2011a) used Non-Dominated Sorting Genetic Algorithm-II (NSGA-II) and multi-objective particle swarm optimization algorithm (MOPSO) for simultaneous optimum selection of design and manufacturing tolerances with alternative process selection. Sivakumar et al. (2011b) used NSGA-II and MOPSO for tolerance allocation for mechanical assemblies with considering alternative manufacturing process selection. Sivakumar, Balamurugan, and Ramabalan (2012) used evolutionary NSGA-II and multi-objective differential evaluation (MODE) optimization algorithms for the simultaneous selection of optimal process tolerances for a wheel assembly. Guofu, Yuege, Ye, and Hu (2013) proposed a method of multi-objective reliability tolerance design for electronic systems. It improved the performance of output response and reduced the operating stresses on components.

It has been observed from the literature review that some of the researchers had used traditional optimization techniques like scatter search method, sequential linear method, Lagrange multiplier method, Monte Carlo simulation, linear programming method, etc. for optimum tolerance design of machine elements. However, these approaches suffer from some drawbacks as following: (i) these traditional techniques may be relevant only for simple cost functions and may not be relevant for discontinuous cost functions. (ii) the traditional techniques may not give global optimum solution. In order to overcome these drawbacks of traditional optimization techniques, few researchers had used the so-called advanced optimization techniques such as GA, SA, NSGA-II, MODE, and MO-PSO algorithms. The advantages of these advanced optimization techniques are: (i) these techniques are relevant for solving all types of complex optimization problems and (ii) these techniques are population-based and hence they may give global optimum solution.

Many researchers, except Sivakumar et al. (2011a, 2011b, 2012) and Guofu et al. (2013), considered only one objective function i.e. minimization of manufacturing cost. But the real-world tolerance design problems have more than one objective function. Some of the other important objective functions are quality loss, stack-up tolerance, stock removal allowance, etc.

The advanced optimization techniques like GA, SA, NSAG-II, MOPSO, MODE, etc. need tuning of algorithm-specific parameters. For example, GA requires crossover probability, mutation probability (Claudio, Jose, & Nestor, 2003) and NSGA-II requires crossover probability, mutation probability, real-parameter SBX parameter, and real-parameter mutation parameter (Daniel, Claudio, & Blas, 2006). SA requires controlling temperature and Boltzmann constant (Luis & Claudio, 2007).

MOPSO requires mutation probability, inertia weight, and social and cognitive parameters. MODE requires scaling factor and crossover constant. The proper tuning of the algorithm-specific parameters is a very essential factor, which affects the performance of those algorithms. The improper tuning of algorithm-specific parameters either increases the computational effort or yields the local optimal solution (Rao & Patel, 2013a). Hence to overcome the problem of tuning of algorithm-specific parameters, we have used a recently developed algorithm-specific parameter-less algorithm known as teaching-learning-based optimization (TLBO) algorithm (Rao, Savsani, & Vakharia, 2011; Rao, Savsani, & Vakharia, 2012; Rao & Savsani, 2012). It requires only common controlling parameters like population size and number of generations for its working (Just like any other population-based optimization algorithms). For more details about the effects of the common controlling parameters on the performance of TLBO algorithm, readers may refer to Rao and Patel (2012, 2013b).

John Holland had developed GAs in the 1960s. Most commonly used genetic operators are reproduction, crossover, and mutation. They are implemented as a computer simulation in which a population of abstract representations (called chromosomes or the genotype of the genome) of candidate solutions (called individuals, creatures, or phenotypes) to an optimization problem evolves toward better solutions (Singh et al., 2005).

NSGA-II algorithm is utilized here as a multi-objective optimization method in order to find the Pareto frontier using the GA. The number of ways NSGA-II works differs from Non-Dominated Sorting Genetic Algorithm (NSGA). It uses an elite-preserving mechanism. Secondly, it uses a fast Non-Dominated Sorting procedure. The solutions on each pareto-front are pareto-optimal with respect to each other (Sivakumar et al., 2011a).

The single-objective PSO algorithm extends to handle multi-objective optimization problems called MOPSO algorithm. The crowding space mechanism together with a mutation operator maintains the diversity of non-dominated solutions in the outer archive. The motivation behind this concept is to attain better convergence to the Pareto-optimal front, while giving sufficient emphasis to the diversity consideration (Sivakumar et al., 2011b).

The case studies presented by Sivakumar et al. (2011b) and Singh et al. (2005) are considered in this work to demonstrate the applicability of the TLBO algorithm for the overrunning clutch assembly, knuckle joint with three arms, and the helical spring design problem to see if any further improvement can be obtained in the results and in the computational time. Out of these three optimization problems, the first two problems are multi-objective problems and the third one is a single-objective problem.

The next section presents the details of the TLBO algorithm.

## 2. Teaching-learning-based optimization algorithm.

TLBO is a teaching-learning process inspired algorithm proposed recently by Rao et al. (2011, 2012) based on the effect of influence of a teacher on the output of learners in a class. The algorithm mimics the teaching-learning ability of teachers and learners in a classroom. Teacher and learners are the two vital components of the algorithm and describe two basic modes of the learning, through teacher (known as teacher phase) and interacting with the other learners (known as learner phase) (Rao et al., 2011, 2012; Rao & Savsani, 2012).

The output in TLBO algorithm is considered in terms of results or grades of the learners which depend on the quality of teacher. So, teacher is usually considered as a highly learned person who trains learners so that they can have better results in terms of their marks or grades. Moreover, learners also learn from the interaction among themselves which also helps in improving their results. TLBO is a population-based method and a group of learners is considered as population and different design variables are considered as different subjects offered to the learners and learners' result is analogous to the 'fitness' value of the optimization problem. In the entire population, the best solution is considered as the teacher. The flowchart of TLBO algorithm is shown in Figure 2 (Rao & Kalyankar, 2012). The working of TLBO is divided into two parts, 'teacher phase' and 'learner phase'. Working of both the phases is explained below.

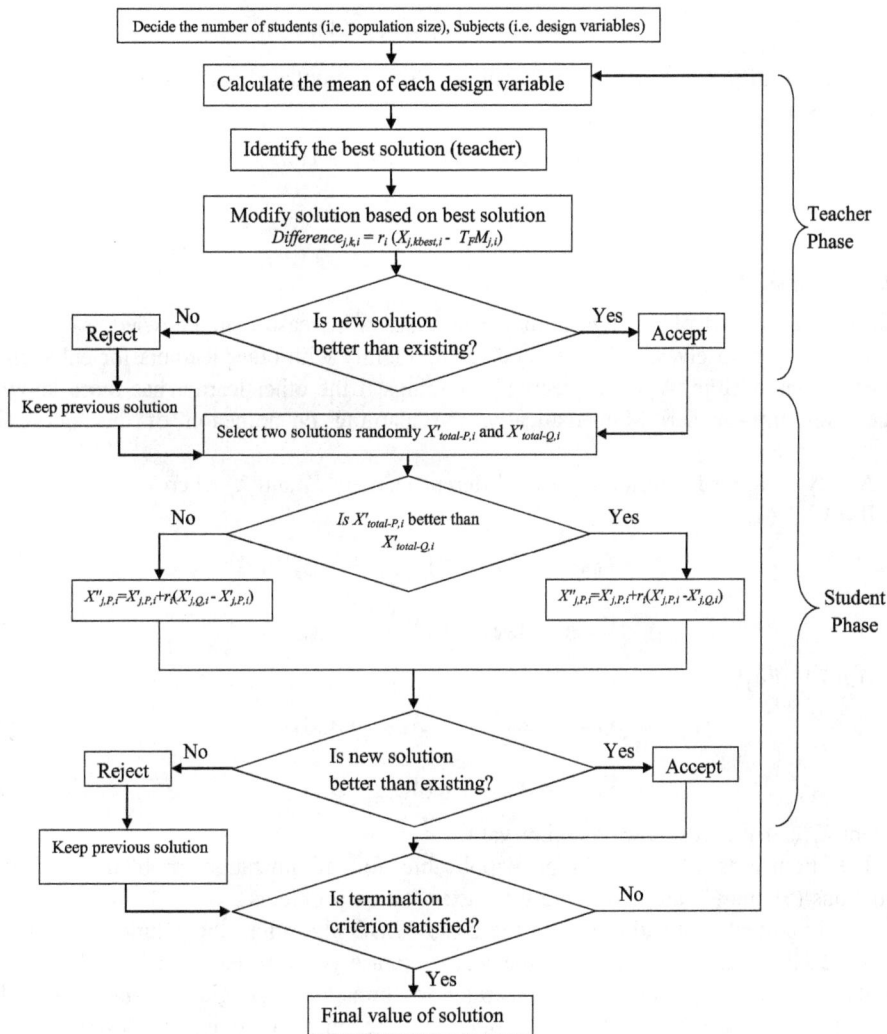

Figure 2.    Flow chart of TLBO algorithm (Rao & Kalyankar, 2012).

## 2.1. Teacher phase

It is the first part of the algorithm where learners learn through the teacher. During this phase, a teacher tries to increase the mean result of the classroom from any value $M_1$ to his or her level (i.e. $T_A$). But practically it is not possible and a teacher can move the mean of the classroom $M_1$ to any other value $M_2$ which is better than $M_1$ depending on his or her capability. Consider $M_j$ be the mean and $T_i$ be the teacher at any iteration $i$. Now $T_i$ will try to improve the existing mean $M_j$ towards him/her so that the new mean will be designated as $M_{new}$ and the difference between the existing mean and new mean is given by,

$$Difference\_Mean_i = r_i(M_{new} - T_f M_j) \tag{1}$$

where $T_f$ is the teaching factor which decides the value of mean to be changed, and $r_i$ is the random number in the range [0,1]. Value of $T_f$ can be either 1 or 2 which is a heuristic step and it is decided randomly with equal probability as,

$$T_f = \text{round}\,[1 + \text{rand}(0,1)] \tag{2}$$

Based on this *Difference_Mean*, the existing solution is updated according to the following expression

$$X_{new,i} = X_{old,i} + Difference\_Mean_i \tag{3}$$

## 2.2. Learner phase

It is the second part of the algorithm where learners increase their knowledge by interaction among themselves. A learner interacts randomly with other learners for enhancing his or her knowledge. A learner learns new things if the other learner has more knowledge than him or her. Mathematically, the learning phenomenon of this phase is expressed below.

At any iteration $i$, considering two different learners $X_i$ and $X_j$ where $i \neq j$
If $f(X_i) < f(X_j)$

$$X_{new,i} = X_{old,i} + r_i(X_{old,i} - X_{old,j}) \tag{4}$$

$$X_{new,i} = X_{old,i} + r_i(X_{old,j} - X_{old,i}) \tag{5}$$

If $f(X_i) > f(X_j)$

$$X_{new,i} = X_{old,i} + r_i(X_{old,j} - X_{old,i}) \tag{6}$$

$$X_{new,i} = X_{old,i} + r_i(X_{old,i} - X_{old,j}) \tag{7}$$

Accept $X_{new}$ if it gives better function value.

The Equations (4) and (6) are applicable for minimization problems and the Equations (5) and (7) are applicable for maximization problems.

It is important to mention that in the basic TLBO algorithm, the solution is updated in the teacher phase as well as in the learner phase (Rao & Patel, 2013a). For more details on TLBO algorithm, one may refer to: https://sites.google.com/site/tlborao. In this paper, the TLBO algorithm is applied for the optimum tolerance design of few machine elements like overrunning clutch assembly, knuckle joint with three arms, and helical spring.

The next section describes the role of design and manufacturing tolerances, stock removal allowances, selection of manufacturing processes, manufacturing cost, and quality loss function.

## 3. Problem definition

### 3.1. Design and manufacturing tolerances

In any product design, the proposed functions and assembly needs are distorted into a set of related tolerances and appropriate dimensions. The dimensions are known as the assembly dimensions and the related tolerances are known as assembly tolerances. The assembly tolerance is sensibly distributed among the part dimensions in particular dimension sequence considering the practical aspects. A tolerance specified in the design stage is further refined to suit the requirements of process planning for producing the constituent dimensions (Singh et al., 2009).

### 3.2. Stock removal allowances

The amount of stock removal allowance has very much effect on the manufacturing tolerance selection. The stock removal allowance is the layer of material to be removed from the surface of a work piece in manufacturing to obtain the required profile accuracy and surface quality through different machining processes. The stock removal allowance is very much affecting the quality and the production efficiency of the manufactured features. A disproportionate stock removal allowance will add to the consumption of material, machining time, tool, and power, and hence raise the manufacturing cost. On the other hand, with an insufficient stock removal allowance, the faulty surface layer caused by the previous operation cannot be rectified. Variation in the stock removal is the sum of manufacturing tolerances in the prior and the current operations. An appropriate stock removal allowance is required for each successful manufacturing operation and the concept is shown in Figure 3 (Haq, Sivakumar, Saravanan, & Muthiah, 2005).

### 3.3. Selection of machining process

The selection of machining process is depending on the equipment precision, machining series, set-up mode, and cutting parameters. The selection of machining process is strongly affected by the tolerance of the part to be machined. So it is vital to do a simultaneous selection of the best machining process while allocating the tolerance (Sivakumar et al., 2011a).

### 3.4. Manufacturing cost

The total expenditure for developing a blueprint dimension is called the manufacturing cost of the dimension. The overall manufacturing cost will constitute both the direct cost and the overheads. For a given manufacturing process, the material, set-up, tool, and inspection costs, in addition to overheads, comprise the fixed cost. The variable cost, which usually depends on the value of tolerance assigned, is mostly because of rework and/or rejection. The actual processing (labour) cost also accounts for the variable cost, though it is not predominant.

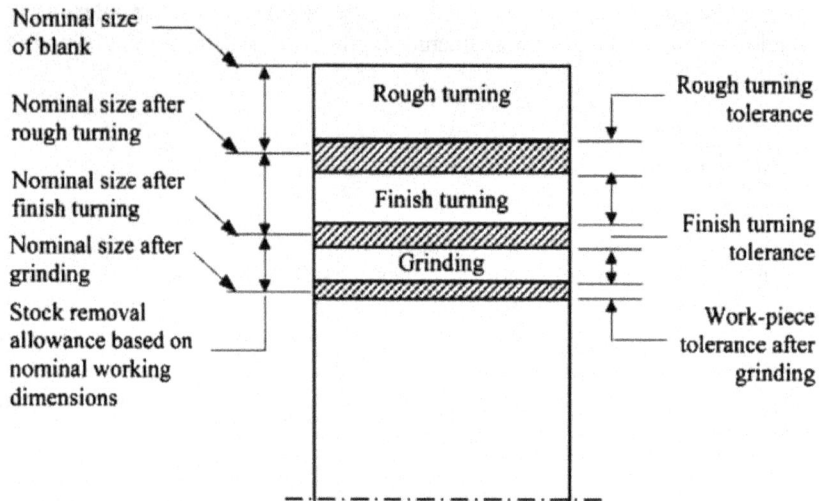

Figure 3.    Machined layers and manufacturing tolerances (Haq et al., 2005).

Manufacturing cost usually increases as the tolerance of quality characteristics in relation to the ideal value is reduced. Due to this, more refined and accurate operations are essential while the suitable ranges of output are reduced. In other hand, large tolerances are having minimum cost to get as they want less specific manufacturing processes; but usually they provide poor result in performance, premature wear, increase in scrap, and part rejection.

The manufacturing cost function for the available manufacturing processes is assumed to be exponential (Equation (16)) (Singh et al., 2005).

$$C = c_0 * e^{-c_1*t} + c_2 \tag{8}$$

where $C$ is manufacturing cost as a function of tolerance; $c_0$, $c_1$, and $c_2$ are cost function values; and $t$ is the design tolerance dimension.

### 3.5.  Quality loss function

Variability in the production process is compulsory due to changeability in tool work piece, material, and process parameters. In this study, it is referred as the quality loss function (Feng & Kusiak, 1997). This loss function is a quadratic expression for measuring the cost in terms of financial loss due to product failure in the eyes of the consumers. The quality loss function (QL) is:

$$QL = \frac{A}{T^2} \sum_{i=1}^{J} \sigma_i^2 \tag{9}$$

$$\sigma_i = \frac{t_i}{3} \tag{10}$$

$$QL = \frac{A}{9T^2} \sum_{i=1}^{J} t_i^2 \tag{11}$$

where $T$ is the tolerance stack-up limit of the dimensional chain, $A$ is the quality loss cost, $i$ is the component tolerance index, $j$ is the process index, $t$ is the design tolerance dimension, and is the standard deviation of dimension $i$.

The next section presents three problems of optimization and the application of the TLBO algorithm for same.

## 4. Examples

The examples of overrunning clutch assembly, knuckle joint assembly with three arms which were solved previously by Sivakumar et al. (2011b) , and the design of helical spring problem, which was solved previously by Singh et al. (2005), are attempted now by using the TLBO algorithm.

### 4.1. Example 1: overrunning clutch assembly (Sivakumar et al. 2011b)

The overrunning clutch assembly as shown in Figure 4 (Sivakumar et al., 2011b) consists of hub, rollers, and cage. The rollers are supplied by vendor hence their having fixed value tolerance and manufacturing cost. So, only for the hub and the cage the tolerances and the manufacturing cost are optimized. The following are the requirements and information for manufacture of the assembly. These data are adopted from (Singh et al., 2008; Sivakumar et al., 2011b; Feng & Kusiak, 1997). Assembly response function is given by

$$Y = a \cos\left(\frac{X_1 + X_2}{X_3 - X_2}\right) \qquad (12)$$

where $X_1$ is hub dimension in mm, $X_2$ is rollers dimension in mm which is supplied by vendor, $X_3$ is cage dimension in mm, $Y$ is the assembly dimension, and $a$ is constant.

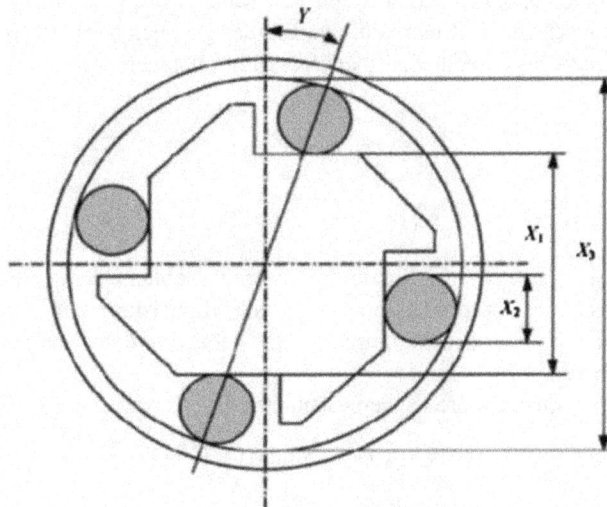

Figure 4.   Overrunning clutch assembly (Sivakumar et al., 2011b).
Note:  ($X_1 = 55.291$ mm;  $X_2 = 22.86 \pm 0.0102$ mm;  $X_3 = 101.6$ mm  and  $Y = 7.0124 \pm 10$ [Sivakumar et al., 2011b; Singh et al. 2009]).

The manufacturing processes for the hub are forging, milling, and grinding; and for the cage are forging, turning, and grinding. Also, there are three machining tolerance parameters for manufacture of each of the hub and the cage $t_{ij}$ ($i = 1$ for the hub, and 3 for the cage; $j = 1$ to 3 for the three manufacturing operations in the respective process plan). For each manufacturing operation, alternative machines are available with the details given in Table 1.

Sivakumar et al. (2011b) used NSGA-II and MOPSO for simultaneous optimum selection of design and manufacturing tolerances with alternative process selection.

The objective functions considered are: minimum tolerances stack-up, minimum total manufacturing cost of the assembly, and minimum quality loss function. The objective functions considered in this paper are same as those considered by Sivakumar et al. (2011b) and these are:

Minimize:

$$Z_1 = \Delta Y = \xi_1 t_{13}|x_{13} + \xi_2 t_2 + \xi_3 t_{33}|x_{33} \tag{13}$$

$$Z_2 = C_{\text{asm}} = \sum_{i=1, i \neq 2}^{3} \sum_{j=1}^{3} (a_0 + a_1 t + a_2 t^2 + a_3 t^3 + a_4 t^4 + a_5 t^5)|_{x_{ij}} \tag{14}$$

$$Z_3 = \text{QL} = \frac{A}{9T^2} \sum_{i=1}^{J} t_{ij}^2 \tag{15}$$

where $Z_1$ is the minimum tolerances stack-up; $Z_2$ is the minimum total manufacturing cost; $Z_3$ is the minimum quality loss function; $C_{\text{asm}}$ is the total assembly manufacturing cost; $\xi_1$, $\xi_2$, $\xi_3$ is the value of sensitivity for component dimensions $X_1$, $X_2$, $X_3$; $t_{ij}$ is the tolerance on the dimension $x_i$ produced by the $j$th process; $a_0$, $a_1$, $a_2$, $a_3$, $a_4$ and $a_5$ is the parameters of cost function.

Constraints on the design tolerances are formulated based on the assumed stack-up criteria. The approach opted by Krishna and Rao (2006) and Sivakumar et al. (2011b) is followed in the present work. Constraints on the design tolerances are formulated based on the assumed stack-up criteria. The manufacturing tolerance constraints can be formulated as follows:

$$\delta_{ij} + \delta_{i(j-1)} \leq \Delta A_{ij} \tag{16}$$

$$\sqrt{\delta_{ij} + \delta_{i(j-1)}} \leq \Delta A_{ij} \tag{17}$$

where $\delta_{ij}$ and $\delta_{i(j-1)}$ are the manufacturing tolerances obtainable in the $j$th and $j-1$th operations, respectively, in production of the $i$th dimension. $\Delta A_{ij}$ is the difference between the nominal and the minimum stock deduction/deformation allowance for manufacturing operation $j$.

The following equations are the representative design constraints:

$$\xi_1 t_{13}|x_{13} + \xi_2 t_2 + \xi_3 t_{33}|x_{33} \leq 0.0349/2 \tag{18}$$

$$\xi_1^2 t_{13}^2|_{x_{13}} + \xi_2^2 t_2^2 + \xi_3^2 t_{33}^2|_{x_{33}} \leq (0.0349/2)^2 \tag{19}$$

The following equations are the representative machining tolerance constraints:

For the hub,

Table 1. Manufacturing process characteristics for the overrunning clutch assembly (Sivakumar et al., 2011b).

| Dimension | Manufacturing operation | Process no. | Parameters of cost function | | | | | | Minimum tolerance (mm) | Maximum tolerance (mm) |
| --- | --- | --- | --- | --- | --- | --- | --- | --- | --- | --- |
| | | | $a_0$ | $a_1$ | $a_2$ | $a_3$ | $a_4$ | $a_5$ | | |
| Hub | Forging | 1 | 304 | −2175 | 6208 | −8620 | 7930 | −5670 | .080 | .500 |
| | | 1 | 220 | −1861 | 5888 | −7985 | 2155 | 3859 | | |
| | Milling | 2 | 208 | −1791 | 5848 | −8065 | 2225 | 3659 | .05 | .350 |
| | | 3 | 199 | −1761 | 5821 | −8062 | 2215 | 3648 | | |
| | Grinding | 4 | 252 | −1952 | 5938 | −8005 | 2315 | 3648 | .010 | .100 |
| | | 1 | 94.60 | −2602.75 | 75,010 | −1,114,190 | 8,219,697 | −23,717,949 | | |
| | | 2 | 101.22 | −2717.55 | 75,108 | −1,108,090 | 8,160,697 | −23,595,949 | | |
| Cage | Forging | 1 | 304 | −2175 | 6208 | −8620 | 7930 | −5670 | .08 | .5 |
| | | 1 | 112.3 | −1060 | 5833 | −15,340 | 18,450 | −8269 | | |
| | Turning | 2 | 106 | −1020 | 5480 | −14,830 | 19,660 | −10,160 | .05 | .5 |
| | | 3 | 110 | −1035 | 5475 | −14,830 | 19,650 | −10,120 | | |
| | Grinding | 4 | 500 | −1035 | 5475 | −14,820 | 19,650 | −10,120 | .010 | .010 |
| | | 1 | 111.28 | −2978.96 | 58,304 | −543,354 | 2,386,396 | −3,966,766 | | |
| | | 2 | 120.67 | −4946.68 | 166,770 | −2,929,392 | 25,171,322 | −83,271,352 | | |

$$t_{11}|x_{11} + t_{12}|x_{12} \leq .7 \tag{20}$$

$$t_{12}|x_{12} + t_{13}|x_{13} \leq .3 \tag{21}$$

For the cage,

$$t_{31}|x_{31} + t_{32}|x_{32} \leq .7 \tag{22}$$

$$t_{32}|x_{32} + t_{33}|x_{33} \leq .3 \tag{23}$$

where $t_{ij}$ is the manufacturing tolerance on the $i$th dimension by $j$th process.

Parameter cost functions and tolerance limits for the overrunning clutch assembly are given in Table 1. The combined objective function considered by Sivakumar et al. (2011b) is used as it is in the present research work to make the comparison of results of optimization meaningful. Sivakumar et al. (2011b) had combined the multiple objectives into scalar objective by using weight vector. Hence, the same approach is used in the present work for comparison purpose. For this problem, the combined objective function ($f_c$) by Sivakumar et al. (2011b) is as follows:

Minimize

$$f_c = W_1 \times Z_1/N_1 + W_2 \times Z_2/N_2 + W_3 \times Z_3/N_3 \tag{24}$$

The values of $N_1 = .1$, $N_2 = 100$, and $N_3 = 10$ are normalizing parameters of the objective functions $Z_1$, $Z_2$, and $Z_3$, and $W_1$, $W_2$, and $W_3$ are the weights given to the objective functions 1, 2, and 3, respectively. $N_1 = .1$ means that it is the minimum value of $Z_1$ if only $Z_1$ is considered as a single objective function (by originally all the other objective function). Similarly, $N_2 = 100$ and $N_3 = 10$ indicate the minimum value of $Z_2$ and $Z_3$ considering them as single objectives (by ignoring the other objective functions). The designer may give any weight to a particular objective function. But the summation of all weight values should be equal to 1. It means that the total weight should be 100%. Sivakumar et al. (2011b) used the weights of $W_1 = W_2 = W_3 = 1/3$ and the same weights are used in the present work for comparison purpose. The calculated values of first, second, and third objective functions using TLBO algorithm are .0046, 233.1067, and 3.2549, respectively. So to get the normalized values, the first, second, and third objective functions are divided by their individual average values (.1, 100, and 10), respectively. The normalized values of first, second, and third objective functions are .046, 2.331067, and .32549, respectively.

### 4.2. Example 2: Knuckle joint with three arms (Sivakumar et al., 2011b)

There are several dimensions in the Knuckle joint assembly as shown in Figure 5 (Sivakumar et al., 2011b). There are three interrelated dimension chains corresponding to the respective design functions giving rise to three constraints. The permissible variation T associated with the respective assembly dimension $Y$ has been assumed to be equal to .2. All similar dimensions are manufactured in same machine, hence the design tolerance associated to $X_{2a}$ and $X_{2b}$ is $t_2$, and that associated to $X_{3a}$ and $X_{3b}$ is $t_3$. For each manufacturing operation, alternative machines are available with the details given in Table 3.

Design functions:

$$Y_1 = X_{2b} - X_1 \tag{25}$$

$$Y_2 = X_{3b} - (2X_{2a} + X_{2b}) \tag{26}$$

$$Y_3 = X_4 - (2X_{3a} + X_{3b} + X_5) \tag{27}$$

where, $Y_1$, $Y_2$, $Y_3$ is design dimensions; $X_1$, $X_{2a}$, $X_{2b}$, $X_{3a}$, $X_{3b}$, $X_4$, $X_5$ are individual dimensions in product assembly.

Sivakumar et al. (2011b) used NSGA-II and MOPSO for simultaneous optimum selection of design and manufacturing tolerances with alternative process selection.

The objective functions considered are: minimum tolerances stack-up, minimum total manufacturing cost of the assembly, and minimum quality loss function. The objective functions considered are same as those considered by Sivakumar et al. (2011b) and these are:

Minimize:

$$Z_1 = \Delta Y_1 = t_{1j} + t_{2j} \tag{28}$$

$$Z_2 = \Delta Y_2 = 3t_{2j} + t_{3j} \tag{29}$$

$$Z_3 = \Delta Y_3 = 3t_{3j} + t_{4j} + t_{5j} \tag{30}$$

$$Z_4 = C_{asm} = Cx_1 + 2Cx_{2a} + Cx_{2b} + 2Cx_{3a} + Cx_{3b} + Cx_4 + Cx_5 \tag{31}$$

$$Z_5 = QL = \frac{A}{9T^2} \sum_{i=1}^{J} t_{ij}^2 \tag{32}$$

where $Z_1$, $Z_2$, and $Z_3$ are the minimum tolerances stack-up; $Z_4$ is the minimum total manufacturing cost and $Z_5$ is the minimum quality loss function.

The following tolerance stack-up constraints are considered:

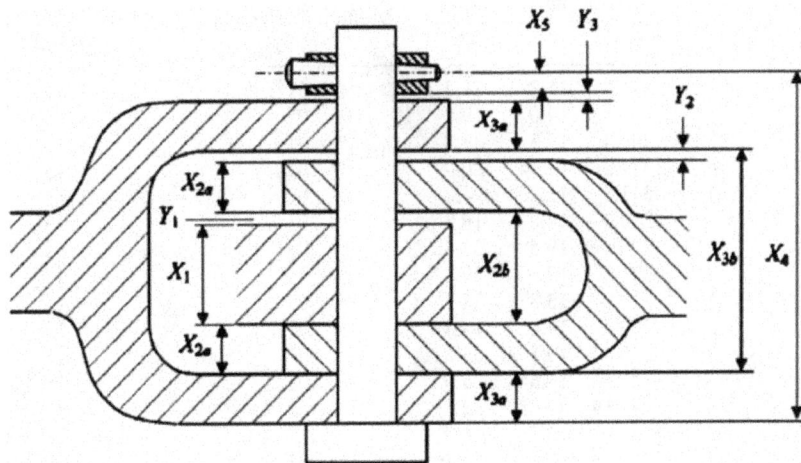

Figure 5.   Knuckle joint assembly with three arms (Sivakumar et al., 2011b).
Note: ($X_i$ is the individual dimensions in product assembly; $Y_1$, $Y_2$, $Y_3$ is design dimensions).

$$t_{1j} + t_{2j} \leq .2 \tag{33}$$

$$3t_{2j} + t_{3j} \leq .2 \tag{34}$$

$$3t_{2j} + t_{4j} + t_{5j} \leq .2 \tag{35}$$

Parameter cost functions and tolerance limits for the knuckle joint assembly with three arms are given in Table 2. Multiple objectives are combined into scalar objective using a weight vector. For this problem, the combined objective function ($f_c$) is defined as follows [18]:

Minimize:

$$f_c = W_1 \times Z_1/N_1 + W_2 \times Z_2/N_2 + W_3 \times Z_3/N_3 + W_4 \times Z_4/N_4 + W_5 \times Z_5/N_5 \tag{36}$$

The values of $N_1 = N_2 = N_3 = N_5 = 1.0$ and $N_4 = 100$ are the normalizing parameters of the objective functions $Z_1$, $Z_2$, $Z_3$, $Z_4$, and $Z_5$ and $W_1$, $W_2$, $W_3$, $W_4$, and $W_5$ are the weights given to the objective functions 1, 2, 3, 4, and 5, respectively. $N_1 = 1.0$ means that it is the minimum value of $Z_1$ if only $Z_1$ is considered as a single objective function (by originally all the other objective function). Similarly, $N_2 = 1.0$, $N_3 = 1.0$, $N_4 = 100$, and $N_5 = 1.0$ indicate the minimum value of $Z_2$, $Z_3$, $Z_4$, and $Z_5$ considering them as single objectives (by ignoring the other objective functions). As mentioned earlier, the designer may give any weight to a particular objective function. But the summation of all weight values should be equal to 1. It means that the total weight should be 100%.

Sivakumar et al. (2011b) used the equal weights of $W_1 = W_2 = W_3 = W_4 = W_5 = .2$ and the same weights are used in the present work for comparison purpose. The calculated values of first, second, third, fourth, and fifth objective functions using TLBO algorithm are .186008, .191812, .196046, 1008.7, and .2466, respectively. So to get the normalized

Table 2. Manufacturing process characteristics for the knuckle joint assembly with three arms (Sivakumar et al., 2011b).

| Dimensions | Process | Parameters of cost function | | | Minimum tolerance (mm) | Maximum tolerance (mm) |
|---|---|---|---|---|---|---|
| | | $C_0$ | $C_1$ | $C_2$ | | |
| $X_1$ | 1 | 311.0 | 15.8 | 24.2 | .01 | .15 |
| | 2 | 280.0 | 14.0 | 19.8 | .01 | .15 |
| | 3 | 296.4 | 19.5 | 23.82 | .01 | .15 |
| | 4 | 331.2 | 17.64 | 20.0 | .01 | .15 |
| $X_{2a}, X_{2b}$ | 1 | 311.0 | 15.8 | 24.2 | .01 | .15 |
| | 2 | 280.0 | 14.0 | 19.8 | .01 | .15 |
| | 3 | 296.4 | 19.5 | 23.82 | .01 | .15 |
| | 4 | 331.5 | 17.64 | 20.1 | .01 | .15 |
| $X_{3a}, X_{3b}$ | 1 | 311.0 | 15.8 | 24.2 | .01 | .15 |
| | 2 | 280.0 | 14.0 | 19.8 | .01 | .15 |
| | 3 | 296.4 | 19.5 | 23.82 | .01 | .15 |
| | 4 | 331.5 | 17.64 | 20.0 | .01 | .15 |
| $X_4$ | 1 | 92.84 | 82.45 | 32.5 | .02 | .20 |
| | 2 | 82.43 | 16.70 | 21.0 | .02 | .20 |
| $X_5$ | 1 | 128.25 | 82.65 | 32.5 | .01 | .10 |
| | 2 | 160.43 | 86.70 | 29.2 | .01 | .10 |
| | 3 | 231.42 | 50.05 | 28.05 | .01 | .10 |
| | 4a | 134.16 | 78.82 | 500.0 | .01 | .10 |

values, the first, second, third, fourth, and fifth objective functions are divided by their individual average values (i.e. 1.0,1.0,1.0100, and 1.0), respectively. The normalized values of the first, second, third, fourth, and fifth objective functions are .186008, .191812, .196046, 10.087, and .2466, respectively.

### 4.3.  Example 3: Helical spring (Singh et al., 2005)

In this example, only one objective function is considered i.e. minimizing the total manufacturing cost (Singh et al., 2005). For the helical spring shown in Figure 6 (Singh et al., 2005), there are three decision variables and Singh et al. (2005) used GA for tolerance design considering interrelated dimension chains and process precision limits. The Design functions are:

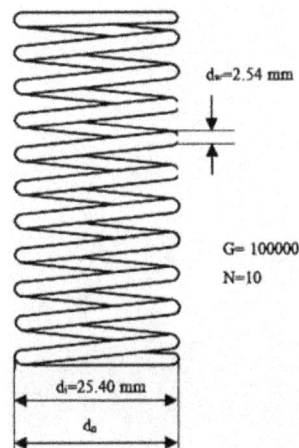

Figure 6.   Helical spring (Singh et al., 2005).

Table 3.   Manufacturing process characteristics for the helical spring (Singh et al. 2005).

| Dimensions | Process | Parameters of cost function | | | Minimum tolerance (mm) | Maximum tolerance (mm) |
|---|---|---|---|---|---|---|
| | | $C_0$ | $C_1$ | $C_2$ | | |
| $d_w$ | 1 | 1.5810 | 7.8735 | 1.44 | .018 | .80 |
| | 2 | 1.4135 | 7.262 | 1.44 | .020 | .82 |
| | 3 | 1.3622 | 6.8138 | 1.52 | .022 | .8 |
| | 4 | 1.6315 | 8.1236 | 1.35 | .018 | .76 |
| $d_i$ | 1 | 7.3964 | 3.9333 | .5 | .218 | 1.2 |
| | 2 | 7.8735 | 3.124 | .55 | .230 | 1.26 |
| | 3 | 6.8138 | 3.522 | .58 | .225 | 1.3 |
| | 4 | 8.8236 | 4.4321 | .625 | .234 | 1.24 |
| $N$ | 1 | 6.5634 | 21.4097 | 1.5 | .02 | .2 |
| | 2 | 6.1138 | 20.682 | 1.5 | .024 | .22 |
| | 3 | 5.8254 | 18.5635 | 1.75 | .022 | .25 |
| | 4 | 7.3425 | 24.22 | 1.855 | .018 | .20 |

Industrial Engineering and Management

Table 4. Optimization results for the knuckle joint with three arms.

| Technique | Dimension | Machine | Tolerance notation | Tolerance value (mm) | $Z_1$ | $Z_2$ | $Z_3$ | $Z_4$ | $Z_5$ | Combined objective function ($f_c$) |
|---|---|---|---|---|---|---|---|---|---|---|
| NSGA-II (Sivakumar et al., 2011b) | $X_1$ | 3 | $tX_1$ | .069665 | .119203 | .194848 | .179988 | 1076.895 | .124079 | 2.2774136 |
| | $X_{2a}$-$X_{2b}$ | 3 | $tX_{2a}tX_{2b}$ | .049538 | | | | | | |
| | $X_{3a}$-$X_{3b}$ | 3 | $tX_{3a}tX_{3b}$ | .046234 | | | | | | |
| | $X_4$ | 2 | $tX_4$ | .020007 | | | | | | |
| | $X_5$ | 2 | $tX_5$ | .021279 | | | | | | |
| MOPSO (Sivakumar et al., 2011b) | $X_1$ | 3 | $tX_1$ | .087427 | .137138 | .197005 | .195297 | 1029.393 | .154244 | 2.1955228 |
| | $X_{2a}$-$X_{2b}$ | 3 | $tX_{2a}tX_{2b}$ | .049711 | | | | | | |
| | $X_{3a}$-$X_{3b}$ | 3 | $tX_{3a}tX_{3b}$ | .047872 | | | | | | |
| | $X_4$ | 2 | $tX_4$ | .027344 | | | | | | |
| | $X_5$ | 2 | $tX_5$ | .024337 | | | | | | |
| TLBO | $X_1$ | 3 | $tX_1$ | .137619 | .186008 | .191812 | .196046 | 1008.7 | .2466 | **2.1814** |
| | $X_{2a}$-$X_{2b}$ | 3 | $tX_{2a}tX_{2b}$ | .048389 | | | | | | |
| | $X_{3a}$-$X_{3b}$ | 3 | $tX_{3a}tX_{3b}$ | .046645 | | | | | | |
| | $X_4$ | 2 | $tX_4$ | .030979 | | | | | | |
| | $X_5$ | 2 | $tX_5$ | .025132 | | | | | | |

Note: The number in bold indicate the better values.

Table 5.  Optimization results for the overrunning clutch assembly.

| Technique | Optimum variable | | | | | | Objective functions | | | Combined objective function ($f_c$) |
|---|---|---|---|---|---|---|---|---|---|---|
| | Manufacture of hub | | | Manufacture of cage | | | $Z_1$ | $Z_2$ | $Z_3$ | |
| | Machine | Tolerance notation | Tolerance value mm) | Machine | Tolerance notation | Tolerance value(mm) | | | | |
| NSGA-II (Sivakumar et al., 2011b) | 1 | $t_{11}$ | .261619 | 1 | $t_{31}$ | .196208 | | | | |
| | 3 | $t_{12}$ | .165386 | 2 | $t_{32}$ | .094096 | .0112 | 288.175 | 13.201 | 1.436501 |
| | 2 | $t_{13}$ | .02747 | 2 | $t_{33}$ | .028307 | | | | |
| MOPSO (Sivakumar et al., 2011b) | 1 | $t_{11}$ | .203162 | 1 | $t_{31}$ | .22952 | | | | |
| | 3 | $t_{12}$ | .200655 | 2 | $t_{32}$ | .11676 | .0127 | 271.952 | 13.702 | 1.404177 |
| | 2 | $t_{13}$ | .023635 | 2 | $t_{33}$ | .042379 | | | | |
| TLBO | 1 | $t_{11}$ | .385848 | 1 | $t_{31}$ | .465393 | | | | |
| | 3 | $t_{12}$ | .24247 | 2 | $t_{32}$ | .121685 | .0046 | 233.1067 | 3.2549 | **.8917** |
| | 2 | $t_{13}$ | .010973 | 2 | $t_{33}$ | .0165781 | | | | |

Note: The number in bold indicate the better values.

$$k = \frac{Gd_w^4}{\{8(d_i + d_w)^3 N\}} \qquad (37)$$

$$d_o = d_i + 2d_w \qquad (38)$$

where $k$ is the spring rate, $G$ is the modulus of rigidity of the spring wire material, $d_w$ is the wire diameter of spring, $d_i$ is the inner diameter of spring, $d_o$ is the outer diameter of spring, and $N$ is the number of active coils in spring.

The objective function considered in this paper is same as that considered by Singh et al. (2005) and these are:

Minimize:

$$Z_1 = C_{\text{asm}} = C_{d_w} + C_{d_i} + C_n \qquad (39)$$

where $C_{\text{asm}}$ is the total assembly manufacturing cost, $C_{dw}$, $C_{di}$, and $C_n$ are manufacturing costs of wire diameter of spring, inner diameter of spring and number of active coils in spring, respectively.

For each manufacturing operation, alternative machines are available with the details given in Table 3.

The following tolerance stack-up constraints are considered:

Table 6.  Optimization results for the helical spring.

|  | Dimension | Process | Individual tolerance value (mm) | Resultant assembly tolerance | Minimum cost |
|---|---|---|---|---|---|
| GA (Singh et al., 2005) | $d_w$ | 2 | .02 | $\Delta K = .04770^*$ | 5.94 |
|  | $D_i$ | 1 | .4645 | $\Delta d_0 = .5045$ |  |
|  | $N$ | 2 | .2054 |  |  |
| TLBO | $d_w$ | 2 | .02007159 | $\Delta K = .08105$ | **5.8365** |
|  | $D_i$ | 1 | .49925285 | $\Delta d_0 = .5392$ |  |
|  | $N$ | 2 | .17866543 |  |  |

$^*$Corrected value is $\Delta K = .08561$.
Note: The number in bold indicates the better value.

Table 7.  Results obtained by using NSGA-II, MOPSO, and TLBO algorithms for the multi-objective optimization problems.

|  | $Z_1$ | $Z_2$ | $Z_3$ | $Z_4$ | $Z_5$ | Combined objective function $(f_c)$ |
|---|---|---|---|---|---|---|
| *Example 1 (Overrunning clutch assembly)* | | | | | | |
| TLBO | .0046 | 233.1067 | 3.2549 | NA | NA | **.8917** |
| NSGA-II | .0112 | 288.175 | 13.201 | NA | NA | 1.436501 |
| MOPSO | .0127 | 271.952 | 13.702 | NA | NA | 1.404177 |
| *Example 2 (Knuckle joint assembly with three arms)* | | | | | | |
| TLBO | .186008 | .191812 | .196046 | 1008.7 | .2466 | **2.1814** |
| NSGA-II | .119203 | .194848 | .179988 | 1076.895 | .124079 | 2.2774136 |
| MOPSO | .137138 | .197005 | .195297 | 1029.393 | .154244 | 2.1955228 |

Note: NA – not applicable.
The numbers in bold indicate the better values.

$$\frac{G(d_w + t_{wj})^4}{8\{(d_i - t_{ij}) + (d_w + t_{wj})\}^3 (N - t_{Nj})} - \frac{G(d_w + t_{wj})^4}{8\{(d_i + t_{ij}) + (d_w - t_{wj})\}^3 (N + t_{Nj})} \leq .477 \quad (40)$$

$$t_{ij} + 2t_{wj} \leq .508 \quad (41)$$

For each manufacturing operation, alternative machines, parameter cost functions, and tolerance limits for the helical spring are given in Table 3.

## 5.  Results and discussion

Tables 4–6 compare the optimum selection of the manufacturing process (machine), allocated tolerance value, the values of objective functions, and the combined objective function value obtained by using various optimization techniques. For optimal result, overrunning clutch assembly, the knuckle joint assembly with three arms and the helical spring, were used population size of 50 and number of generations of 20, 60, and 40, respectively, for overrunning clutch assembly, the knuckle joint assembly with three arms, and the helical spring. As the evaluation is done in the both the teacher and student phases, the number of function evaluation used in TLBO is calculated as, Number of function evaluation = Population size * number of generations * 2. Thus the function evaluations are 2000, 6000, and 4000, respectively, in the case of for overrunning clutch assembly, the knuckle joint assembly with three arms, and the helical spring. It is observed that TLBO algorithm gives better results than GA, NSGA-II, and MOPSO algorithms. The TLBO algorithm is found superior to GA, NSGA-II, and MOPSO for

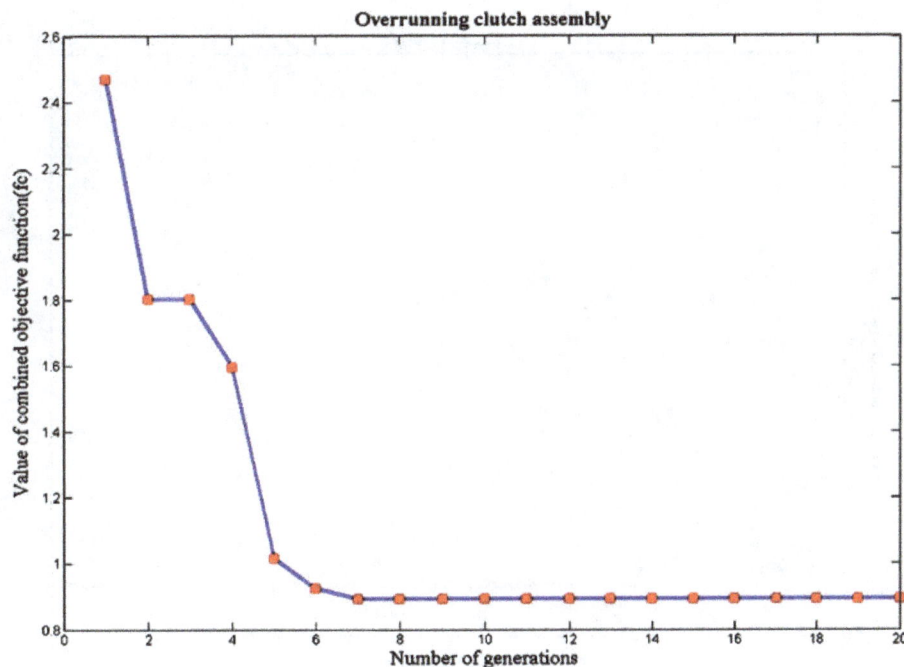

Figure 7.    Convergence of combined objective function obtained by TLBO algorithm for Overrunning clutch assembly.

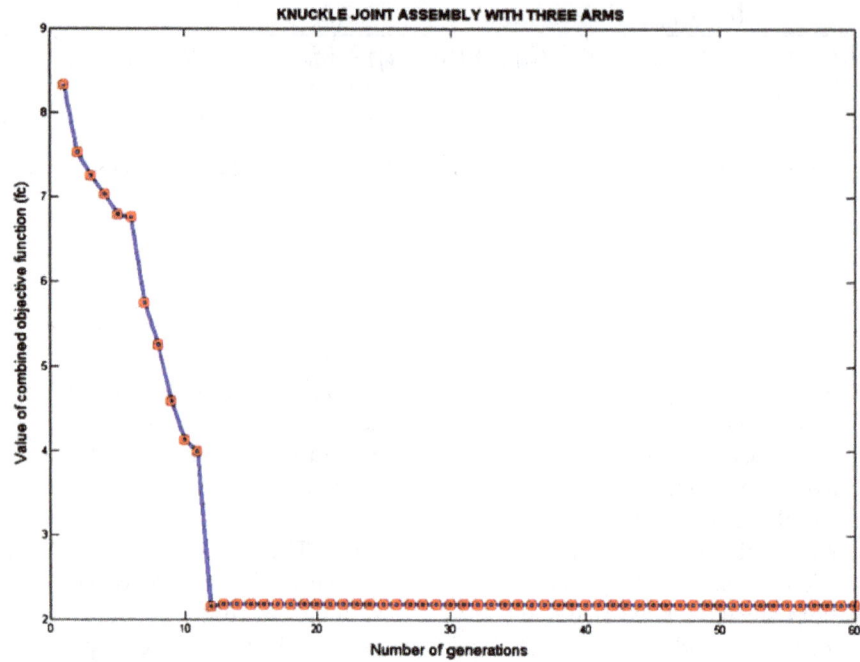

Figure 8.    Convergence of combined objective function obtained by TLBO algorithm for Knuckle joint assembly with three arms.

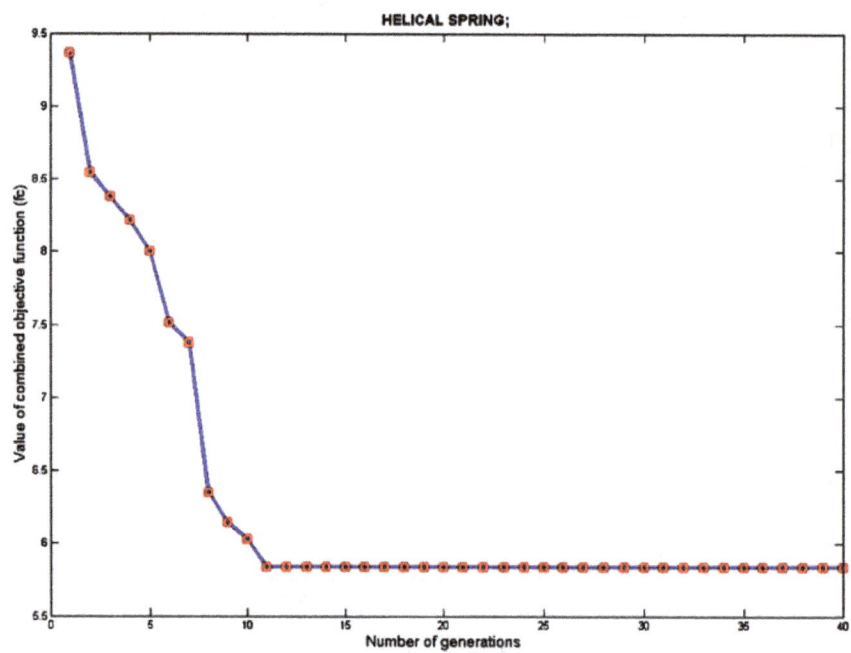

Figure 9.    Convergence of combined objective function obtained by TLBO algorithm for Helical spring.

all the three case studies considered, i.e. overrunning clutch assembly, the knuckle joint assembly with three arms, and the helical spring. Table 7 shows that the results obtained using NSGA-II, MOPSO, and TLBO algorithms for the multi-objective optimization problems. Each problem is run 30 times. Standard deviation for clutch assembly is .00156, for knuckle joint assembly is .0246, and for spring design is .0894.

Figures 7–9 show the convergence rate of the TLBO algorithm for the three examples. In the example of overrunning clutch assembly convergence takes place after 7th iteration, in knuckle joint assembly with three arms convergence takes place after 12th iteration, and in design of helical spring the convergence takes place after 11th iteration of the TLBO algorithm. It is noted that the TLBO algorithm gives the best results (i.e. minimum values of $Z_1$, and $Z_3$ for overrunning clutch assembly and minimum values of $Z_1$, $Z_2$, $Z_3$, and $Z_5$ for knuckle joint assembly) and the computational time to find the optimum solutions by TLBO algorithm is less than that of GA, NSGA-II, and MOPSO algorithms which implies that the TLBO algorithm is faster than the GA, NSGAII, and MOPSO algorithms. Figure 10 shows the optimal solution trade-offs obtained from NSGA-II, MOPSO, and TLBO algorithms for the knuckle joint assembly. Figure 11 shows the optimal solution trade-offs obtained from NSGA-II, MOPSO, and TLBO algorithms for the overrunning clutch assembly.

In overrunning clutch assembly, TLBO gives 38 and 36.5% improvements in the combined objective function as compared to the results given by the NSGA-II and MOPSO algorithms, respectively. In the case of knuckle joint with three arms assembly, TLBO gives 4.2% and .62% better value of the combined objective function than those

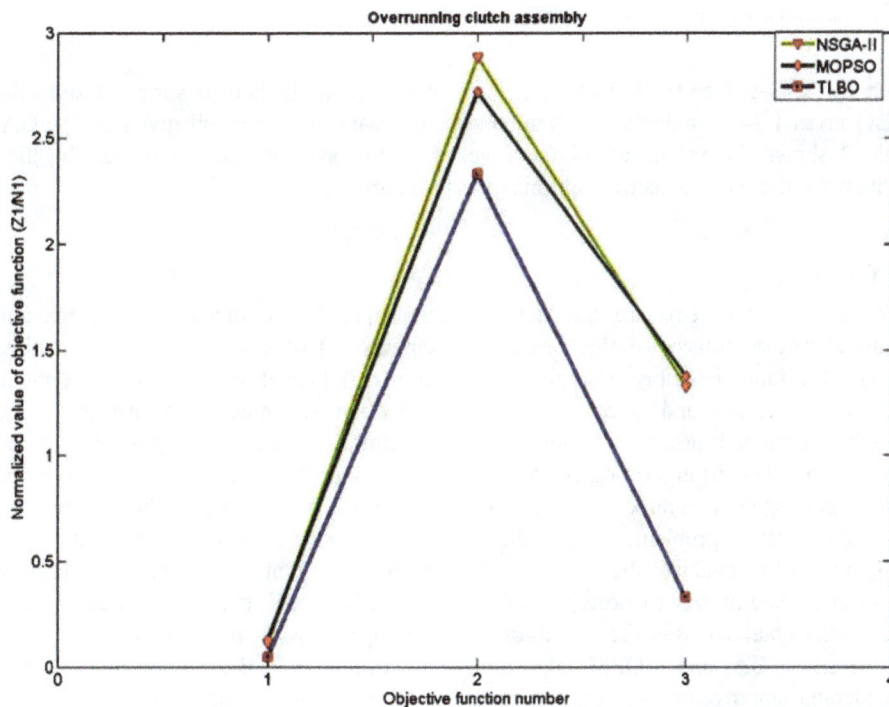

Figure 10.   Solution trade-offs obtained from TLBO, NSGA-II, and MOPSO algorithms for the overrunning clutch assembly.

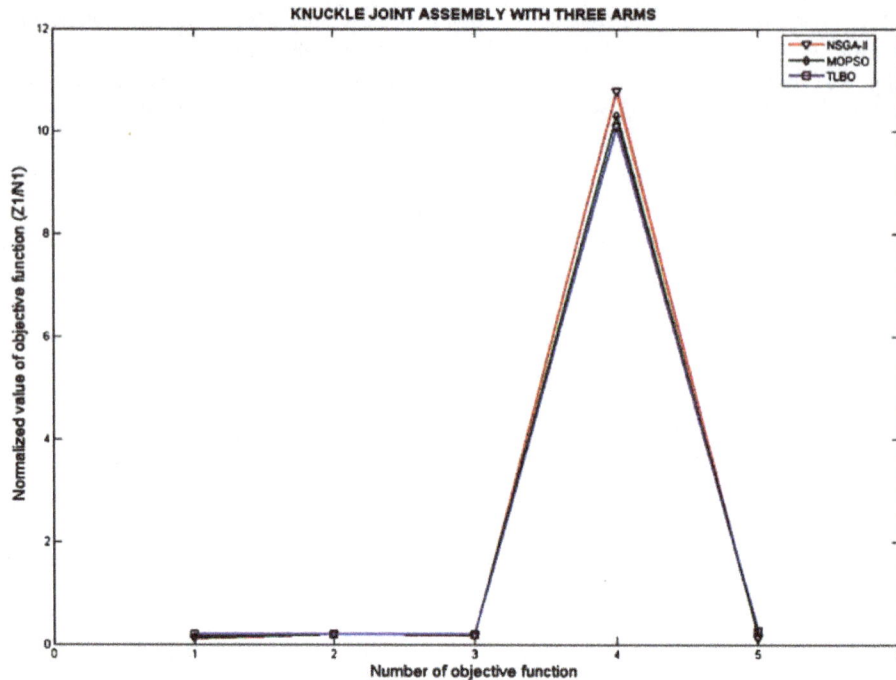

Figure 11.   Best solution trade-offs obtained from TLBO, NSGA-II, and MOPSO algorithms for the Knuckle joint with three arms assembly.

given by NSGA-II and MOPSO algorithms, respectively. In helical spring design, the TLBO gives 1.742% reduction in total cost as compared to the result given by the GA. Table 7 shows the values of all the objective functions and the combined objective function for the multi-objective optimization problems.

## 6.   Conclusions

Three different problems are considered in this paper for optimization of design and manufacturing tolerances of three machine elements: (i) overrunning clutch assembly, (ii) knuckle joint assembly with three arms, and (iii) helical spring. For the optimal tolerance allocation and alternative process selection for mechanical assemblies, a recently developed advanced optimization algorithm, called teaching-learning-based optimization (TLBO) algorithm, is considered. This algorithm does not require selection of algorithm-specific parameters. Out of the three problems considered, the first two are the multi-objective problems and the third is a single-objective problem and these problems are used to evaluate the strength of the TLBO algorithm. The same models were earlier attempted by other researchers using GA, SA, NSGA-II, and MOPSO algorithms. The results obtained using TLBO algorithm is compared with those obtained by using the NSGA-II, GA, and MOPSO algorithms. However, the TLBO algorithm has given considerable improvements in the results and in the convergence rate.

The TLBO algorithm has shown its ability in solving multi-objective optimization problems by using normalizing weighting objective function. The convergence behaviour of the TLBO algorithm to a near global solution has been observed to be more effective than that of GA, NSA-II, and MOPSO algorithms. The results indicate that the TLBO technique gives better results than GA, NSGA-II, and MOPSO algorithms. Hence, the TLBO algorithm is proved better than the other optimization algorithms in terms of results and the convergence. The TLBO algorithm may be conveniently used for the optimal tolerance design of the other machine elements also.

## References

Claudio, M. R., Jose, A. M., & Nestor, C. (2003). Robust design using a hybrid-cellular-evolutionary and interval-arithmetic approach: A reliability application. *Reliability Engineering and System Safety, 79*, 149–159.

Daniel, S., Claudio, M. R., & Blas, J. G. (2006). Optimization of constrained multiple-objective reliability problems using evolutionary algorithms. *Reliability Engineering and System Safety, 91*, 1057–1070.

Feng, C.-X., & Kusiak, A. (1997). Robust tolerance design with the integer programming approach. *Transaction of ASME Journal Manufacturing Science Engineering, 119*, 603–610.

Forouraghi, B. (2002). Worst-case tolerance design and quality assurance via genetic algorithms. *Journal of Optimization Theory and Applications, 113*, 251–268.

Forouraghi, B. (2009). Optimal tolerance allocation using a multiobjective particle swarm optimizer. *International Journal of Advanced Manufacturing Technology, 44*, 710–724.

Gonzalez, I., & Sanchez, I. (2009). Statistical tolerance synthesis with correlated variables. *Mechanism and Machine Theory, 44*, 1097–1107.

Guofu, Z., Yuege, Z., Ye, X., & Hu, B. (2013). A method of multi-objective reliability tolerance design for electronic circuits. *Chinese Journal of Aeronautics, 26*, 161–170.

Haq, A. N., Sivakumar, K., Saravanan, R., & Muthiah, V. (2005). Tolerance design optimization of machine elements using genetic algorithm. *International Journal of Advanced Manufacturing Technology, 25*, 385–391.

Huang, Y. M., & Shiau, C.-S. (2006). Optimal tolerance allocation for a sliding vane compressor. *Journal of Mechanical Design, 128*, 98–107.

Huang, M. F., & Zhong, Y. R. (2007). Optimized sequential design of two-dimensional tolerances. *International Journal of Advanced Manufacturing Technology, 33*, 579–593.

Huang, M., & Zhong, Y. (2008). Dimensional and geometrical tolerance balancing in concurrent design. *International Journal of Advanced Manufacturing Technology, 35*, 723–735.

Krishna, G., Rao, K. M. (2006). Simultaneous optimal selection of design and manufacturing tolerances with different stack-up conditions using scatter search. *International Journal of Advanced Manufacturing Technology, 30*, 328–333.

Muthu, P., Dhanalakshmi, V., & Sankaranarayanasamy, K. (2009). Optimal tolerance design of assembly for minimum quality loss and manufacturing cost using metaheuristic algorithms. *International Journal of Advanced Manufacturing Technology, 44*, 1154–1164.

Pierluissi, L., & Rocco, C. M. (2007). Optimal design centring through a hybrid approach based on evolutionary algorithms and monte carlo simulation. ICANNGA'07, 8th International Conference Adapt Nat Comput Algo. Warsaw, Polonia, Lecture Notes in Computer Science (LNCS), series by Springer Verlag, 31–38.

Prabhaharan, G., Asokan, P., & Rajendran, S. (2005). Sensitivity-based conceptual design and tolerance allocation using the continuous Ants colony algorithm (CACO). *International Journal of Advanced Manufacturing Technology, 25*, 516–526.

Rao, R. V., & Kalyankar, V. D. (2012). Parameter optimization of machining processes using a new optimization algorithm. *Materials and Manufacturing Processes, 27*, 978–985.

Rao, R. V., & Patel, V. (2012). Comparative performance of an elitist teaching-learning-based optimization algorithm for solving unconstrained optimization problems. *International Journal of Industrial Engineering Computations, 4*, 29–50.

Rao, R. V., & Patel, V. (2013a). Multi-objective optimization of two stage thermo electric cooler using a modified teaching–learning-based optimization algorithm. *Engineering Applications of Artificial Intelligence, 26*, 430–445.

Rao, R. V., & Patel, V. (2013b). An elitist teaching-learning-based optimization algorithm for solving complex constrained optimization problems. *International Journal of Industrial Engineering Computations, 3*, 535–560.

Rao, R. V., & Savsani, V. J. (2012). *Mechanical design optimization using advanced optimization techniques*. London: Springer-Verlag.

Rao, R. V., Savsani, V. J., & Vakharia, D. P. (2011). Teaching–learning-based optimization: A novel method for constrained mechanical design optimization problems. *Computer-Aided Design, 43*, 303–315.

Rao, R. V., Savsani, V. J., & Vakharia, D. P. (2012). Teaching–learning-based optimization: An optimization method for continuous non-linear large scale problem. *Information Sciences, 3*, 1–15.

Rocco, C. M., Moreno, J., & Carrasquero, N. (2003). Robust design using a hybrid-cellular-evolutionary and interval-arithmetic approach: A reliability application. *Reliability Engineering and System Safety, 79*, 149–159.

Salazar, D., Rocco, C. M., & Galvan, B. (2006). Optimization of constrained multiple-objective reliability problems using evolutionary algorithms. *Reliability Engineering and System Safety, 91*, 1057–1070.

Singh, P. K., Jain, S. C., & Jain, P. K. (2003). Tolerance allocation with alternative manufacturing processes-suitability of genetic algorithm. *International Journal of Modeling and Simulation, 2*, 22–34.

Singh, P. K., Jain, S. C., & Jain, P. K. (2004). A GA based solution to optimum tolerance synthesis of mechanical assemblies with alternate manufacturing processes: Focus on complex tolerancing problems. *International Journal of Simulation and Modelling, 42*, 5185–5215.

Singh, P. K., Jain, P. K., & Jain, S. C. (2005). Advanced optimal tolerance design of mechanical assemblies with interrelated dimension chain and process precision limits. *Computers in Industry, 56*, 179–194.

Singh, P. K., Jain, P. K., & Jain, S. C. (2008). Optimal tolerance design of mechanical assemblies for economical manufacturing in the presence of alternative machines- a genetic algorithm-based hybrid methodology. *Proceeding Institution of Mechanical Engineers Journal of Engineering Manufacture, 222B*, 591–604.

Singh, P. K., Jain, P. K., & Jain, S. C. (2009). Important issues in tolerance design of mechanical assemblies. Part-2: Tolerance synthesis. *Proceedings IMechE Journal of Engineering Manufacture, 223B*, 1249–1287.

Sivakumar, K., Balamurugan, C., & Ramabalan, S. (2011a). Concurrent multi-objective tolerance allocation of mechanical assemblies considering alternative manufacturing process selection. *International Journal of Advanced Manufacturing Technology, 53*, 711–732.

Sivakumar, K., Balamurugan, C., & Ramabalan, S. (2011b). Simultaneous optimal selection of design and manufacturing tolerances with alternative manufacturing process selection. *Computer-Aided Design, 43*, 207–218.

Sivakumar, K., Balamurugan, C., & Ramabalan, S. (2012). Evolutionary multi-objective concurrent maximisation of process tolerances. *International Journal of Production Research, 50*, 3172–3191.

Sivakumar, M., Kannan, S. M., & Jayabalan, V. (2009). A new algorithm for optimum tolerance allocation of complex assemblies with alternative processes selection. *International Journal of Advanced Manufacturing Technology, 40*, 819–836.

Sivakumar, M., & Stalin, B. (2009). Optimum tolerance synthesis for complex assembly with alternative process selection using Lagrange multiplier method. *International Journal of Advanced Manufacturing Technology, 44*, 405–411.

Wu, F., Jean-Yves, D., Alain, E., Ali, S., & Patrick, M. (2009). Improved algorithm for tolerance allocation based on Monte Carlo simulation and discrete optimization. *Computers and Industrial Engineering, 56*, 1402–1413.

Ye, B., & Salustri, F. A. (2003). Simultaneous tolerance synthesis for manufacturing and quality. *Research in Engineering Design, 14*, 98–106.

# 12

# Stabilizing wave attenuation effects in turning process

Chigbogu Godwin Ozoegwu*

*Department of Mechanical Engineering, Nnamdi Azikiwe University, PMB 5025, Awka, Nigeria*

Efforts have been made and continue to be made to understand the self-excited vibrations of machine tools called regenerative chatter. Theory of stabilizing wave attenuation effects in machining process presented in another work is expanded in this work to postulate the behaviour of turning stability lobes to changes in material, process and structural parameters. The wave attenuation theory simply stated that rise in attenuating forces suppresses chatter instability. Analysis in this work showed that rise in tool natural frequency $\omega_n$, damping ratio $\xi$ or feed speed $v$ causes a rise in attenuating forces thus suppressing chatter. As expected from the wave attenuation theory harder workpiece materials with higher cutting coefficients $C$ will better resist wave attenuation, and thus exhibit more chatter instability. Numerical verification of postulations is given.

**Keywords:** turning process; wave attenuation; chatter; turning point critical force; stability

## 1. Introduction

Regenerative vibrations are the self-excited vibrations in machine tools that are referred to as secondary chatter in literature (Wiercigroch & Budak, 2001). Regenerative chatter is a much more detrimental problem to the machining industry than the primary chatter phenomena that stem from frictional, thermo-mechanical and mode-coupling effects. Some of the unwanted effects of regenerative chatter are noise pollution in the work environment, poor surface quality, aggravated tool wear, reduced productivity and waste of materials and energy. Two of the most important reviews of machine tool chatter research (Quintana & Ciurana, 2011; Siddhpura & Paurobally, 2012) noted that regenerative chatter is not yet fully understood even though it has received most attention of all chatter phenomena studied over a century. Regenerative chatter is caused by Regenerative effects which are effects of waviness on a machined surface that are in turn caused by response of the machine tool system to disturbances. The waves of consecutive turns are normally out of phase causing cutting force variation that excites the tool. If cutting process parameters are stable, then cutting force variation decays because phase difference of consecutive turns die out with time. The opposite is the case when cutting process parameters constitute an unstable combination. Before regenerative chatter was distinguished and understood as stemming from self-excitation effects it was originally thought by Arnold (1946) as stemming from negative damping. Some of the standard early works (Merritt, 1965; Tobias, 1961; Tobias & Fishwick, 1958) on regenerative chatter considered a single degree of freedom orthogonal turning process vibrating in

---

*Email: chigbogug@yahoo.com

the feed direction only. Analysis of non-linear regenerative chatter was introduced by Hanna and Tobias (1974). It is known that the so-called Hopf bifurcation (lose of stability) in turning operation occurs when a pair of complex conjugate characteristic roots crosses from the left-half plane to right-half plane of the complex plane. More recently non-linear analysis has been strengthened by the experiment of Shi and Tobias (1984) and analysis of Stepan and Kalmar-Nagy (1997) by revealing that Hopf bifurcation of turning operation is subcritical. In addition to chatter frequencies of Hopf bifurcation type, damped natural frequency can also be spotted in the frequency response spectrum of high-speed turning tool due to the effects of tool-workpiece contact loss. Damped natural frequency has been experimentally confirmed to exist in the frequency response spectrum of high-speed milling (Insperger, Stepan, Bayly, & Mann, 2003). Experimental data of chatter and other vibrations in the turning and milling processes have been characterized using multi-scale methods (Litak, Syta, & Rusinek, 2011; Sen, Litak, Syta, & Rusinek, 2013) and flicker-noise spectroscopy (Litak, Polyakov, Timashev, & Rusinek, 2013). The theory of stabilizing wave attenuation effects in machining process was presented by Ozoegwu and Omenyi (2013). The theory posits that some of the force of interaction between the tool and workpiece acts normal to surface waves, thus depressing or flattening them in a way as to attenuate the chatter–causing regenerative effects. The attenuating force is either compressive when there is a continuous cutting edge-workpiece engagement or impulsive when the active pass of the cutting edge is highly interrupted as in high-speed machining. The theory further posits that any variable of the attenuating force that varies in a way as to increase the amplitude of the attenuating force will suppress chatter instability. In Ozoegwu and Omenyi (2013), the wave attenuation theory was able to explain why chatter stability rises with rise in feed, fall in number of teeth and fall in hardness of workpiece material. The theory of stabilizing wave attenuation effects is being expanded in this work to cover the turning process.

## 2.  The cutting force, dynamics and stability of regenerative turning

Figure 1 represents an orthogonal turning process in which the cutting edge of the tool is perpendicular to the feed motion. In the physical configuration the workpiece is clamped in the chuck of a rotating spindle while the tool is fed in the workpiece. Regenerative chatter is the self-excited vibration in machining due to the effects of waviness (regenerative effects) on a machined surface resulting from disturbed dynamic interaction between the tool and the workpiece. The present turn of the workpiece has waviness that is not in phase with those of the last turn resulting chip thickness variation, hence cutting force variation that powers vibration which subsequently builds up to chatter if cutting parameter combination is unfavourable. As shown in Figure 1, a tool with modal parameters; $m$ the mass, $c$ the equivalent viscous damping coefficient and $k$ the stiffness is fed into the workpiece at a speed $v$ in $x-$ direction. The response of the tool $x(t)$ to the regenerative cutting force variation is seen from Figure 1 to be governed by the differential equation

$$m\ddot{x}(t) + c[\dot{x}(t) - v] + k[x(t) - vt] + F_x(t) = 0 \qquad (1)$$

The cutting force $F_x(t)$ has the empirical form (Insperger, 2002; Stepan, Szalai, & Inspeger, 2004; Tlusty, 2000);

$$F_x(t) = Cwf_a^\gamma \qquad (2)$$

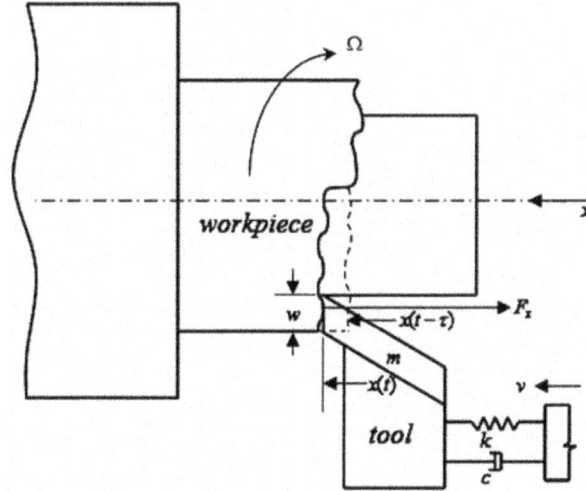

Figure 1.   Mechanical model of orthogonal turning.

where $C$ is the cutting coefficient, $w$ is the depth of cut, $f_a$ is the actual feed rate defined as difference of present and one period delayed position of tool; $f_a = x(t) - x(t - \tau)$ and $\gamma$ is the feed exponent of the cutting force that has typical values 1, .8 and 3/4. With the use of Equation (2), Equation (1) is re-arranged to give

$$m\ddot{x} + c\dot{x} + kx = cv + kvt - Cw(x - x_\tau)^\gamma. \tag{3}$$

where $x(t) = x$ and $x(t - \tau) = x_\tau$. The response of the tool is a linear combination of prescribed feed motion $vt$, static deflection $x_t(t) = -(Cwf^\gamma)/k = -[Cw(vt)^\gamma]/k$ and perturbation $z(t)$ such that Equation (3) becomes.

$$m\ddot{z} + c\dot{z} + kz = Cw(v\tau)^\gamma - Cw[v\tau + (z - z_\tau)]^\gamma. \tag{4}$$

The linearized Taylor's series expansion Equation (4) about $v\tau$ gives

$$m\ddot{z} + c\dot{z} + kz = -hw(z - z_\tau). \tag{5}$$

where $h = C\gamma(v\tau)^{\gamma-1}$. Recognizing that the natural frequency and damping ratio of the tool are given in terms of modal parameters $k, m$ and $c$, respectively as $\omega_n = \sqrt{k/m}$ and $\xi = c/2\sqrt{mk}$ gives the modal form

$$\ddot{z} + 2\xi\omega_n\dot{z} + \left(\omega_n^2 + \frac{hw}{m}\right)z = \frac{hw}{m}z_\tau. \tag{6}$$

Equation (6) is the linear dynamical model for the regenerative turning tool. With the substitutions $y_1 = z$ and $y_2 = \dot{z}$ made, Equation (6) could be put in state differential equation form

$$\underline{y} = \mathbf{A}y + \mathbf{B}y_\tau \tag{7}$$

where $\mathbf{A} = \begin{bmatrix} 0 & 1 \\ -\left(\omega_n^2 + \frac{hw}{m}\right) & -2\xi\omega_n \end{bmatrix}$, $\mathbf{B} = \begin{bmatrix} 0 & 0 \\ \frac{hw}{m} & 0 \end{bmatrix}$, $y = \begin{Bmatrix} y_1 \\ y_2 \end{Bmatrix}$, $y_\tau = \begin{Bmatrix} y_{1,\tau} \\ y_{2,\tau} \end{Bmatrix}$ and $y_{i,\tau} = y_i(t - \tau)$ for $i = 1$ and 2. A trial solution of form $y(t) = Ke^{\lambda t}$ put in Equation (7) to give the characteristic equation

$$|\lambda \mathbf{I} - \mathbf{A} - \mathbf{B} e^{-\lambda \tau}| = 0. \tag{8}$$

On the stability lobes of turning process the pair of maximum-magnitude eigenvalues of Equation (8) are on the imaginary axis; $\lambda = \pm i\omega$ given the pair of equations

$$-\omega^2 + \omega_n^2 + \frac{hw}{m}(1 - \cos\omega\tau) = 0 \tag{9}$$

$$2\xi\omega_n\omega + \frac{hw}{m}\sin\omega\tau = 0. \tag{10}$$

These are solved simultaneously to give the critical parameter combinations

$$\Omega_c = \frac{60}{\tau_c} = \frac{30\omega}{j\pi - tan^{-1}\left(\dfrac{\omega^2 - \omega_n^2}{2\xi\omega_n\omega}\right)}, \qquad j = 1, 2, 3, \tag{11}$$

$$w_c = \frac{m}{2h}\left\{\frac{(\omega^2 - \omega_n^2)^2 + 4\xi^2\omega_n^2\omega^2}{\omega^2 - \omega_n^2}\right\}. \tag{12}$$

The critical speed $\Omega_c$ and critical depth of cut $w_c$ constitute the pair of stability equations for turning process. The details of deriving Equations (11) and (12) from Equations (9) and (10) are seen in other works (Ozoegwu, 2011; Ozoegwu & Omenyi, 2012).

## 3. The stabilizing wave attenuation effects in turning process

What follows is geared towards establishing the wave attenuation theory for orthogonal turning process. The cutting force model used in this work is seen from Figure 1 to be normal to the machined surface waves and thus, it is reasonably true that the magnitude of attenuating force will be proportional to cutting force $F_x(t)$ in the manner; $|F_{at}| = \alpha|F_x(t)|^\beta$ where $\alpha$ and $\beta$ are both positive real numbers. Equations (11) and (12) that jointly describe the stability lobes of turning process also defines the critical cutting force $F_c$. Being that $\alpha, \beta \geq 0$ the values $\alpha = 1$ and $\beta = 1$ can be adopted without loss of generality such that $F_c$ becomes used as a measure of attenuating force on a stability lobe. This is possible because sign of the partial derivative $\frac{\partial}{\partial\varepsilon}|F_{at}| = \alpha\beta|F_x(t)|^{\beta-1}\frac{\partial}{\partial\varepsilon}|F_x(t)|$ is solely determined by the sign of the partial derivative $\frac{\partial}{\partial\varepsilon}|F_x(t)|$ where $\varepsilon$ is a parameter of $F_{at}$. Since a turning process on stability lobe is stable in the Liapunov sense, then the critical cutting force $F_c$ is seen from Equation (2) to be given by

$$F_c = Cv^\gamma w_c\tau_c^\gamma \tag{13}$$

where in light the specific force variation $h = C\gamma(v\tau)^{\gamma-1}$ and Equations (11) and (12), $\tau_c$ and $w_c$ are as given as

$$\begin{aligned}\tau_c &= \frac{60}{\Omega_c} = \frac{2}{\omega}\left\{j\pi - tan^{-1}\left(\frac{\omega^2 - \omega_n^2}{2\xi\omega_n\omega}\right)\right\} \\ w_c &= \frac{mv^{1-\gamma}}{2C\gamma}\tau_c^{1-\gamma}\left\{\frac{(\omega^2 - \omega_n^2)^2 + 4\xi^2\omega_n^2\omega^2}{\omega^2 - \omega_n^2}\right\}\end{aligned} \tag{14}$$

It is of interest here to investigate the effect of variation of the passive parameters $\omega_n$ and $\xi$ and the prescription parameter $v$ on $F_c$ and interpret the result in light of the wave attenuation theory. The most mathematically amenable is the critical cutting force at the turning points of the stability lobes. The stability lobes generated using Equation (14) are normally given as $w_c$ as a function of $\Omega_c$. The turning point (minimum point) for the $j$th lobe is derived in details in (Ozoegwu, Omenyi, Achebe & Uzoh, 2012) the non-dimensionalized cutting parameter space to be located at

$$\tilde{\Omega}_{ct}^{j} = \frac{\pi \Omega_{ct}^{j}}{30\omega_n} = \frac{\pi\sqrt{2\xi+1}}{\{j\pi - \tan^{-1}(1/\sqrt{2\xi+1})\}}$$

$$\tilde{W}_{ct} = \frac{w_{ct}h}{m\omega_n^2} = 2\xi(\xi+1) \qquad (15)$$

It is clear from Equations (14) and (15) in light of the specific force variation $h = C\gamma(v\tau)^{\gamma-1}$ that the critical cutting parameters at the turning point of $j$th lobe is at

$$\tau_{ct}^{j} = \frac{2}{\pi\sqrt{2\xi+1}}\left\{j\pi - \tan^{-1}\left(\frac{1}{\sqrt{2\xi+1}}\right)\right\} \qquad (16)$$

$$w_{ct} = \frac{2^{2-\gamma}m\omega_n^{\gamma+1}}{\gamma C v^{\gamma-1}}(\xi^2 + \xi)(2\xi+1)^{\frac{1}{2}(\gamma-1)}\left\{j\pi - \tan^{-1}\left(\frac{1}{\sqrt{2\xi+1}}\right)\right\}^{(1-\gamma)} \qquad (17)$$

When Equations (16) and (17) are inserted in Equation (13) the turning point critical force reads

$$F_{ct} = \frac{4mv\omega_n}{\gamma}(\xi^2 + \xi)(2\xi+1)^{-\frac{1}{2}}\left\{j\pi - \tan^{-1}\left(\frac{1}{\sqrt{2\xi+1}}\right)\right\} \qquad (18)$$

It is seen from Equation (18) that the magnitude of the turning point critical force $F_{ct}$ will rise with rise in either feed $v$ or natural frequency $\omega_n$ since both partial derivatives; $\frac{\partial F_{ct}}{\partial v}$ and $\frac{\partial F_{ct}}{\partial \omega_n}$ are positive. This causes a rise in the attenuating forces which by the wave attenuation theory means more flattening of the regenerative effects and suppression of chatter instability. This point is verified by generating stability charts of 15 lobes ($j = 1$ to 15) for a system with $m = .0431\,\text{kg}$, $\gamma = .75$, $C = 3.5 \times 10^7\,\text{Nm}^{-7/4}$ and damping ratio $\xi = .02$. The natural frequencies $\omega_n = 1000, 3000, 5700$ and $7000\,\text{rad s}^{-1}$ are considered at fixed feed $v = .0025\,\text{ms}^{-1}$ in generating Figure 2. The feed speeds $= .0008333, .0025, .08$ and $.6\,\text{ms}^{-1}$ are considered for a turning tool with natural frequency $\omega_n = 5700\,\text{rad s}^{-1}$ in generating Figure 3.

The effect of variation of damping ratio $\xi$ on the turning point critical force $F_{ct}$ is not obvious from Equation (18) except on further investigation. This further investigation involves determining and checking the sign of the derivative $\frac{\partial F_{ct}}{\partial \xi}$. The derivative is

$$\frac{\partial F_{ct}}{\partial \xi} = \frac{4mv\omega_n}{\gamma(2\xi+1)}\left\{\frac{\xi^2+\xi}{2\xi+2} + \frac{3\xi^2+3\xi+1}{\sqrt{2\xi+1}}\left[j\pi - \tan^{-1}\left(\frac{1}{\sqrt{2\xi+1}}\right)\right]\right\} \qquad (19)$$

Since machine tools usually have low damping ratio that lie in the range $\xi \approx .005 - .02$ (Insperger, 2002), the derivative $\frac{\partial F_{ct}}{\partial \xi}$ as given in Equation (19) is positive meaning that $F_{ct}$ rises with rise in damping ratio $\xi$. Based on the wave attenuation theory a rise in $\xi$ will suppress chatter being that rise in $F_{ct}$ which is a measure of the attenuating force improves flattening of the waves. To verify this point damping ratios $\xi = .005, .01, .015$ and $.02$ are considered at fixed natural frequency $\omega_n = 5700\,\text{rad s}^{-1}$ and feed speed $v = .0025\,\text{ms}^{-1}$ in generating Figure 4. It is visible from Figure 4 that rise in $\xi$ only improves the domain of delay independent stability of the turning stability chart. This

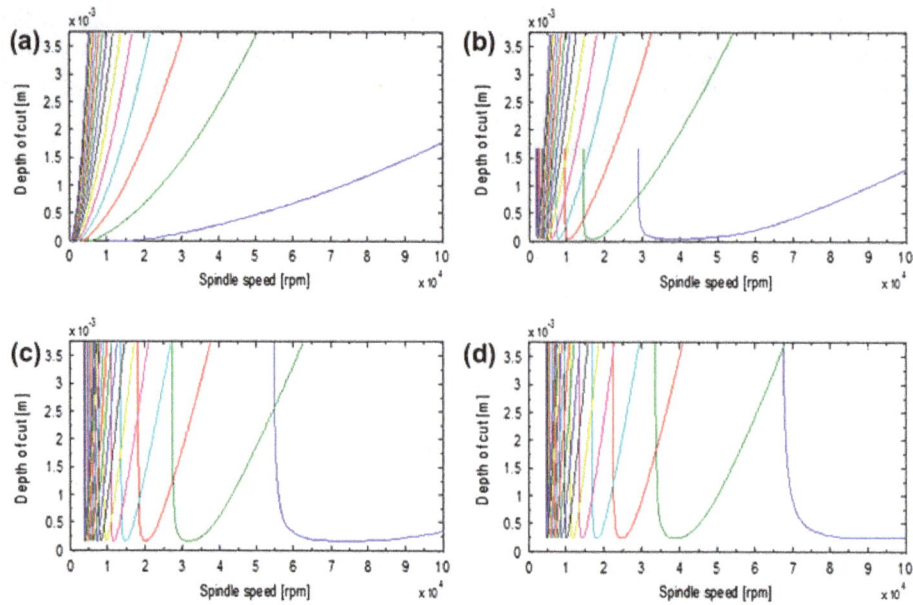

Figure 2.  Stability charts of turning tools with same $m = .0431$, $\gamma = .75$, $C = 3.5 \times 10^7 \, \text{Nm}^{-7/4}$, $\zeta = .02$ and $v = .0025 \, \text{ms}^{-1}$ when natural frequencies are (a) 1000 rpm, (b) 3000 rpm, (c) 5700 rpm and (d) 7000 rpm. It is seen that in conformity with the wave attenuation theory that the chatter instability (union of all domain above the stability lobes) is suppressed with rise in natural frequency.

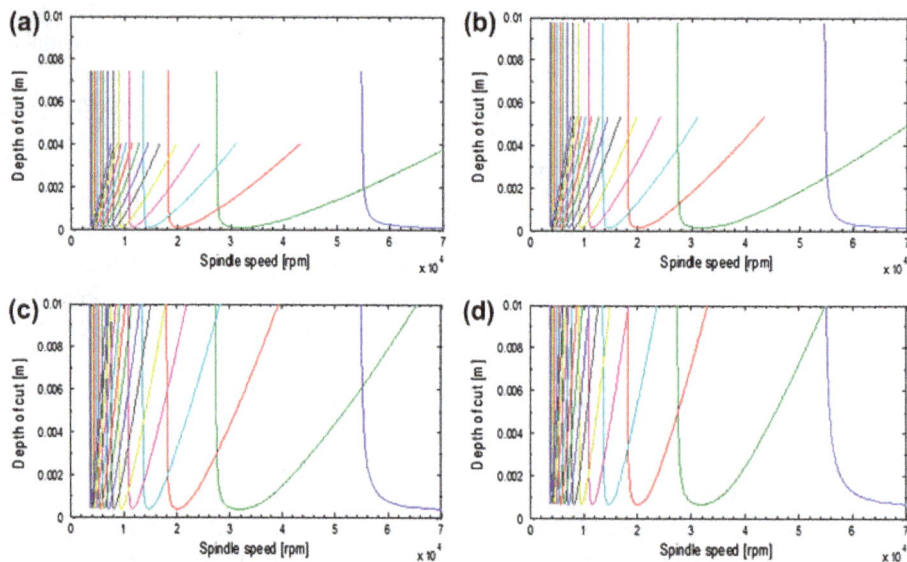

Figure 3.  Stability charts of a turning tool with $m = .0431$, $\gamma = .75$, $C = 3.5 \times 10^7 \, \text{Nm}^{-7/4}$, $\zeta = .02$ and $\omega_n = 5700$ rpm when feed speeds are (a) $.0008333 \, \text{ms}^{-1}$ (b) $.0025 \, \text{ms}^{-1}$ (c) $.08 \, \text{ms}^{-1}$ (d) $.6 \, \text{ms}^{-1}$. It is seen that in conformity with the wave attenuation theory that the chatter instability (union of all domain above the stability lobes) is suppressed with rise in feed speed.

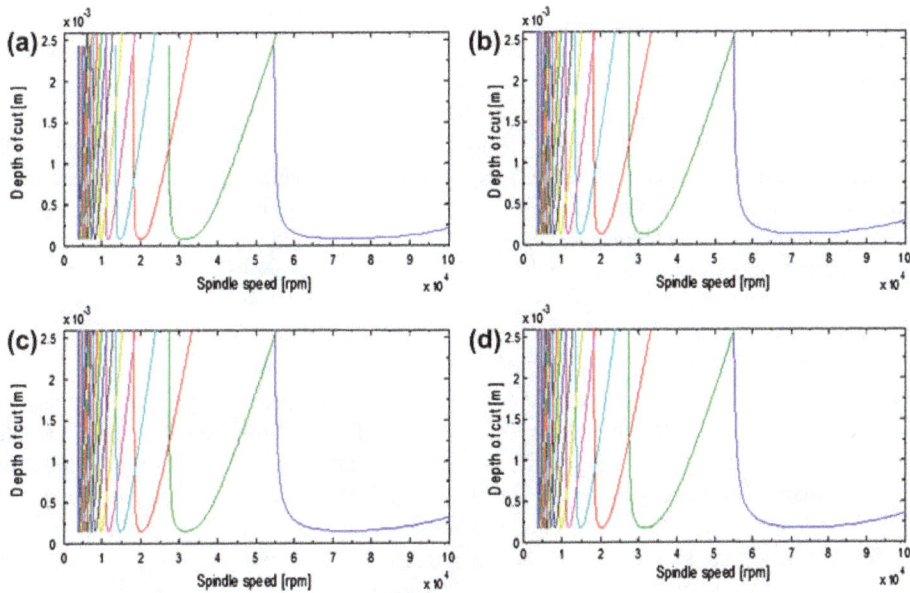

Figure 4. Stability charts of turning processes with $\omega_n = 5700$, $v = .0025\,\text{ms}^{-1}$ and $C = 3.5 \times 10^7\,\text{Nm}^{-7/4}$ and damping ratios (a) .005, (b) .01, (c) $\xi = .015$ and (d) $\xi = .02$. The domain of delay-independent stability (the domain in which a fixed depth of cut line does not cross any of the lobes) rises with rise in $\xi$.

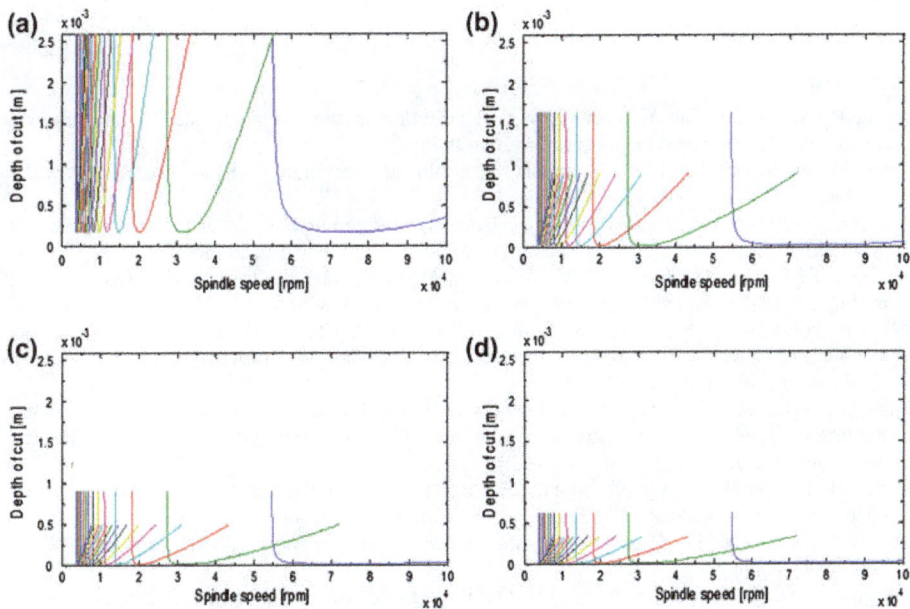

Figure 5. Stability charts of turning with $m = .0431$, $\gamma = .75$, $\xi = .02$ and $v = .0025\,\text{ms}^{-1}$ and $\omega_n = 5700$ for the cutting coefficients (a) $C = 3.5 \times 10^7\,\text{Nm}^{-7/4}$ (b) $C = 2.0667 \times 10^8\,\text{Nm}^{-7/4}$ (c) $C = 3.7834 \times 10^8\,\text{Nm}^{-7/4}$ (d) $C = 5.5 \times 10^8\,\text{Nm}^{-7/4}$. It is seen that size of stable domain falls with rise in $C$ suggesting that stabilizing effects of wave attenuation is resisted more at higher $C$.

point was noted in Ozoegwu et al. (2012). One major consequence of the wave attenuation theory pointed in Ozoegwu and Omenyi (2013) is that the waves of harder and more heat resistant materials will better resist the attenuating forces, and thus enhance chatter instability. Thus, wave attenuation effect is expected to favour chatter stability in cutting of some light materials like composites and magnesium but this favour will considerably diminish when machining materials of hard machinability like stainless steel and inconel. Numerical verification of this point is presented in Figure 5 for a turning tool with parameters; $m = .0431\,\text{kg}$, $\gamma = .75$, $\xi = .02$ and $v = .0025\,\text{ms}^{-1}$ and $\omega_n = 5700\,\text{rad}$ $\text{s}^{-1}$ machining four workpiece materials with different cutting coefficients $C$.

## 4.  Conclusion

The theory of stabilizing wave attenuation effects is developed for the turning process. The theory generally states that some of the force of interaction between the tool and workpiece is flattening on regenerative waves thus attenuating and depopulating the chatter –causing regenerative effects. The flattening force is called the attenuating force. In this work a measure for the attenuating force in turning process is established as a function of tool modal parameters; natural frequency $\omega_n$ and damping ratio $\xi$ and turning process parameter; feed speed $v$. It is shown that the attenuating force rises in response to rise of each of $\omega_n$, $\xi$, $v$. This by the wave attenuation theory means more flattening of the regenerative effects and suppression of chatter instability. One major consequence of the wave attenuation theory is noted; the waves of harder and more heat-resistant materials will better resist the attenuating forces thus retaining chatter instability. All the above points deriving from the wave attenuation are verified numerically.

## References

Arnold, R. N. (1946). The Mechanism of tool vibration in the cutting of steel. *Proceedings of institution of mechanical engineers, 54*, 261–284.

Hanna, N., & Tobias, S. (1974). A theory of nonlinear regenerative chatter. *ASME Journal of Engineering for Industry, 96*, 247–255.

Insperger, T. (2002). *Stability analysis of periodic delay-differential equations modelling machine tool chatter*. Budapest: Budapest University of Technology and Economics.

Insperger, T., Stepan, G., Bayly, P. V., & Mann, B. P. (2003). Multiple chatter frequencies in milling processes. *Journal of Sound and Vibration, 262*, 333–345.

Litak, G., Polyakov, Y. S., Timashev, S. F., & Rusinek, R. (2013). Dynamics of stainless steel turning: Analysis by flicker-noise spectroscopy. *Physica A: Statistical Mechanics and its Applications, 392*, 6052–6063.

Litak, G., Syta, A., & Rusinek, R. (2011). Dynamical changes during composite milling: Recurrence and multiscale entropy analysis. *The International Journal of Advanced Manufacturing Technology, 56*, 445–453.

Merritt, H. E. (1965). Theory of self-excited machine-tool chatter: Contribution to machine-tool chatter research – 1. *Journal of Engineering for Industry, 87*, 447–454.

Ozoegwu, C. G. (2011). *Chatter of plastic milling CNC machine*. Awka: Nnamdi Azikiwe University Awka.

Ozoegwu, C. G., & Omenyi, S. N. (2012). Stability characterization of a turning process. *Journal of Engineering and Applied Sciences, 8*, 68–75.

Ozoegwu, C. G., & Omenyi, S. N. (2013). Wave attenuation effects on the chatter instability of end-milling. *Noise Control Engineering Journal, 61*, 436–444.

Ozoegwu, C. G., Omenyi, S. N., Achebe, C. H., & Uzoh, C. F. (2012). Effect of modal parameters on both delay-independent and global stability of turning process. *Journal of Mechanical Engineering and Automation, 2*, 159–168.

Quintana, G., & Ciurana, J. (2011). Chatter in machining processes: A review. *International Journal of Machine tools and manufacture, 51*, 363–376.

Sen, A. K., Litak, G., Syta, A., & Rusinek, R. (2013). Intermittency and multiscale dynamics in milling of fiber reinforced composites. *Meccanica, 48*, 738–789.

Shi, H. M., & Tobias, S. A. (1984). Theory of finite amplitude machine tool instability. *International Journal of Machine Tool Design and Research, 24*, 45–69.

Siddhpura, M., & Paurobally, R. (2012). A review of chatter vibration research in turning. *International Journal of Machine tools and manufacture, 61*, 27–47.

Stepan, G., & Kalmar-Nagy, T. (1997). *Nonlinear regenerative machine tool vibrations*. Proceedings of DETC'97 1997 ASME Design Engineering Technical Conferences, Sacramento, CA, September 14–17.

Stepan, G., Szalai, R., & Inspeger, T. (2004). This is a chapter. In R. Radons & R. Neugebauer (Eds.), *Nonlinear dynamics of production systems: Nonlinear dynamics of high-speed milling subjected to regenerative effect* (pp. 111–127). Weinheim: Wiley-VCH.

Tlusty, J. (2000). *Manufacturing processes and equipment*. Abingdon: Prentice Hall, New Jersey.

Tobias, S. A. (1961). Machine tool vibration research. *International Journal of Machine Tool Design and Research, 1*, 1–14.

Tobias, S. A., & Fishwick, W. (1958). The chatter of lathe tools under orthogonal cutting conditions. *Transactions of ASME, 80*, 1079–1088.

Wiercigroch, M., & Budak, E. (2001). Sources of nonlinearities, chatter generation and suppression in metal cutting. *Philosophical Transactions of the Royal Society A: Mathematical, Physical and Engineering Sciences, 359*, 663–693.

# Permissions

# List of Contributors

**Hasan Hosseini Nasab, Mahdi Tavana and Mohsen Yousefi**
Industrial Engineering Department, Faculty of Engineering, Yazd University, Yazd, Islamic Republic of Iran

**M. Tucci, F. De Carlo, O. Borgia and N. Fanciullacci**
IBIS Lab, Department of Industrial Engineering, University of Florence, Viale Morgagni 40, Florence 50134, Italy

**Mahdi Bagheripoor and Hosein Bisadi**
Department of Mechanical Engineering, Iran University of Science and Technology, Tehran, Iran

**Suha Karim Shihab, Zahid A. Khan, Aas Mohammad and Arshad Noor Siddiquee**
Department of Mechanical Engineering, Jamia Millia Islamia (A Central University), MMA Jauhar Marg, New Delhi 110025, India

**Ehsan Afshar Bakeshlu**
Department of Industrial Engineering, College of Engineering, University of Kharazmi, Alborz, Iran

**Javad Sadeghi, Tahereh Poorbagheri and Mahziar Taghizadeh**
Faculty of Industrial and Mechanical Engineering, Qazvin Branch, Islamic Azad University, Qazvin, Iran

**Rastee D. Koyee, Uwe Heisel and Rocco Eisseler**
Institute for Machine Tools, University of Stuttgart, Holzgartenstr. 17, D-70174 Stuttgart, Germany

**Siegfried Schmauder**
IMWF, University of Stuttgart, Pfaffenwaldring 32, D-70569 Stuttgart, Germany

**Reina Angkiriwang, I Nyoman Pujawan and Budi Santosa**
Department of Industrial Engineering, Sepuluh Nopember Institute of Technology, Kampus ITS Sukolilo, Surabaya 60111, Indonesia

**Victor Chukwunweike Nwokocha and Iganatius Ani Madu**
Department of Geography, University of Nigeria Nsukka, Nigeria

**Jalel Ben Hmida, Jim Lee and Xinchun Wang**
Systems Engineering, University of Louisiana at Lafayette, P.O. Box 44170, Rougeou, Hall Room # 244, Lafayette, LA 70504, USA;

**Fathi Boukadi**
Department of Petroleum Engineering, University of Louisiana at Lafayette, P.O. Box 44408, Madison Hall, Room # 134, Lafayette, LA 70504, USA

**Joakim Wikner**
Department of Management and Engineering, Linköping University, Linköping, Sweden
School of Engineering, Jönköping University, Jönköping, Sweden

**R.V. Rao and K.C. More**
Department of Mechanical Engineering, S.V. National Institute of Technology, Ichchanath, Surat, Gujarat 395 007, India

**Chigbogu Godwin Ozoegwu**
Department of Mechanical Engineering, Nnamdi Azikiwe University, PMB 5025, Awka, Nigeria

www.ingramcontent.com/pod-product-compliance
Lightning Source LLC
Chambersburg PA
CBHW080459200326

41458CB00012B/4034